Jack Alterman

About the Author

Edward Ball was born in Georgia, raised in the South, and worked in New York as an art critic. His first book, *Slaves in the Family*, told the story of his search for the descendants of his ancestors' slaves. He lives in Litchfield, Connecticut, with his wife, Elizabeth.

Also by Edward Ball

Slaves in the Family

THE SWEET HELL INSIDE

The Rise of an Elite Black Family in the Segregated South

EDWARD BALL

Perennial
An Imprint of HarperCollins*Publishers*

A hardcover edition of this book was published in 2001 by William Morrow, an imprint of HarperCollins Publishers.

HarperCollins books may be purchased for educational, business, or sales promotional use. For information please write: Special Markets Department, HarperCollins Publishers Inc., 10 East 53rd Street, New York, NY 10022.

First Perennial edition published 2002.

Designed by Bernard Klein

The Library of Congress has catalogued the hardcover edition as follows:

Ball, Edward.

The sweet hell inside : a family history / Edward Ball. —1st ed.

p. cm.

ISBN 0-688-16840-X (acid-free paper)

1. Harleston family. 2. African American families—South Carolina—Charleston Region—Biography. 3. Interracial marriage—South Carolina—Charleston Region—History. 4. Slaves—South Carolina—Charleston Region—Biography. 5. African Americans—South Carolina—Charleston Region—Biography. 6. Slaveholders—South Carolina—Charleston region—Biography. 7. Charleston Region (S.C.)—Biography. 8. Charleston Region (S.C.)—Race Relations. 9. African American families—Biography. 10. Racially mixed people—United States—Biography. I. Title.

F279.C453 H37 2001
975.7'915'0099—dc21

2001030880

ISBN 0-06-050590-7 (pbk.)

07 08 09 ❖/RRD 10 9 8 7 6 5 4 3 2

By the sea under the yellow and sagging moon,
The messenger there arous'd, the fire, the sweet hell within,
The unknown want, the destiny of me.

—*Walt Whitman*

Contents

The Harleston Family

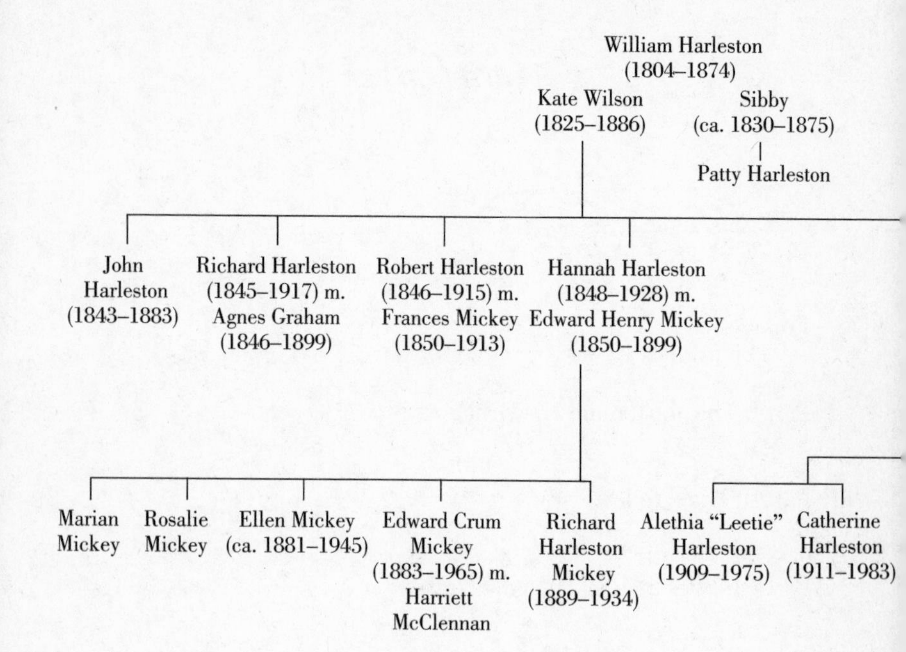

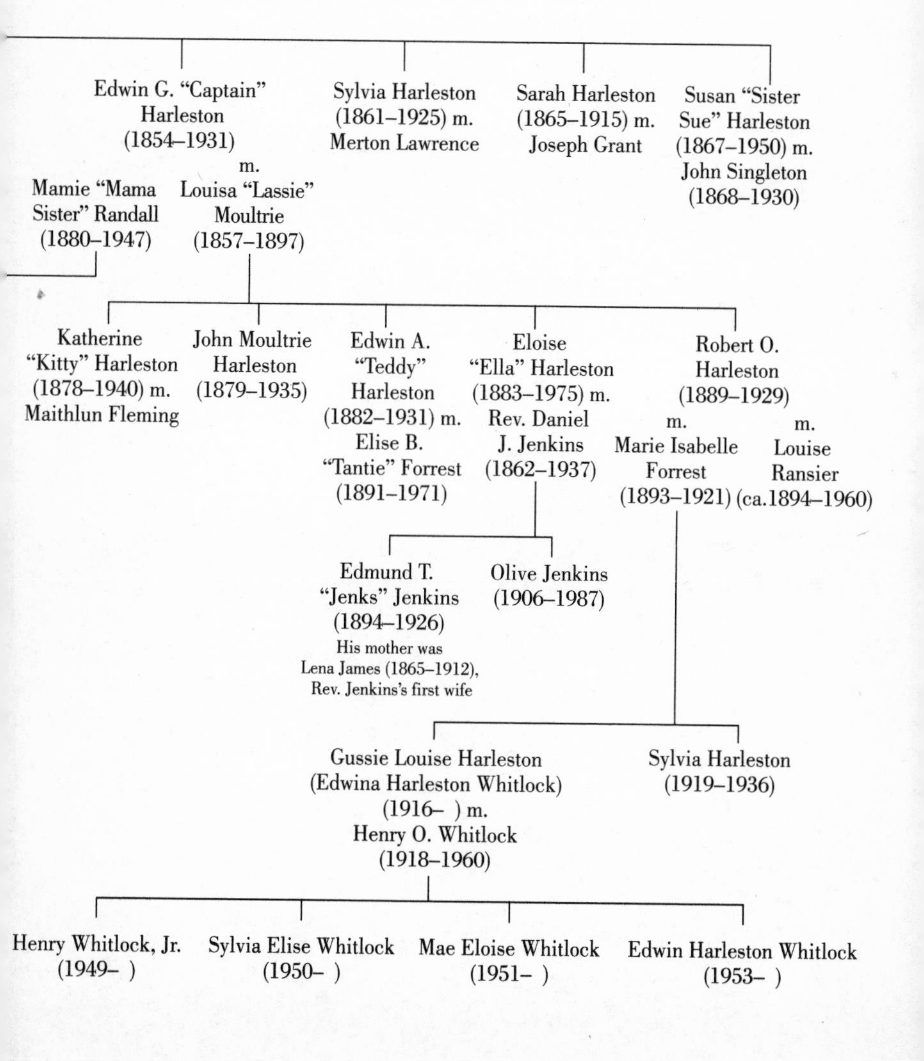

Edwin G. "Captain" Harleston (1854–1931)
m.
Mamie "Mama Sister" Randall (1880–1947)
Louisa "Lassie" Moultrie (1857–1897)
Sylvia Harleston (1861–1925) m. Merton Lawrence
Sarah Harleston (1865–1915) m. Joseph Grant
Susan "Sister Sue" Harleston (1867–1950) m. John Singleton (1868–1930)
Katherine "Kitty" Harleston (1878–1940) m. Maithlun Fleming
John Moultrie Harleston (1879–1935)
Edwin A. "Teddy" Harleston (1882–1931) m. Elise B. "Tantie" Forrest (1891–1971)
Eloise "Ella" Harleston (1883–1975) m. Rev. Daniel J. Jenkins (1862–1937)
Robert O. Harleston (1889–1929)
m.
Marie Isabelle Forrest (1893–1921)
m.
Louise Ransier (ca.1894–1960)
Edmund T. "Jenks" Jenkins (1894–1926)
His mother was Lena James (1865–1912), Rev. Jenkins's first wife
Olive Jenkins (1906–1987)
Gussie Louise Harleston (Edwina Harleston Whitlock) (1916–) m. Henry O. Whitlock (1918–1960)
Sylvia Harleston (1919–1936)
Henry Whitlock, Jr. (1949–)
Sylvia Elise Whitlock (1950–)
Mae Eloise Whitlock (1951–)
Edwin Harleston Whitlock (1953–)

Preface

EDWINA Harleston Whitlock rose from the table where we were sitting in her living room in Atlanta and said, "I want to show you something." The elderly woman led me to the rear of her house, where I had never gone, pushed open the door to a small office, and there, in heaps, was the most chaotic mess of papers I had ever seen. Every chair and tabletop was layered with documents. Filing cabinets were jammed so full their contents spilled to the floor. Sepia-colored photographs lay among piles of stained folders. From window to window, old papers stood in tilting stacks.

"I've been collecting information about my family" was her explanation.

For days, I had been listening to stories about Edwina's relatives, each more intriguing than the last, and now she was showing me her sources. I like old things, and the office and its ancient files were to my eyes like a ship to the eyes of a castaway. At that moment, I could picture myself spending a year sifting through the hoard.

The first time I spoke with the collector of that jumble of papers was on a Saturday afternoon several years ago. I picked up the phone, and on the other end was an unusually pleasing voice, which sounded like it belonged to a woman who was in her later years. She told me her name, Edwina Harleston Whitlock, and abruptly announced that we were cousins. I was living in Charleston, South Carolina, and the possibility that I might be related to someone I had never met didn't surprise me. In the South, to discover a cousin is nothing unusual, because Southerners claim to be related to almost everybody, practically to the trees. The one distinct exception is that white people never admit they might be related to blacks.

At the time, I was in the midst of writing my first book, *Slaves in the Family*. The book told the story of my family's long history of slave ownership in the American South, and the story of my search for and meetings with descendants of my ancestors' slaves. Edwina Harleston Whitlock said she had heard about my research into the lives of my family's former slaves, and she wanted me to know that she was one of the nonwhite descendants of the Ball family. She said she was the great-granddaughter of a slave owner and (in her word) his slave "consort." Put another way, we were cousins, and she was a person of color.

In the book I was writing, I didn't intend to skirt the subject of sex between slave owners and slaves, and I thought I might receive a phone call like this one. But the idea of having a cousin of color was still jarring. There were no black Americans I thought of in the same way I think of my immediate family, who are white. As a boy, growing up in the South in the late 1960s and 1970s, I knew of the Harleston family, some of whom had married cousins of my father. I knew that various Harleston relatives were buried next to the graves of my father's people in a small cemetery near Charleston. But I didn't know that some of our Harleston relatives had been black. As I put down the phone, I sensed my life had suddenly and irreversibly changed.

Although it occurred to me that the caller might be mistaken, I eventually found the wills, probate records, diaries, and oral tradition that documented our kinship. We were, in fact, very distant relatives: Edwina Harleston Whitlock's great-great-great-great-grandfather had been the brother of my great-great-great-great-great-great-grandmother. The connection was labyrinthine, but no less real. Two of my forebears, Elizabeth Harleston (ca. 1678–1719) and Elias Ball (1676–1751), were married in South Carolina in 1700; Elizabeth Harleston was the sister of

Edwina Whitlock's forebear John Harleston (ca. 1675–1738). In 1707, John Harleston married Elizabeth Willis (ca. 1680–1754), and that couple formed the initial white roots of Edwina Whitlock's family, which later became African American by means of the previously mentioned "consort." By the rules of kinship, Edwina and I were sixth cousins, two times removed.

After our introduction, I would sometimes make the drive from Charleston to Atlanta to spend a day or two with her. She stood five feet seven, had a beautiful smile, and at eighty-something was successfully fighting her age. She painted her fingernails, wore silk, and walked as erect as a student in a posture class. Her skin was not much darker than mine, its texture smooth from decades of creams and care, and her eyes were penetrating when she uttered a devious remark, which was often.

We began by tracing the small blood we had in common and starting to come to terms with our connection. But it soon became apparent that our kinship would take second place to another preoccupation in our relationship, and that was storytelling. My newfound cousin was good with a tale, especially when it concerned her family's past, and her stories made me interested in her life. She was ironic and made fun of things. We enjoyed one another, despite the anxiety we both felt about being related by an act of (probably not consensual) interracial sex.

Edwina Harleston Whitlock is the descendant of a white slave owner and his young black concubine, whose names were William Harleston and Kate Wilson, who had lived in the 1840s on an estate in South Carolina. William Harleston never married a white woman, and he had eight children with Kate Wilson, including Edwina's grandfather Edwin G. Harleston.

In the early 1900s, Edwina's family, the Harlestons, played a little-known but fascinating role in the American national saga. For years, Edwina had wanted to put down some of their experiences, and she had gotten as far as compiling the documentation. Although she could describe ancient scenes in vivid detail and remember fragments of conversations from her childhood in the 1920s, finding a written voice had eluded her.

When the white William Harleston died, his children, who had previously lived in some comfort, were pushed into black society and into destitution by their white relatives. But Edwina's grandfather saved the family from poverty by establishing a funeral business, and the Harlestons eventually landed in prosperous and secure lives.

At this point, with the generation of Edwina's grandfather Edwin, the story of the Harlestons becomes truly special. Edwin Harleston's own five children, born between 1878 and 1889, were American cultural pioneers. One became an artist who opened a studio with his wife, a photographer, and painted portraits of the likes of industrialist Pierre Du Pont; that couple, in the early 1900s, were two of only a handful of nonwhites in the country who could say that art was their career. Another of Edwin's children helped to run an unusual orphanage, some of whose young black inmates became prodigies in a group of child musicians called the Jenkins Orphanage Band, whose pubescent entertainers share credit with the bands from New Orleans for the birth and spread of jazz music. Other personalities in the family included a gay undertaker and hypochondriac who believed he was always on the verge of death; a classically trained black composer who haunted the nightclubs of 1920s Paris; and a Harleston family mistress who doubled as an abortionist. From their beginnings as the discarded children of a case of "miscegenation," the Harleston family rose to play strange and wonderful roles in the American pageant, with scenes that were acted out in Chicago, New York, Boston, and beyond, in London and Paris.

Edwina's family thrived even during some of America's cruelest years. From the 1890s until about 1960, black and "mixed-race" Americans lived nearly as aliens in the wider society—deprived of a choice of work, denied the ability to educate their children, as well as the freedom to select a home, and subjected to random violence—while governments looked away, and white citizens remained indifferent. But the Harlestons shrugged off all these troubles as though they were so much bad weather and went about their business building a cultural dynasty.

The Harlestons had an unusually good time of it in part because of who they were. Edwina Harleston Whitlock's family belonged to a tiny group of light-skinned blacks at the top of the nonwhite world, an educated, powerful "colored elite," most of whom were the children of white men and their black lovers. The Harlestons were members of this in-between class—a group set apart both from white society, which shunned them, and from poorer black families, who often envied them.

But despite their riches in life, the Harlestons carried a deep pain inside. They were neither white nor black, and they wondered who they might be for themselves, as well as in the eyes of others. Parents passed on this anxiety to their children, and it shadowed the family for generations.

While I was visiting Edwina and listening to her recollections, it became apparent to me that my newfound cousin filled up with pleasure and desire whenever she spoke of her family. Her emotion was pure, and I found the Harleston family's story to be so rare and so moving that I longed to tell it in its entirety. Eventually, with Edwina's consent, I agreed to try to write it.

We began by arranging a meeting schedule. I would visit, and the two of us would talk only about family history. Next, I put in order Edwina's disorganized papers. They amounted to some two thousand pages of correspondence, scrapbooks, diaries, photographs, and notebooks, many of them handed down in her family from the early 1900s. I studied the papers for clues and followed the leads of Edwina's memories and experience. After I was engorged, I sketched an outline and went home to my desk to write.

The Sweet Hell Inside is the biography of one family, from their origins as the offspring of an interracial sexual "arrangement," straight down to the present. The story of the Harlestons is a tale of black and white sex in America, and its latter-day harvest. Although members of Edwina's family are my distant cousins, at this late hour the connection is too remote to let me claim their experiences as my own. And so their extraordinary journey, and its satisfying culmination, belongs only to them.

Part I

~

The Master and His Orphans

Chapter One

WILLIAM Harleston was born in 1804 at The Hut, a plantation owned by his parents, twenty-five miles north of Charleston, South Carolina. The Hut was one of several assets in a Harleston dynasty of land and slaves. (Other family estates were named Rice Hope and Richmond.) Few records have survived from William Harleston's youth. Even his birth date is in question—the year is known, but there is no record of the day—as though it had been clipped from memory. A genealogy states that he was one of five children, with three older sisters, Hannah, Sarah, and Constantia, and one younger brother, John. As a boy growing up on The Hut, a one-thousand-acre rice plantation with about sixty slaves, William would have passed his youth in ways appropriate to his class and sex: he hunted deer and fished, was schooled in classics and mathematics by private tutors, and attended the glittering society balls held every winter in nearby Charleston. The Harleston family's holdings made his life indolent and luxurious.

Despite its many comforts, William's childhood was strained by high expectations, because he carried the unusual burden of a heroic family legacy handed down by his father. In the 1770s, a quarter of a century before William's birth, the Harlestons had fought bravely during the Revolutionary War. William's father,

William Harleston Sr., and his uncle Isaac had both risked everything to win American independence. Only twenty when the war began, William Senior served as an infantryman before returning to the family lands to help supply food to American soldiers; but Isaac, then in his thirties, passed through an extraordinary war. Isaac Harleston had left his plantation, which was known as Irishtown, to become a captain in one of the earliest fights of the conflict, the Battle of Fort Sullivan, on June 28, 1776, a deadly assignment during which a few hundred American patriots repulsed an attack on Charleston by a flotilla of British warships. After that American victory, the British left the South alone for two years. In the meantime, Isaac loaned the revolutionary movement sixty-five hundred pounds, putting muscle in the phrase of the Declaration of Independence, "our lives, our fortunes, and our sacred honor." Six months after the loan, in 1778, the British returned, and Isaac was elevated to the rank of major in command of the Sixth Regiment of the Continental army. He fought for eighteen months in various campaigns, but in May 1780 Isaac was in Charleston when it fell to the British, and he was taken prisoner. While many rebels now swore allegiance to the Crown in order to escape punishment, Major Harleston stayed true to the cause. Eventually Isaac was released to a victor's welcome and returned to quiet citizenship. He died in 1798.

To be an heir to such patriotism gave young William Harleston of The Hut the aura of an American prince.

As he grew up, William seemed ready to take his natural place in the elite circle of slaveholding landlords. But his own father foreshadowed another outcome when, adding to his white children, the elder William had a mixed-race son, born to one of his slaves. The name of the boy's mother has not survived, but the child was given the name Isaac, after the war hero. Although there was a taboo against interracial sex, in reality ruling-class men often had brown children with their black slaves. As long as the children were kept away from white relatives, and the taboo received lip service, silence about the offspring could be maintained. But this was not the case with Isaac.

Family tradition describes Isaac as a short brown man. Born a slave in the 1780s, the illicit Isaac grew up at The Hut and at some point was evidently given his freedom. However, instead of leaving the vicinity, a frequent outcome in such cases, Isaac stayed right in the thick of family business. He became a steward on a riverboat that his Harleston relatives used, then married and fathered three children of his own, the first of whom (his name was Edward) settled on the Harleston plantation known as Rice Hope. Rice Hope stood about a mile from The Hut, which meant that the young William, as an impressionable child, had ample time to see and speak to the living fruit of his father's wayward sexuality.

In South Carolina, a few rich families like the Harlestons lived comfortably, while the majority of people, black slaves, were consigned to a form of living hell. South Carolina covers a relatively small area, thirty-one thousand square miles, about a third less land than Virginia. But in William Harleston's day, Charleston, a large and queenly port city, set a worldly tone. Charleston ranked fourth in size in cities in the United States, after New York, Boston, and Philadelphia. Its port shipped out homegrown necessities—mainly rice and cotton tilled by black hands—and received a stream of luxuries in return from the Northern states and from Europe. The tons of silver, imported clothing, and fine furniture that came back all floated up to the top tier of society. The crumbs fell to a poor white working class, and the dregs dribbled to the slaves.

Although the United States had banned the import of slaves in 1808, West Africans who had previously arrived in Charleston in chains filled the state. In 1820, blacks in South Carolina outnumbered whites 265,000 to 237,000. (The state was one of only two with a black majority, the other being Louisiana.) William's family was among the lucky few. With their houses in Charleston and plantations outside town, the Harlestons were within the wealthiest one percent of all Americans.

Census records show that during America's slaveholding years, there were more white bachelors living in the South than in the rest of the country. This anomaly was probably not the result of a surplus of gay men. Homosexual men often married and had children while also carrying on affairs with same-sex lovers. The prevalence of single men in the South expressed the reality that white men did not have to marry white women in order to have sex.

In 1802, a Virginia newspaper, the *Richmond Recorder*, accused President Thomas Jefferson of having an enslaved lover, Sally Hemings, who lived at Jefferson's mountaintop plantation, Monticello. Jefferson had been widowed for many years and remained unmarried. The *Recorder* named the children the president had fathered with Hemings, a woman thirty years younger than he. "It is well known that the man [Jefferson], whom it delighteth the people to honor, keeps, and for many years past has kept, as his concubine, one of his own slaves," said the paper. "Her name is SALLY." The *Connecticut Courant* and Boston's *Gazette of the United States* soon picked up on the story. A cartoon in one broadsheet poked fun at Jefferson's reputation as a sage who had written the Declaration of Independence: the drawing showed Jefferson and Hemings disguised as a rooster and a hen, above the caption "A Philosophic Cock." The scandal did not punish Jefferson: two years after the story surfaced, the president won reelection.

No paintings or photographs of William Harleston survive. But other branches of the Harlestons, like other families of William's generation, left behind daguerreotypes or paintings of themselves. If William resembled his cousins, he probably grew to perhaps five feet eight inches, with straight, dark hair. His forehead receded a little from the eyebrows to the hairline, and he had a straight nose, high cheekbones, and a square jaw. In all, he appeared moderately refined, but not quite the aristocratic ideal.

While he was still in his teens, William's easy coming-of-age came to a shuddering end. In 1816, when he was twelve, his father died; and when he turned seventeen, his mother, Sarah, followed her husband to the grave. Just when he most needed parental help, William became an orphan. In her will, Sarah Harleston appointed her husband's brother as a guardian of her remaining minor children, who included William and his younger brother, John. But this brother-in-law (Edward Harleston, the boys' uncle) was a man of sixty and busy with his own large family, so he was not often available to help, and it appears the children were largely on their own.

After the death of his parents, William Harleston became increasingly private and withdrew from society. Unlike his famous uncle, Isaac, he had no interest in the military. He didn't study art, although he had both time and money; and he didn't marry, when this was expected. William eventually took charge of The Hut, his father's rice plantation. Although there is no evidence about how well he did as a plantation master, old families like the Harlestons often had business managers who could step in to prevent ruin. The heir retreated to the extent that little can be said about the decade or so after his mother's death. William's young adulthood is unmarked, a doldrums.

In fall 1835, John, William's younger brother, acted to bring an end to this isolation. John Harleston, then thirty and unmarried, asked William, thirty-one, to move with him to a new estate John had bought, known as Elwood. John had purchased the 602-acre tract in partnership with a brother-in-law (the husband of John's sister Hannah), who had invested in Elwood but lived elsewhere. John wanted companionship, so he asked William to set up house with him.

Elwood, three miles north of The Hut, was one of the smaller and newer slave farms on the rivers near Charleston. It had been settled a generation earlier on a wedge of land squeezed between two older plantations. Perhaps fifty black slaves were attached to the land—field hands, craftsmen, and house servants—many of whom had served the previous owner. John and William moved into the vacant mansion in the winter of 1836 and began introducing themselves around the neighborhood, though their famous name hardly needed mentioning.

On a list of slaves attached to Elwood, the name "Katey" appears. Katey, or Kate, was the only person with that name on the property. She had been born on December 4, 1825, the daughter of the house cook, Sarah, and a man named Anthony Wilson, who was probably white. Kate also had a sister, Chloe, and two brothers, Richard and Anthony. But little is recorded about Kate's early life.

Slaveholders rarely wrote down biographical facts about their workers, and slaves themselves were kept away from books. When she got older, Kate used her father's name, Wilson, although to do so was unusual for slaves. Harleston family tradition says something more about Kate's mother, Sarah. In a district with many black cooks, Sarah had a reputation as a notable chef. She became famous for her banquets, parties talked about by guests for weeks. The children of cooks usually became their helpers, and it's likely that Kate Wilson followed her mother into the kitchen.

Although no photographs or drawings of Kate Wilson survive, oral tradition from her descendants says she was pretty and soft-spoken, with a light brown complexion. (Her white father and black mother made her a tan "mulatto.") When William and John Harleston moved to Elwood, Kate was just ten, and as a kitchen worker, she would have been much in their sight. It's likely that when the new masters of the farm ate their meals, Kate squeezed between the chairs, delivering dishes and clearing from the table. Though direct evidence is slim, Kate and William may have met in the dining room.

In the slave days, it was not unusual for white adult men to force sexual relationships on young nonwhite women. Thomas Jefferson turned forty-four before he began his affair with fourteen-year-old Sally Hemings. Kate's freshness may have captured the attention of William, who by then was nearly forty. Like Sally Hemings, Kate was less dark-skinned than most field hands, a trait that seemed to appeal to many men.

William Harleston's life had been frozen in youth, when he lost his parents, and he had withdrawn from social life. He was unmarried, lonely, and one can assume he felt lust. (One can also assume that Kate was probably not the first slave on whom his lust had settled.) William's motives lay on top of each other like tangled sheets. But whatever his mix of urges, after several years in Kate's company, William decided he wanted her. Sometime in 1842, William began to have sex with Kate. He was thirty-eight, and she was sixteen.

Chapter Two

THE first child of William Harleston and Kate Wilson, John, was born on August 5, 1843. Their second, Richard, came in 1845, and the third, Robert, was born in 1846. Elwood plantation began to be known in the neighborhood as the home of the Harleston brothers, and "Will's black brood."

It's safe to say that their initial lovemaking was not consensual. Kate's master most likely forced her to have intercourse, and she became his concubine. It's impossible to know what they talked about, but easy to imagine how things looked. A middle-aged man, pale and bony, suspended himself over a slight, brown girl.

By later codes of sexual behavior, the start of the relationship would have to be described as rape. It may have been a violent act, or it may have been quiet: slave women gave in to their masters in different ways, but perhaps after their second or third child, William's heart began to soften, and he became emotionally attached to Kate. Some of the babies would be given the names of William's own white siblings: the firstborn took the name of William's brother and housemate, John Harleston, while two others, Hannah and Sarah, shared names with two of William's sisters. This seems an unlikely outcome if their father looked on their mother purely as a partner who would give him an orgasm.

One must be careful when it comes to describing the feelings of Kate Wilson. But in present terms, Kate could be seen as a victim of child sexual abuse. She arguably suffered emotional and physical wounds. The circumstances of slave mistresses, however, if they survived in the role, often improved. Kate may have moved into nicer rooms, or gotten relief in the kitchen for her aging mother, Sarah. She may even have grown accustomed to the touch of her master.

Tradition reports that William was not faithful to Kate. He is said to have had another black lover, named Sibby, who was also his slave. Sibby had a daughter called Patty, who was known as an "outside child," because she was born outside the core couple, William and Kate.

Far from diminishing their bond, William's infidelity with Sibby made his relationship with Kate resemble a traditional marriage. Evidence suggests that after several years, William began to see Kate as his life companion, and their children as his only family. Kate's contact with William differed in one important way from the customary sex between white men and black women: rather than ending after a week or a season, their relationship survived nearly thirty-five years. They became, in the quaint term of the era, common-law husband and wife; and eventually, the couple had eight children. Still, the sex Kate Wilson had with William cannot be called consensual, because it was too far from a pairing of equals. But just as hostages often grow attached to their captors, Kate could have become fond of William, if only to protect her mind from the shock of frequent assault. Their relationship seems to have started with force and ended with resignation, or even comfort. And the idea cannot be dismissed that after many years, they developed an affection for each other. It was not possible for the two to marry, but William never bent to the pressure from his large and influential family to quit "that nigger concubine" and marry a white woman.

When William and Kate began sleeping together, his family tried to pull him back into the pale of white life. He remained a dinner guest at other family estates. He went to church and courted ladies of his own station, though he never proposed to them. But when it became apparent that Kate Wilson had become his partner, William seems to have been ostracized. His name cannot be found in the customary places where people of his class came together: the member lists of private clubs, the agricultural societies, the church vestry rolls.

In black society, Kate faced similar judgment. Throughout the slave South, on the dirt streets of a thousand plantations, black people pitied the women who were raped by whites and shunned those who bargained sex for better treatment. When she became a concubine, Kate was William's prey. To try to refuse a master would have been dangerous or impossible. But as awful as it sounds, for many black women, sex with whites was the only way up and out. The longer Kate stayed with him, the more it looked like she was bartering her body. Whenever her master treated her to some favor—new shoes or a cheap necklace—other slaves isolated her. Kate's children benefited from their mother's edge in getting extras, but they also carried the onus of her sexual commerce.

The living arrangements at Elwood plantation have faded from memory. William and John Harleston almost certainly shared the mansion, a two-story wooden Carolina-style I-house, with a central hall ("as long as the letter 'I'") flanked by symmetrical rooms. But Kate and the children probably lived apart. House slaves occupied cabins near the master's dwelling. After all, the Harleston brothers had to entertain white business contacts, and any respectable white person would have shriveled at the sight of black "family."

When a slave owner was inclined to have sex with a black woman, little, apart from his own scruples, could stop him. A taboo stood in the way, but the tradition ran deep. Slaves could not testify against whites, so no legal obstacles slowed the rape of black women, unless a white person of conscience saw it and brought charges—and this never happened. In reality, it was a rite of passage for a young man to "get a colored girl" before marriage. The ban made it more alluring, especially to the teenage sons of rich parents. Upper-class boys in adolescence, like predators on the family lands, cornered black women for sex. After several years, they married a proper white woman and pretended it had all been in fun.

Most memoirs written by women who were once in slavery refer to the rape and sexual bartering of black women. Harriet Jacobs's autobiography, *Incidents in the Life of a Slave Girl,* in 1861, and Elizabeth Keckley's memoir, *Behind the Scenes; or, Thirty Years a Slave and Four Years in the White House* (published in 1868 by a former housekeeper for Abraham and Mary Todd Lincoln), both make explicit that rape often went with enslavement. Years later, when it was easier to speak, former slave women said more. One black woman, looking back from the 1930s, said to a government interviewer: "I can tell you that a white man laid a nigger gal whenever he wanted her. Seems like some of them had a plumb craving for the other color. Leastways they wanted to start themselves out on the nigger women."

Throughout the South during the 1800s, elaborate customs were put in place to enable interracial sex. In New Orleans, the former French colony, the placage (placement) system regulated the supply of mixed-race women to wealthy white men. Placage began when a man showed his desire for a young woman, usually a light-skinned "quadroon" girl (who was supposedly one-quarter black). They had often met at a so-called quadroon ball, or formal dance, put on for the purpose of black and white matchmaking. If she accepted him, the woman referred the man to her mother, who then acted as her agent. The mother received promises that the seducer would support her daughter in high style; and if he broke off the affair, he would give her money for each child they may have. The bargain struck, the concubine would be set up in a high-ceilinged apartment, ready for her lover.

For less theater, but a similar outcome, New Orleans men could go to the posh St. Louis Hotel to inspect the "fancy girls." These were young slave women advertised and sold as household servants, though blacks and whites widely understood them to be sex slaves. At the St. Louis Hotel, pretty quadroons were dressed up in gowns and sold on a platform under the rotunda of the marble-clad lobby. The auctioneer muttered sexual banter during the sale, and the women were sometimes undressed for closer examination. A less flagrant market in "fancy girls" was available in Charleston. Bachelor heirs frequently bought the women, who commanded the peak price men would pay for a warm body, higher even than the cost of a profitable young field hand.

Kate and William's children grew up. The boys probably spent time out on the creeks, fishing, riding horses, hunting duck and deer. The girls practiced their needlework and kitchen chores, but nothing too strenuous, because less fortunate slaves would have done the real work.

It was against the law for the children to attend school. In 1834, the South Carolina legislature had outlawed schools for nonwhites: "If any person shall hereafter teach any slave to read or write . . . such person if a free white person upon conviction thereof shall for each and every offense against this act be fined not exceeding $100 and imprisoned not more than six months." Still, circumstantial evidence shows that Kate's children got some instruction. Years later, as adults, two of the siblings (Hannah and Edwin) would run a business together. Their account books would be scrupulous—which means the children were probably tutored in their youth. And although no letters written by Edwin Harleston survive, there are several notes from family members about Captain Harleston reading his mail. The instruction the children received would have had to involve either William Harleston himself or a private teacher.

Pictures of the family of Kate Wilson's son Edwin have been found. Their skin is neither white nor black, but tan. The faces of the women tend to an oval shape, while the men are more square-jawed. There is pride in the way they hold themselves, but in their eyes there is a gleam of insecurity, as though something about life isn't right. The Harleston family had just escaped being born into the cruel jail of lifetime subservience.

In 1850, the Federal Bureau of the Census introduced a new racial category: "mulatto." It was to be added to the already existing classes of "white," "black," and "Indian." The agency did not define mulatto, which had been in use for centuries, but everyone took the word to mean a person with one white and one black parent. When the 1850 census was completed, the government had counted 3,639,000 Negroes and 406,000 mulattoes living in the United States. Twenty years later, census enumerators again had to distinguish mulattoes from other Americans, and this time, the Census Bureau supplied a confusing definition. The category of mulatto was now said to include "quadroons, octoroons and all persons having any perceptible trace of African blood."

The word "mulatto" derives from the Spanish *mula,* meaning mule. A mule is a cross between a mare and a jackass. It cannot reproduce: it has no forebears that are mules and can have no offspring. Because of its beastly connotations, the term "mulatto," five hundred years old, has lately fallen out of use; but the idea of a mixed-race group, an in-between society flanked on either side by whites and blacks, was once a given of American life. Mulattoes were thought to be a link between two pronounced races. Because of their access to education, they became leaders in black society, the home class of ministers, politicians, and businesspeople. Whites alternately looked down on them and (because they were more likely to own property or be self-employed) approved of their ambitiousness. Darker and poorer blacks both resented the half-white skin color and relative richness of mulatto families and acknowledged them as the elite of the nonwhite world. Pulled between shifting signals of contempt and praise, the mixed-race minority in the middle went through life with more than its share of anxiety and wounded pride.

When Kate Wilson was having children, nine out of ten mulattoes lived in the South—states south of Pennsylvania and east of Texas. They clustered heavily in two areas: southern Louisiana (New Orleans) and eastern South Carolina (Charleston). In Louisiana, with its French and libertine heritage, white fathers of mixed-race children felt little shame about their families of color. Kate's children, had they been from Louisiana, would have been part of a large group of in-between families known as Creoles. But in South Carolina, settled by especially conservative English colonists, sexual neurosis was the rule, and many whites felt physical repulsion for blacks. Whereas in Louisiana, white fathers often provided for their "colored" children, in South Carolina, the pattern grew among white fathers to discard them and let them fend for themselves.

The desertion took one of two forms. Usually, fathers allowed the children to remain slaves, which was the supreme form of neglect. But sometimes fathers wrote their children's freedom into a last will, leaving the surviving white family the job of abandonment: pushing their nonwhite siblings out of the dead man's house.

This group of freed mulattoes usually moved off the plantation and to the city. In the rural South, more than three-quarters of mixed-race people were counted as slaves; but in Charleston, two-thirds of them were free. Law required that they wear a metal tag, which read "City of Charleston, FREE," with a registration number stamped on it. Mixed-race former slaves became so numerous that the government coined a name for them: "free people of color," or "f.p.c.'s."

Kate's children were not lucky enough to have even received that status. By the time they came along, their father could no longer free them. In 1800, the legislature made it illegal for a slave owner to manumit a slave (that is, give freedom to him or her) without court approval. In 1820, the ban was tightened to require the consent of the state legislature itself; and in 1841, deathbed wills that granted manumission were voided. No records show that William ever approached lawmakers to request f.p.c. status for his children, but neither did most other white fathers of mixed-race offspring. They feared public ridicule, which was the real purpose of the law in the first place.

In the late 1850s, William Harleston decided to return to The Hut, the plantation where he had been born, and he took Kate and the children with him. Perhaps his brother John had grown tired of bumping into dark nieces and nephews and asked them all to leave. The Hut stood deep in the swamp, behind several walls of trees, and was more private than Elwood. The isolation brought fewer white visitors, which proved a blessing. William and Kate may have finally been able to share a house.

The move may also have been an attempt by the couple to go into hiding. In the South during the 1850s, anti-Northern sentiment ran high, and some whites aimed their anger at mulattoes, whose frequent literacy and ability to gain privileges looked like a challenge to slavery. The census counted nearly ten thousand f.p.c.'s in South Carolina, who differed from enslaved blacks in an important way: they could own property and accumulate wealth. As a result, f.p.c.'s were often prosperous and competed for jobs with working-class whites. At various times, the South Carolina legislature considered bills that would expel them from the state, because any advantages among nonwhites were hard to tolerate. Though Kate and the children were nominal slaves, they thought of themselves as part of the community of free people of color. The public talk of "getting even with the coloreds" made things untenable. There had been reports of mob attacks on white-black couples that had previously lived openly together, and no one could say where it would end.

As it turned out, William and Kate retreated to The Hut just in time, because a crackdown came in 1860. That summer, the governments in the Southern states contemplated what they might do if Abraham Lincoln, a known opponent of slavery, were elected president. The path taken by officials in South Carolina was not to await the presidential election in November, but to strengthen attacks on mulattoes as soon as possible, because f.p.c.'s were the enemy within. In August 1860, eight months before the start of the Civil War, with the consent of the mayor, Charleston marshals began going door-to-door to the houses of mixed-race families demanding evidence of their status. Some f.p.c.'s had kept their freedom papers current, paying a special annual tax on themselves and safeguarding an actual manumission document that proved their status. But others had lost their papers, or the "free" badge, or had let the tax slide. Those who could not prove they were free were arrested and sold into slavery.

William Harleston was frightened for his family, because he knew interracial couples could be targeted next. Joining with his brother John, William circulated a petition and sent it up to the state legislature, demanding that the police leave the decent mulattoes alone:

> The petition of the undersigned citizens of Charleston respectfully showeth that they have seen with regret in the papers the draught of a bill proposing to drive our free coloured people from

the State under the heaviest penalties. Your petitioners can find no reason for such severity; on the contrary, they believe the project one of wrong and injustice to a class of our inhabitants who ought to be objects of our care and protection.

[signed]
John Harleston
William Harleston
et al.

Nearly a hundred men, most of them slaveholders and friends of mulatto citizens, signed the petition. But it was too late for nearly a thousand people of color, who fled Charleston for the North or were sold into slavery.

In 1861, the Civil War intervened. Both John, fifty-five, and William, fifty-seven, were too old to join the rebel armies. Further, men who owned more than twenty slaves (as the Harlestons did) could be exempt from service. But as loyal Southerners, the brothers enlisted in a home guard company called the Etiwan Rangers. Named after a Native American tribe that once lived in the vicinity, the Etiwan Rangers was the service of choice for old men and big shots who wanted out of the line of fire. There is no evidence about William's views of the war. Although he had a large African American family, it would have been impossible for him to back the Union cause without being branded a traitor. And the possibility is real that in the war fever, stronger in South Carolina than anywhere else, William may have supported the rebel cause, believing it was separate from the status of his family. In either case, circumstances protected him from having to take a dangerous public position.

Life for Kate and her children must have been worse during the war years than before. The family had to hide to an extent and rarely leave The Hut, at the risk of encountering a rebel zealot who disliked mulattoes. Things worsened when, soon after the fighting began, the South Carolina legislature took up a bill that would have required the capture and sale into slavery of all free people of color after January 1, 1862. Interestingly, a cousin of William Harleston's by marriage, John Harleston Read, was at that time chair of the legislature's Committee on the Colored Population. Read opposed the bill, arguing that f.p.c.'s were "good citizens, and patterns of industry, sobriety and irreproachable conduct." Read told the *Baltimore Sun,* "Whilst we are battling for our rights, liberties, and institutions, can we expect the smiles and countenance of the Arbiter of all events when we make war upon the impotent and unprotected, enslave them against all justice?" When the bill was defeated, Kate must have been relieved.

In the Etiwan Rangers, both John and William had the lowest rank, private, and little was expected of them. The occasional drills resembled get-togethers, and now and then they gave assurances their guns at home were cleaned and stocked with shot. But for four years, as hundreds of thousands died on distant battlefields, William and Kate were relatively safe. In February 1865, Union troops arrived in the area, bringing an end to Kate and William's impossible situation. At last, after the truce, William no longer held legal title to his family.

At the end of the war, William was sixty-one, and Kate was thirty-nine. The couple had their last child two years after the Civil War, in 1867. The children ranged in age from infancy to their early twenties, though most were still minors.

When the war ended, thousands of freed slaves left the plantations of their long imprisonment and moved to nearby Charleston. The Civil Rights Act of 1866 and the Reconstruction Acts of 1867 offered freed people new rights and services: black men could vote, and within a short time hundreds of nonwhites took office, many of them of mixed race. A greater promise of freedom was education. In 1868, the state of South Carolina drafted a new constitution that guaranteed free schools for all. Kate Wilson had probably had no schooling, and she may have wanted it for her children, but nothing was available to them, because despite the law, schools were not being built in rural areas. While the Harleston family continued to live at The Hut, far from city life, several schools for nonwhites were opening in Charleston. The first was the Avery Institute, founded just after the end of the truce, in 1865; named for a minister from Pennsylvania, Charles Avery, who gave money for the building, the Avery Institute quickly became the school of choice for the nonwhite elite. The white Episcopal Church started another. An Episcopal clergyman, Rev. Anthony Toomer Porter, opened a school for black students, using contributions from the North; the teachers were white Charleston women. Other schools were Shaw Memorial Institute (named for Robert Gould Shaw, a white Civil War colonel who had died with most of his black company in an assault on the outskirts of Charleston) and the Morris Street School, on a road of that name through a black neighborhood.

Kate Wilson may have urged her "husband" to move the family to town, so the children could take advantage of one of the new schools. But there was another reason for them to leave the countryside: the growing menace of the Ku Klux Klan. The Ku Klux Klan first took root in Tennessee in 1866, a year after the Confederate surrender. In 1868, factions of the Klan wrapped themselves in sheets and began riding at night, dispensing beatings and torture to blacks who tested their new rights. Klan violence was at its worst in South Carolina after an election in 1870, when masked riders went out almost every night for several months. The districts near Charleston were relatively peaceful, though inland counties a hundred miles away ran with the blood of Klan victims. Many attacks fell on (frequently mulatto) Republicans and on white men who supported black rights. In a typical incident in 1870, according to court records, Klan members set upon the home of a "respectable mulatto" and member of the Republican Party named Amzi Rainey, whipping him, raping his daughter, beating his wife, and forcing Rainey to run for his life while gunmen fired after him. During a twelve-month period, eleven black citizens were killed and six hundred whipped by South Carolina Klansmen.

Whether William and Kate wanted education for their children or were fleeing the prospect of Klan terror, in May 1870, William and his "wife" went into Charleston to buy a house. The place they chose stood at 28 Laurel Street, not far from the Morris Street School. William bought the house from a man named Thomas Nestor, for $1,050. It was a simple two-story wooden structure on a block lined with similar single-family houses. It had four rooms, plus another building in the back that contained two rooms, on a lot measuring thirty feet by one hundred feet. (Both the street and number have since changed: the same house is now 50 Ashe Street.) In addition to being closer to the schools, Laurel Street had the advantage of putting Kate near her family. At some point, Kate's son Richard had moved four doors down, to number twenty. And the block was full of people of Kate's kind: nearly half the families were mixed-race people who had been free before the Civil War.

Charleston, with its population of forty-nine thousand, was 46 percent white, 54 percent nonwhite. But nonwhite society in the city was split along old lines. At the top were people like Kate and her children: those with connections to white families, with lighter skin, and with homes they owned themselves. This "colored elite" numbered about four thousand. Below them lay the majority of blacks, some twenty-one thousand, who were darker, destitute, and without education. In the city, Kate and her children befriended other families that shared their identity, but kept their distance from the mass of Negroes. A cultural gap magnified the gap in class status. Poorer blacks attended Baptist and Methodist churches, held down servants' jobs, and mocked their richer brethren for their skin color and airs. Richer families took pride in their light skin, read books for pleasure, and went to Congregational, Presbyterian, or Episcopal churches.

Although Kate Wilson and her younger children moved from The Hut to their new home, it seems William stayed behind and began to divide his time between the farmland and Charleston. It's not clear how he stayed with Kate when he was in town, because despite the new freedoms, a white man could not live openly with a colored woman. If William had gone to the trouble of buying his family a house, it stands to reason he continued to support them with infusions of money. But Kate may also have gotten work in Charleston, as a dressmaker or a baker, since none of the colored elite were truly wealthy, especially after the devastation of the war.

Racial identities flourished after the Civil War. Black people coined a batch of names for whites. They were "ofay," from pig latin for "foe." Whites were also "daps," "pinktails," "Celts," and "buckra." The last term was said to be an Americanization of a West African word.

Members of the colored elite were called "high yellow" for their shade of skin. They were also "dicty," "hincty," and "arnchy"—all adjectives meaning snobbish. "Muckety-mucks" were people who had arrived in the class, "strivers" wanted to arrive. Perhaps the weirdest coinage was "mariny," meaning of questionable race, a word that seems to come from the same root as "marine" and "marinated." A mariny person could go both ways, passing as either white or black.

In 1872, the South Carolina legislature surprised many by passing a law that recognized interracial relationships. The law said that the "children of white fathers and Negro mothers may inherit from the father if he did not marry another woman, but continued to live with their mother." Because such couples could not marry (the ostracism was too fierce), it was a backhanded acknowledgment. But it was relevant and meaningful to hundreds of families.

The following year, on July 17, 1873, William wrote out his will. The short document publicly claimed his family and dealt carefully with their needs:

> I, William Harleston . . . desire that all my debts shall be paid. I give to the colored woman Kate (formerly my slave) my bed and all my bedding.
>
> The rest and residue of my estate which I now or may hereafter possess I give and bequeath to my brother John Harleston and to the colored woman Kate (formerly my slave) to be equally divided between them, share and share alike, and for their sole and separate use forever.

The will singled out the "marital" bed. If the bed followed local style, it was probably a four-poster with a canopy and layers of feathering on the mattress. (The bed has since been lost.) William then divided his assets between Kate and his brother John; and with his signature, that was all. No apologies, nothing hidden. It was almost unheard-of for a white man to leave even a small inheritance to a nonwhite person.

About a year later, at age seventy, William Harleston died. Neither the date nor the cause of death is known—for some reason, his death certificate has disappeared from state records—but his age was considered advanced. He was buried in a graveyard near Charleston, close to other members of the white Harleston family. Little is known about the funeral. If Kate and the children attended, the white Harlestons probably stayed away, except his brother John. John Harleston had seen William grow old with Kate, and he would have paid her the respect of his presence.

According to an inventory, at his death, William Harleston was practically a poor man compared with the large assets his family had possessed before the Civil War. He had only $2,000 in cash, $1,000 in railroad bonds, and $2,210 in stock. The war had shrunk his riches, as it had those of his peers. For most of William's life, human property made up nearly half of the family capital. When freedom came, half of the fortunes of the rich vanished. Land values had also fallen since the war, to a quarter of their previous appraisal, leaving most families with a fraction of their former wealth in acreage. When he died, William owned no land, having evidently transferred The Hut to the hands of his white relatives (while maintaining his home there) to pay for his and Kate's growing expenses. Still, the estate he left was large enough to support a family in style for years.

When the will was proved, in January 1875, William's inclusion of Kate in his inheritance stunned his relatives. The white Harlestons had known all along about the sad case of their cousin, whose illicit lover disgraced their name. But they never dreamed he would humiliate them from the grave by blackening the family money.

William had squandered the fame of his forebears, heroes of the Revolution, and hurt the Harleston reputation. The family had tried to overlook his offensive behavior. Now, they hated him. With William out of reach, rejection from memory was the only reproach available.

Some years later, a local journal published a genealogy of the white Harleston family. The essay contained descriptions of family achievements and praise for Harleston connections, but the terse entry on William Harleston revealed his status as "black sheep." William's name warranted no birth date, no childhood stories—

only this brief reference to the relative who had settled down with a slave:

"William Harleston. Died at 'The Hut,' in 1874, unmarried."

The news of the inheritance must have struck like a storm in the house on Laurel Street, filling Kate with wonder. Only ten years earlier, the colored Harlestons were slaves, and now they were heirs to upper-class whites. When William died, half of Kate's eight children—John, Richard, Robert, and Hannah—were grown, with work and married lives of their own. The money would improve their situation, but William's bequest would especially benefit the younger four—Edwin, Sylvia, Sarah, and Susan. The lucky Harleston children would be protected by their father's money and his famous name. Or at least, this is what they wanted very much to believe.

For the white Harlestons, however, things had not gone well in the ten years since the Civil War. Several family members had been killed fighting for the Confederate cause; others had lost their plantations to taxes and economic decline. For those whose land still remained intact, the new system of sharecropping, which meant dividing profits with black workers, was proving far less lucrative than slavery had been.

The white family's world, in many ways, had been wheeled around backward. Charleston, wrecked by wartime artillery and fires, stood half in ruins, a city in poverty and decay. Reconstruction had been an experiment with black leadership, a disgusting idea to educated whites. While Negroes and mulattoes exercised power in state government, their authority was enforced by the presence of Yankee soldiers, who had never left the state. Under the heel of defeat, the white Harlestons and their peers longed for a return of the old days.

When William Harleston died, many knew Kate Wilson to have been his life companion. The ex-concubine and her children were a stain on the white family's honor, and the news that Kate would actually have a bit of money came as a further insult. Not only would there be a house of colored cousins, they would be living almost comfortably.

William's three sisters were dead, leaving John as the only surviving sibling and everything in his hands. Soon after William's funeral, John, executor of the will, began to pay Kate support from the estate. By law, the distribution of the inheritance had to wait until an inventory and appraisal. Months went by, then a year. For whatever reason, the probate court would not give the order to distribute; so every few months, John approved small outlays for Kate's expenses. March 19, 1875: $13.35. May 3: $24.17 "in groceries." August 9: $15 cash.

According to family memory, the white Harlestons had not been idle during the long delay. They had no intention of allowing Kate to take possession of her wealth, modest as it was. Though no papers survive that would prove it, tradition says that William Harleston's nephew, Benjamin Frost Huger, decided to "take care of" Kate and her money.

The son of William's sister, Sarah, Benjamin Huger was a polished and influential attorney. At thirty-eight, he had known his uncle William all through life. By his father, Huger had a French background: the name Huger was that of an old Huguenot family. (It had a local pronunciation, *Hew-gee,* which, though not quite French, was prettier than its guttural English appearance.) Benjamin Huger cut a wide swath in legal circles: in a few years, he would be appointed postmaster of Charleston. Huger must have liked his uncle, because around the time the old man died, the lawyer had a son, whom he named William. But he disliked intensely his black "aunt," Kate.

Benjamin Huger was a resentful man, and to him, the times looked awful. With his wife, the former Anna May Walker, he had five young children, whom he loved, but who presented a heavy burden. Though his legal practice was sound, the economy had been shrinking for years, and fees were down. What's more, the spread of black rights had been galling, so much so that it was not unusual to see black men, many of them ex-slaves, walking around town in top hats, putting on airs and enjoying their new freedom. Huger could not stand these displays, which he considered part of the harvest of emancipation.

The story goes that one day in the winter of 1875, Benjamin Huger asked Robert Harleston, Kate's third son, if he would take

Huger to visit Robert's mother. Huger may have gone at the urging of his Harleston relatives, or perhaps he acted alone. Robert agreed (though if he had known Huger's plan he probably would have refused), and the two went to see Kate. Huger had his eye on Kate's money and knew she was vulnerable. Her world had changed, but she had not: she couldn't read and she had lived much of her life dependent on the behavior of whites. Huger, the smooth attorney, persuaded Kate that with her limited financial skills, she would be incapable of handling her estate. He presented her with documents—a power of attorney, statements of gift—that would take her worries away. Huger, worldly and expert with accounts, would take care of things. There may have been blandishments, threats. "William would have wanted it this way." Kate signed Huger's documents with an "X."

Kate immediately realized her mistake. She turned to her only white friend, John Harleston, bringing him copies of the documents she had signed. Kate described Huger's visit, and John could not believe what had happened. She begged him to undo the damage. John examined Huger's handiwork, and was crestfallen. According to family tradition, John turned to Kate and said, "It's too late. The cloth has been cut."

The inventory was finally ready. On August 12, 1875, the court issued a lump-sum payment of Kate's whole inheritance. Kate's descendants say Huger took the money for himself.

The word "miscegenation," from the Latin *miscere* (mix) and *genus* (race), came into American English during the Civil War. It was coined in an anonymous pamphlet, "Miscegenation: The Theory of the Blending of the Races," published during the election in 1864. The pamphlet had been written by Democratic Party ideologues and suggested that Republican candidates favored interracial sex. Although it was hoped this would help the Democrats, the Republican, Abraham Lincoln, was reelected.

Interracial sex was said to be a violation of both natural and divine law, as it produced a "mixed" people previously not seen on earth and also unsanctioned by God. Whites were to have sex with whites, and blacks were to have sex with other blacks. Otherwise, there would be a risk of decaying the species. The most frequent users of the term "miscegenation" came from the ranks of white supremacists. Initially, they argued for the preservation of slavery; when that failed, they turned their sizable energies to promoting the separation of blacks from whites.

Until 1879, no law on the books in South Carolina banned interracial sex or interracial marriage. But at the end of Reconstruction, the movement to divide the races gained strength. White legislators chose this moment to pass such a bill. The resulting law threw a wide net: "It shall be unlawful for any white man to intermarry with any woman of either the Indian or negro races, or any mulatto, mestizo or half-breed, or for any white woman to intermarry with any person other than a white man, or for any mulatto, half-breed, Indian, Negro, or mestizo to intermarry with a white woman; and any such marriage, or attempted marriage, shall be utterly null and void and of no effect." Punishment consisted of a fine (five hundred dollars) or imprisonment for at least a year.

The frequency of black and white sex had in fact declined after the Civil War. Black women found it easier to refuse white men, and the men came under growing social pressure to leave them alone. But Kate Harleston may have at least smiled on hearing of the ban against what she had been through.

The legislation took aim once again at the mixed-race Harlestons. Kate and her family were members of a special social class, the offspring of former slave owners and former slaves. Their light skin gave them status, but it was an uncomfortable cachet, like wearing broken jewelry. Their pale skin carried the memory of the black and white sex that brought them into the world and marked them forever.

When John Harleston saw what Benjamin Huger had done, he decided to retaliate. He would use the only weapon available to him: embarrassment. Years before, John had written his own will, which had designated a piece of his estate—five times larger than his brother's—to his nephew. In February 1875, John went to a lawyer and added a codicil to his will, writing Benjamin Huger out of his bequest: "I will and direct that the share bequeathed to my nephew Benjamin F. Huger shall not be paid to him. . . . In witness whereof I, John Harleston, have hereunto set my hand."

John's symbolic retort to his nephew (he left the money to a niece instead) was the last thing he would do for his brother and Kate. He died two years later, on August 26, 1877, at seventy-one.

Benjamin Huger's deft removal of money from Kate's hands came at a time when whites in Southern society were reasserting their power over blacks. The election of 1876 brought the overthrow of "black Reconstruction" and the withdrawal of federal troops that enforced it. In South Carolina, a former Confederate general named Wade Hampton took the governor's seat and appointed other rebel veterans as his aides. With the occupying army gone, the process of ejecting blacks from the voting rolls began. In 1876 ninety-one thousand black Carolinians voted; in 1888 only fourteen thousand did. It was the beginning of a decline for black citizens that would later give way to worse.

Kate and her children had been turned out of their white family, and the ranks had closed behind them.

There was a single exception to the complete loss of Kate's property: the house at 28 Laurel Street. William had left the house in trust to his eight children, "to permit and suffer" them "to use and occupy and enjoy the said premises during their natural lives."

Within a few years, Kate became seriously ill. According to the census of 1880, she was an "invalid" and unable to leave home. In 1886, twelve years after the death of her common-law husband, Kate died in her bed at 28 Laurel Street. She was sixty years old. The cause of her death, on November 14, was given by the attending physician as "pulmonary edema," with a secondary cause being "paralysis." She had possibly suffered a stroke. The death certificate gave her name as "Kate Harleston" and described her as a "widow."

The Harleston children gathered to console one another. Most of them had long since grown and had children of their own. Their mother had done well for herself, people said. She was born a slave and was illiterate her whole life—but by Southern standards, she had risen rather high. For the funeral, the children followed their mother's coffin in the procession to the graveyard. There was no chance she would be buried next to William, even if this had been her wish. Mulattoes had separate burial grounds from whites and also from darker Negroes. The body was placed in a cemetery just north of Charleston, among the graves of other colored elite families, even though the Harlestons were now destitute.

Years later, Kate's children would place a large granite marker over her grave. The stone would show her name, Kate Wilson Harleston, and the date of her death, but nothing else. It would make no mention of her "husband," William, or of his family. About their white roots, the Harleston children could be silent as the tomb.

Part II

~

High Yellow

Chapter Three

EDWIN Gaillard Harleston was born on December 23, 1854. His middle name (which came from a white family related to his father) was pronounced *Gill-yard*. The fifth child of Kate and William, Edwin spent his first years at Elwood plantation. As a boy, he moved with his parents to The Hut plantation, where he grew up along the tidal creeks of the Carolina swamp.

According to a relative who knew him, Edwin had piercing eyes. He often narrowed his gaze to show he was not to be trifled with. He exuded confidence, was difficult to get along with, and was quite controlling.

In 1870, when Kate moved from the countryside to Charleston, young Edwin, then fifteen, probably stayed behind with his father. A note in old family files states that when his father died, Edwin became a small-scale rice planter, at age nineteen. It's not likely he would have been able to do this except after a period of apprenticeship with William Harleston on plantation land. For four years, between 1874 and 1878, he cultivated the staple crop with paid black laborers. The land probably came from William (a few acres of The Hut), as did the workers (ex-slaves to the Harleston family).

Family tradition says that in 1875, when Kate's inheritance vanished, which effectively kicked support out from under him, Edwin hardened. Betrayed by circumstance, and then by a white man, he became wounded. While still a child, Edwin had seen the wealth of his Harleston cousins, but he had come up with nothing. The boy grew into a handsome, proud, and driven man. In his mind, Edwin could see the dim outline of the old Harleston family dynasty. He wanted to get his hands on it—or, knowing that to be impossible, something like it—and he was willing to do all he needed to get there.

Edwin's first move (in 1877, when he was twenty-two) was to make an advantageous marriage. He courted and proposed to a prosperous nineteen-year-old named Louisa Moultrie. With his white roots and fair skin, Edwin already stood near the top of colored society, but there was one group above him: the former free people of color. Louisa Moultrie, whom Edwin called Lassie, came from a light-skinned family that had been free for more than fifty years before the Emancipation Proclamation. (Lassie's great-grandmother Lucy Wilkinson had bought herself out of slavery in 1804 and then purchased freedom for her daughter Flora.) After the defeat of the South, the f.p.c. label no longer had legal meaning; but because such families had been able to save money and run businesses before anyone else, they knew about commerce and were the richest of black Americans. Louisa Moultrie may have brought money to the marriage. Edwin Harleston married Lassie in April 1877, and after Edwin gave up rice farming, they settled in Charleston. The marriage shows Edwin to have been a social climber: Lassie's family would widen his contacts and help him to get on with making money.

Edwin looked around for other ways to earn a living and eventually went to work on a cargo boat. In those days, fleets of cargo schooners plied the Atlantic coast, running goods from Charleston, south to Florida, up to North Carolina, and back. In 1881, four years after Edwin's marriage, a Charleston city directory listed Edwin as a "boatman" and living at 20 Laurel Street, with his brother Richard. Edwin was reliable as a sailor, and he knew how to talk to white people. But a deckhand could only earn a lowly wage.

Not long after the boatman listing, Edwin went into the cargo business as a sole proprietor, perhaps using money from his wife.

He bought a schooner called the *Dorothea,* took the helm, and hired himself out. The *Dorothea* carried rice and cotton two hundred miles north, along the shore of North Carolina, and two hundred miles south, down the Georgia coast. Edwin's directory listings evolved: from "boatman," to "seaman," he became, finally, "captain of sloop." He bought increasingly large boats, eventually taking possession of a twenty-seven-foot sloop called the *Beulah Benton*. People started calling him "Captain" Harleston, and the name stuck.

Ninety percent of Afro-Americans were propertyless farmhands and still worked in gangs, but Captain Harleston's schooner business made him into an entrepreneur. As the business grew, Edwin became a swashbuckler on the waves. By the late 1880s, he was comfortable at home, having moved into his mother's old house at 28 Laurel Street. He had a little money, an attentive wife, and some status.

Edwin stayed in the cargo business for nearly twenty years. But one night in the late 1890s, the *Beulah Benton* collided with a steamer, doing damage to the steamer and frightening its passengers. Edwin wasn't on board at the time, having stayed on shore following a bout with malaria. A negligence trial ensued in which he was named as a defendant. The hired skipper who had been piloting Captain Harleston's boat offered conflicting testimony, with the result that Edwin lost both the case and the *Beulah Benton*. Thrown out of the shipping trade, stripped of his property for the second time in his adult years, Edwin must have been frustrated and furious. Scrambling for a new way to make a living, at the end of the 1890s, he opened a fruit and vegetable stand, where he bided his time until he could work up a better business. "Captain Harleston, the vegetable seller" might have sounded like an incongruous sales pitch, but Edwin liked it. Though he had given up the sea, the name Captain would stay with him for the rest of his life.

The years after Reconstruction gave way to a slow-moving war between whites and blacks. A fragile truce between the races that had come out of the Northern victory vanished. In 1880, South Carolina had the highest proportion of black residents in the nation: nearly 61 percent, and added to them were the mulattoes. Each side maneuvered for advantage, with whites attempting to "corner the coloreds" into submission. The return of white power to the government meant this might be possible, but even the most anxious white supremacists knew it would take time.

It was an incongruous period. State appropriations for education were abysmally low, but about equal for whites and blacks. Black men could still vote, and South Carolina elected nonwhite representatives to the U.S. Congress. But in the 1880s, the Democratic Party—at the time, the Democrats were the party of white conservatives—used violence and money to control the turnout. Ballot boxes were stuffed as a way of driving out the Negro politicians. Black candidates for office often came home to find their houses on fire.

The strange career of Jim Crow had begun. Jim Crow is sometimes understood to mean the disconnection of the races in schools and on public transportation, but its deeper meaning was the separation of nonwhites from good jobs and their removal to special neighborhoods. It also meant required demonstrations of subservience by all colored people to all whites. The name Jim Crow is said to have come from the traveling variety shows, or minstrel acts, of the mid-1800s. Two black character types tended to appear in minstrel theater: Jim Crow and Jim Dandy. Jim Crow was portrayed as crude, poorly educated, and rustic, whereas actors played Jim Dandy as an effeminate and urbane mixed-race Negro. The customs and later the laws of race discrimination came to be known as Jim Crow because they pushed all blacks into the lowest possible status. In time, the new order would be politely termed "segregation."

Lynching had also begun in earnest, as a way of backing up white authority. According to Tuskegee Institute, the historically black college in Alabama, which kept count, thirty-six lynchings occurred in South Carolina in the 1880s, all of them involving white mobs and black victims.

Captain Harleston had seven brothers and sisters: John, Richard, Robert, Hannah, Sylvia, Sarah, and Susan. Though each was comfortable, none did as well in coping with the shifting sands of Jim Crow as did he.

The eldest, John Harleston, had worked as a butler, but died at age thirty-nine in 1883, of asthma.

The next in line, Richard, became a house painter in Charleston. In 1865, at twenty, he married a nineteen-year-old woman named Agnes Graham. Family memory is divided about Agnes Graham. Some say that she was a servant from Europe, and white, while others say that she was a free person of color. The "color" in Agnes's status, if there was any, seems theoretical: she was so light as to be able to pass for white, although she lived with Richard as a black woman. Richard and Agnes settled in Agnes's hometown of Red Bank, South Carolina, where they had five daughters. The last two girls were called Kate and Hannah. Agnes Harleston died in 1899, when her youngest was still a toddler, leaving Richard to raise the girls alone. Kate and Hannah Harleston grew up to marry two brothers, John and George Shokes, from a small fishing village called McClellanville. The story goes that Kate and George were seeing each other first; then George met Hannah, and he married her instead. After that, Kate married George's brother John. Kate apparently never spoke to her former boyfriend again, even though they lived around the corner from each other for much of their lives. The village of McClellanville stood on the coast, and the sisters moved to their husbands' hometown, bringing their father, Richard Harleston, with them. The Shokes branch of the family became keepers of Richard's legacy, and many progeny of Kate and William were born at McClellanville.

Robert Harleston, the third son, apprenticed to become a tailor. Because no nonwhites were allowed to rise into the professions, to be a tailor put one firmly in the colored elite. (The mixed-race middle class was in this way unlike its white counterpart: most of its members had craftsmen's jobs, such as tailor, dressmaker, carpenter, or bricklayer.) Robert Harleston's shop stood at 265 Meeting Street, near the edge of the downtown business district. Robert married a woman named Frances Mickey, who lived three houses from his business, after which the couple settled on Laurel Street, next door to the old family homestead, and had two sons.

The youngest of Kate's children, Susan, was born in 1867. At twenty-two, she married a ship carpenter named John Singleton, and moved to his hometown, the port town of Beaufort, South Carolina, a daylong steamboat ride to the south. Susan was known in the family as Sister Sue, an endearment shrunk down by repetition to Sis' Sue. Sister Sue, according to one of her nieces, was about five feet four inches and light-skinned. She had beautiful black hair, which she wore in a bun on the back of her head. Sister Sue Harleston and John Singleton had no children of their own, and as a result, Susan often came up from Beaufort to Charleston to see her family and entertain her siblings' young ones.

Sylvia, born in 1861, was Kate's second daughter. As a young girl, Sylvia attended Plymouth Congregational Church, one of a handful of churches where the colored middle class went to worship. Most of the Harleston family also began attending Plymouth Congregational, but Sylvia must have liked the order of service more than her brothers and sisters, because in the 1880s, when she was in her twenties, she married the organist. His name was Merton Lawrence, one of the most capable choirmasters to walk the streets. Sylvia and her husband moved to a block near Captain Harleston, where they had nine children. Their teeming house became something of a musical colony, with the parents as well as hired tutors training and coaching the legions of young Lawrence children in their scales and vocal exercises.

Sarah, the seventh of Kate's eight, born in 1865, married a man named Joseph Grant and had two sons and two daughters.

The final sibling, Hannah Harleston, was born in 1848. Like the others, Hannah was raised first on the farm. At twenty-nine, she married a twenty-seven-year-old man named Edward H. Mickey. Edward Mickey was already part of the family: he was the brother of Frances Mickey, the wife of Hannah's brother Robert. The Mickeys, having been free before the Civil War, were from an even higher colored elite than the Harlestons. Edward and Frances were the grandchildren of a white slave owner, Morris Mickey, and his consort, Ellen, a free woman of color. The patriarch of the Mickey family, Edward Mickey Sr., was a fifty-six-year-old tailor and a minister in the African Methodist Episcopal Church who had gone into politics during Reconstruction, serving in the South Carolina legislature between 1868 and 1872. According to bank records, the

Mickey family lived comfortably at 259 Meeting Street and had a well-tended savings account.

Of the eight Harleston siblings, Hannah had chosen a spouse whose influence would overshadow the whole family. Edward H. Mickey had followed his father into the ministry and then pioneered a profitable new service for black families: undertaking. As a sideline to preaching over the dead, the younger Reverend Mr. Mickey began to handle the actual bodies of the faithful when they passed away. Previously, black families had buried their own; now, with their growing wallets, a few had begun to hire out the job. The Reverend Mr. Mickey cleaned and dressed the deceased, placed the body in a crude coffin, and brought it to the service. He was helped by the reality that white coffin makers would not touch a nonwhite body. And naturally, the pastor charged a fee.

Hannah and Edward Mickey lived on a nice block with other comfortable families. But after the Reverend Mr. Mickey began his funeral business, Hannah had to adjust to a frequent surprise. Though she always had money to shop for groceries, very often when Hannah returned, she found a dead body waiting for her in the kitchen.

In 1890, the Bureau of the Census decided to distinguish more than in any previous count among the various classes of "colored." According to state census data for that year (the federal census was lost in a fire), the United States population numbered 62,117,000, which included approximately 6,338,000 "blacks," 957,000 "mulattoes," 105,000 "quadroons," and 70,000 "octoroons." Each citizen had been classed by an enumerator—usually a twentyish white man—and not by self-description. Around the South, the fresh young head counters squinted at the faces of homemakers who answered the door, trying to guess whether they had one-fourth or one-eighth "black blood." Occasionally, the census manuscripts show a "W" that has been scratched out (~~W~~) and replaced with a "Q" or an "O," evidently after a mixed-race respondent politely corrected her visitor.

The census takers were not the only ones who studied complexion, as nonwhites developed a whole vocabulary of skin tones. In colored society, a person was not merely "black." Skin could be "chestnut," "ebony," "cream," "chocolate," "coffee," "yellow," "blue-black," or "redbone."

White people had developed strong views about the differences between darker Negroes and people with both black and white roots.

Black people, it was said, were dirty and careless with their clothing, but mulattoes were prim and neatly dressed.

Blacks were loud and boisterous, while lighter people were effete and quiet.

Black people had no sense of time and were always late. Mixed-race people were punctual.

Blacks were religious, but their religion was all emotion. The colored elite worshiped thoughtfully, mindful of the nuances of faith.

Blacks had an uncontrolled sex drive, and slept with any willing partner. Mulattoes stayed with their own kind: they "married light, and never dark."

The Harlestons often reflected the stereotypes. They indeed became scrupulous with their clothes. They attended a distinctly quiet church. They spoke softly and laughed lightly. Overall, the Harleston family knew themselves to be special. They were founding members of an original class—a new black bourgeoisie.

The rise of Jim Crow coincided with the growth of the mixed-race middle class. In Mississippi in 1890, state election officials designed a scheme to take the vote away from nonwhites once and for all. The method combined a poll tax and a verbal examination for voters, the "understand and explain" test. Applicants to vote first had to pay a tax. Then, a white registrar would read aloud from a section of the voting law and ask the applicant to discuss it. The registrars had discretion to decide which applicants had been able to "understand and explain" the law. According to the script, registrars would accept the answers of illiterate whites (a large percentage of the applicants) but fail to understand even the most educated Negroes. The scheme worked, and within a few years, most black voters were stripped from the rolls.

In 1894, Benjamin "Pitchfork Ben" Tillman, the governor of South Carolina, won support for a new constitutional convention that would recast the state charter, which had been written largely by mixed-race politicians thirty years before. In his inaugural speech as governor, Tillman had denounced the "niggerdom of Beaufort" (where many colored politicians lived) and called his own election a "triumph of democracy and white supremacy over mongrelism and anarchy, of civilization over barbarism." The purpose of the constitutional convention was at some level to ratify these sentiments.

The caucus took place, and the resulting document applied both a poll tax and an "understand and explain" test, following the Mississippi model. But in addition, buried deep in the text at article III, section 33, there was this strike at interracial families: "The marriage of a white person with a negro or mulatto, or person who shall have one-eighth or more negro blood, shall be unlawful and void." The provision restated the already existing law banning interracial marriage and made it a constitutional offense. In South Carolina, the fight against miscegenation had grown more vehement.

While the draft constitution was being debated, one of the few nonwhite legislators remaining in the state assembly, Robert Smalls of Beaufort, created a quandary for white lawmakers. Believing the anti-mulatto provision would pass, Smalls proposed an amendment that would bar from public office any white man found to be consorting with a black woman and provide that the children from the relationship had the right to their father's name as well as his property. The amendment failed.

The constitution of 1895, passed by a white electorate, solidified the rule of Jim Crow. The new law required separate schools for whites and blacks. Various instruments barred all but a few nonwhites from public office and from voting itself. Similar changes swept through every state in the South. In 1896, the U.S. Supreme Court, in *Plessy v. Ferguson,* affirmed the change and ratified the new age. That case upheld the right of Louisiana to require different public amenities for different "races." Apartheid had come to America, as laws of separation spread from marriage to railroad cars, restaurants, and hotels.

* * *

The week before Christmas, in December 1896, Captain Harleston's wife, Louisa, went into labor. The baby was stillborn, and after the delivery, Lassie remained ill. During the pregnancy, she had developed a kidney disease known as albuminuria, a condition that poses a risk to life for new mothers. Medical care for Southerners was famously inadequate, and on February 28, 1897, she died at age thirty-nine.

Lassie's sudden exit left her husband, at age forty-two, with five children. At that time, men did not nurture children: fathers either remarried immediately or called on female relatives to help them. Captain Harleston did the latter. He asked his sister in the town of Beaufort to come to his aid: this was Susan, or Sister Sue. Thirty years old and childless, Sister Sue was married to ship carpenter John Singleton. When her brother appealed to her for help, Sister Sue moved back to Charleston, leaving her husband in Beaufort, to become a substitute mother to her nieces and nephews. Now the house at 28 Laurel Street was a somewhat unusual home, with a brother and sister at the head of a sizable family.

Widowed, and anxious to get out of selling vegetables, Captain Harleston looked around for other businesses he might try. As it happened, his sister Hannah had entrée to a good one. Hannah and Edward Mickey held a virtual monopoly on the funerals and burials of Charleston's colored elite. At Hannah's invitation, Captain Harleston began stopping in at the Mickey funeral office to see what he could learn.

The Captain set about to study the fine art of embalming. Then, from his brother-in-law, a minister, he also learned the importance of befriending the local preachers. Clergy were often the first ones to know about a death, and the Mickey family taught Captain Harleston that one had to flatter preachers, so they would reveal the names of the sick and dying. Captain Harleston made himself valuable to the Mickeys. After several months of apprenticeship, he felt he had learned enough and wanted to try one or two funerals himself.

Living next door, at number 30 Laurel Street, was the Captain's brother, Robert Harleston. Robert was the tailor married to Edward Mickey's sister Frances. Captain Harleston wanted to conduct a funeral without arousing the jealousy of the Mickey family and asked his brother for a hand. Though in his fifties and making a good living with his sewing machine, Robert was obliging and the two decided to try their first burials together. Captain Harleston could do the embalming, while Robert could sew a suit for a corpse before it went into the coffin.

In a brochure printed for customers, Captain Harleston advertised his experience: "After a business career as a rice planter on the Cooper River, and later for twenty years as a sea captain, the proprietor of the Harleston Funeral Establishment entered the undertaking business with his brother . . . as the firm of Harleston Brothers in 1901."

The business operated out of the two side-by-side houses on Laurel Street. It was cramped, and the brothers' methods were crude, but revenue started to come in. No sooner had the Harleston brothers set up shop, however, than their competitor, Edward Mickey, suddenly died. Mickey's widow, Hannah Harleston, was left with five children and a business to run.

For Captain Harleston, this particular death came at the right time, because it wasn't long before Hannah approached her brothers to help her continue with her husband's company. The funeral business, she thought, needed to have a male figurehead. The Harleston Brothers merged with the Mickey family operation. A year later, Captain Harleston bought out his brother Robert's share, and the undertaking business became solely the Captain and Hannah's domain.

The brother and sister moved the business into a rented location on Meeting Street, one of the main commercial roads. With his forceful personality pushing out ahead, Captain Harleston increasingly took charge of the company. He began dressing in three-piece suits and carried a wooden walking stick with a silver tip. He handled clients with expert dispatch. He exuded charisma but fell quiet when necessary, affecting a convincing sadness in the presence of widows. Within a short time, Hannah was merely a silent partner. Under pressure from her brother, she even agreed to change the name of the business: by 1903, it had become the Mickey and Harleston Funeral Home.

* * *

Edwin Harleston had finally found the instrument that could take back a little of his father's standing. He would build a dynasty of mortuaries, comparable to a dynasty in plantations, he thought. His children, once they grew up, would help him run the business. When he was gone, they would embalm him and then carry on his legacy.

Chapter Four

AFTER his wife, Louisa, died in 1897, Captain Harleston acquired a mistress by the name of Mamie Blanchard. Captain Harleston probably met her through dealings with her husband, because account books show she was married to a stable hand, Fred Blanchard, who worked with horses belonging to the Mickey and Harleston firm. Family tradition says the Captain took his afternoon meals with Mamie. About 2:00 P.M., he would put on his hat and make his way down the sidewalk, tapping out his step with the silver-tipped cane. After a toss on the mattress with his girlfriend, the Captain sometimes came back late to the office. Though the trysts were common knowledge, no one said anything, and the pretense of "a late supper" was maintained.

Mamie Blanchard, according to a family member, was short, had a happy face and a big laugh. She and her husband, Fred, lived in a wooden cottage without electricity. Mamie worked as a domestic servant, yet she managed to get off every afternoon to prepare meals for her lover, and her husband never seemed to come home during the day. The opinion has come down in the family that Fred did not really mind being a cuckold. He is said actually to have been proud that his wife had coupled with a rich, "high yellow" Negro like Captain Harleston. Fred judged the affair to be a mark of his own upward mobility.

Captain Harleston appreciated the favor of Fred's silence, and one year offered to show his gratitude. Inquiring as to the couple's needs, the Captain arranged to install electricity in their cottage. Afterward, Mamie and Fred invited neighbors for an inspection. All were impressed by the lightbulbs hanging naked from the ceiling, the exposed wires, and the big switches. Fred's silence had paid off. The Blanchard house became the first on the block with electric lights, and the only way Fred could have afforded such a luxury was by lending out his wife.

The Mickey and Harleston firm had no white clients, and white morticians had none who were black. When the industry's first professional group, the National Funeral Directors Association, was formed in 1882, nonwhite undertakers were barred from membership. Jim Crow customs made the Harlestons resident aliens in their own hometown, but segregation also helped profits. Captain Harleston and Hannah built their franchise without competition from whites.

At the start of the 1900s, black undertakers were strictly an urban phenomenon, because families in the rural black villages still prepared their dead without hired help. The majority of country people died at home, in their own beds. When news of a death spread, mourners immediately began to arrive at the house of the deceased. Women family members laid out the body, washed it, and wrapped it in a sheet. Neighbors and friends stayed up all night with the chief mourners, in a wake and grieving session, telling stories about the dead, singing, and praying. The next day, a male friend made a pine coffin, and interment took place swiftly, usually within twenty-four hours. (In the semitropical heat of South Carolina, waiting was out of the question.) Grave markers consisted of cypress planks.

A century had lapsed since the slave trade, and most West African death rituals had faded. But a few African customs survived in the remote plantation districts. One was the practice of placing "grave goods," mementos and keepsakes, on fresh burial plots. Many families piled a variety of objects and symbols on graves. Things that had belonged to the deceased came first: smoking pipes, clothing, and jewelry. Smashed goods also appeared, including broken dishes, and later, broken lamps, mirrors, and clocks. These had been ruined to signal the end of life and were often turned upside down on the ground. (The afterlife was said to be a mirror, or reverse, image of that of the living, hence the need for an inverted clock.) Another West African holdover was the custom of passing infant children over the body. During the funeral, adults would carry babies up to the coffin and hand them back and forth over the deceased. This was said to protect infants from future fear of the dead and to dispel any specters that might be lurking.

But when black Americans in the South moved from the farmland into cities, customs shifted. In Charleston, the colored elite were the first to discard slave funeral rites and bury "like white folks." Most Mickey and Harleston clients came from the mixed-race class and wanted none of the taint of the plantation. Instead of broken plates, grieving families placed flowers on graves, and they bought their coffins instead of making them. Because they had more money, mourners put up tombstones rather than wood boards. And, as whites had been doing for fifty years, they paid outside handlers, such as the firm of Mickey and Harleston, to prepare the body.

Despite wariness of old habits, mixed-race families held on to one custom handed down from the slave years: the "burial society." In West Africa, many cultures had developed special cliques to take care of grieving relatives and carry out burials. When a person in the clique died, all other members were required to attend the funeral (ensuring a good turnout) and to contribute to the deceased's family (donations that acted as a kind of life insurance). In America, black Southerners set up similar private clubs connected to death. The first appeared in Charleston in 1790 and was known as the Brown Fellowship Society. A men's guild, the society collected dues to pay for the funerals of its members. The guild also bought land and maintained a private cemetery; and when one of its fellows died, all the members came to the funeral. The wife of a member was entitled to the same treatment given to her husband. Members consisted only of free people of color, and not slaves: the name reminded everyone that these men were not black but mixed.

Similar groups formed in the 1800s—the Unity and Friendship Society, the Humane Brotherhood—until nearly every "respectable" colored male belonged to at least one. Rev. Edward Mickey Sr., Hannah Harleston's father-in-law, had been a founding member of the Humane Brotherhood.

The Mickey and Harleston firm made money from these cash-rich cliques, because each guild functioned like an insurance company, collecting premiums and paying a "death benefit" to the family. The payout ranged from fifty to two hundred dollars and went to settle debts. But mourners also used the money to finance rich funerals. A conspicuous burial was a family's way of showing its status, while saying a decent good-bye to the dead. A lot of families joined the most expensive burial society they could afford, and spent the payment on a big farewell.

When, a few days after the service, Captain Harleston and Hannah Mickey presented their bill, the burial society often paid them directly. Discussion of fees was almost unnecessary.

The word "undertaker" has a curious etymology. It appears in English about the year 1400, with the same meaning as "underwriter," as in a financial backer or provider of insurance. An undertaker supported people in an emergency. Eventually, the term was formalized to describe one who offers needed help with burials, and "undertakes" the funeral.

Funeral undertaking emerged as an American occupation in the early 1800s. Most often, a carpenter who made coffins began to sell burial services as a sideline. In Charleston in 1835, a white man named William B. Purves advertised himself as "Cabinet Maker, Upholsterer, and Undertaker." Purves built coffins and hired carriages for the procession to the graveyard.

Seventy-five years later, Mickey and Harleston became a clearinghouse for death-related services. An account book shows the variety and prices of items the Captain and Hannah offered. The firm paid physicians five to ten dollars for deathbed care (and sometimes a finder's fee for sending the Captain a body). Hannah and her brother also bought cemetery plots. (Seven dollars for a plot in the Centenary Baptist Church graveyard; ten dollars at Zion

Presbyterian Church.) Their clients were charged for the "slumber robes" in which bodies were dressed before going into the coffin. Florists were paid for standard arrangements, which were set up in the church and later moved to the gravesite. (A floral "broken heart" cost four dollars; a "wheel of life" went for eight; and a gaudy "gates of heaven ajar" sold for ten.) There was sometimes a line item for a "female funeral attendant." The attendant, assigned to the chief mourners, consoled the widow of the deceased, accompanied her to the cemetery, held her up if she collapsed, and fanned her face when she wept. Finally, there was the livery end of the business: five dollars for the horse-drawn hearse, and another five for each carriage that hauled the mourners.

The cost for a prominent burial hovered around thirty dollars, at a time when many Southerners, white and black, earned no more than ten dollars per month. To maintain good word of mouth, the Mickey and Harleston firm buried children at a loss—charging about four dollars, or well below cost. This was a considerable risk, because one in four black infants died before the age of twelve months; but the business could be made up during a death boom, which usually came in August and September. In those months, the heat took a heavy toll. So did pneumonia and tuberculosis, which claimed a disproportionate number of black adults. Business records show that Mickey and Harleston usually performed one or two funerals per week, but the pace increased to three or four burials a week in the late summer and early fall.

The most deluxe item, and the one that allowed the biggest markup, was the coffin, which could cost twelve dollars, a workingman's monthly wage. Accounts show the Captain and Hannah bought them mainly from manufacturers in Florida and Tennessee.

In the early 1900s, a coffined burial was a basic amenity, expected by both whites and blacks, but interment in a box was never a universal practice. The earliest Hebrews tied the body in the fetal position and wrapped it in a shroud for burial. During the Middle Ages in England, only the rich were buried in coffins: the poor went into the ground in winding sheets. (Because fabric was handwoven, and thus expensive, the shrouds of the poor were often too short, exposing the feet and giving rise to the expression "short shrift.")

It was only in colonial America that the octagonal wooden coffin became the standard burial case. Narrow at the feet, wide at the shoulder, and narrow at the head, the shape varied little and became a familiar icon. Eighteenth-century cabinetmakers sometimes displayed coffins outside their shops, leaning them on the storefront wall. The more discreet carpenters advertised their skill by making doll-size coffins, each a foot long, and hanging them five or six in a row from a beam over the sidewalk, like wind chimes.

With the spread of factory production, coffin makers branched out. In 1848, an inventor named Almond Fisk patented the Fisk Metallic Burial Case, the first radical departure from the wooden box. Form-fit to the body, like an Egyptian mummy case, the metal coffin had a pair of sculpted arms crossed over the chest and featured a glass plate over the face, which allowed viewing of the head until the first shovelful of dirt. When John C. Calhoun, the famous South Carolina politician and slavery advocate, was buried in a Fisk case in 1850, the popularity of mass-produced coffins in Charleston soared.

Before Captain Harleston and Hannah opened their shop, manufactured burial cases had already been through a period of weird experiment. Glass coffins, entirely see-through, had come briefly on the market, and wicker coffins had made a fleeting appearance, appealing to unpretentious families. The most curious innovation had been the "life signal" coffin. At a time when a coma often went undiagnosed, and seemingly dead patients sometimes "came back to life," ordinary people had a strong fear of being buried alive. The life signal coffin served this market. A typical device fixed one end of a rope to an aboveground bell and ran the other end down a chute into the coffin, where it was wrapped around the hands of the deceased. A person stirring from a coma could clang his awakening.

In the late 1800s, the traditional octagonal shape of the wooden coffin began to offend more prim tastes, because its form reminded one of the body lying within. This led, finally, to the rise of a distinctly American burial box, the rectangular "casket," whose innovator seems to have been Samuel Stein, a cabinetmaker in Rochester, New York. In the 1870s, the Stein Patent Burial Casket popularized both the new word and the object. (The word "coffin" comes from the Greek *kofinos,* meaning heavy chest or coffer; the word "casket" was first floated because it suggested a jewelry case.)

The advantage of the casket, from a business point of view, was its extra size, which allowed for elaborate padding and decoration; this led to value differentiation, luxury pricing, and a higher return. Samuel Stein's company, merging with its competitor, the National Casket Company, in 1890, went on to build a factory that produced some six hundred cloth-covered caskets per week. These were shipped by rail to funeral directors across the United States. Mickey and Harleston records show that they were good customers of the National Casket Company.

Of all the funeral services, the only one representative of an actual specialty was the mortician's art of embalming. Most of the rest of the work consisted of hiring help: florists, drivers, gravediggers. Expertise at embalming gave the Mickey and Harleston firm rightful claim to a professional skill.

Itemized bills show that Captain Harleston used two kinds of embalming fluid: the Champion brand, made in Springfield, Ohio, and Durfee Embalming Fluid, from Grand Rapids, Michigan. (Earlier chemical embalming fluids had high concentrations of arsenic, to kill microbes, and were dangerous to handle; but by 1905, these had been outlawed, and the Captain used less toxic compounds.)

To embalm a body is to attempt to stave off the decay of its flesh. The perfect embalming operation preserved the body in its natural form. The word "embalm" dates from the Middle Ages, borrowed from the French *embaumer,* itself traceable to Latin (*balsamum,* "balsam") and Greek (*balsamon*). In the ancient world, to embalm meant to preserve with balsamic resin or oil.

Chemical embalming was a rare thing in the United States until the mid-1800s, when Dr. Richard Harlan, a professor of anatomy at the Philadelphia Museum, traveled to Europe to study disease control. Harlan came back enamored of the sophisticated embalming processes that had become widespread in France. Translating an obscure work, J. N. Gannal's *History of Embalming and of Preparations in Anatomy, Pathology, and Natural History,* from French into English, Harlan aroused interest in the subject among a small group of scientists. In the United States, anatomists had already been preserving human specimens with poisonous chemicals, such as bichloride of mercury, zinc chloride, and arsenic-based compounds. These fairly recondite methods would now be made available to any mortician.

Embalming methods seemed new in America, but they had reached a high level of articulation several thousand years prior to the nineteenth century. Famously, the early Egyptians refined the technique into a discipline. In the fifth millennium B.C., rich families in Egypt hired preservation specialists, who worked on a corpse for several months. For clients of more modest income, the embalmer injected the body with cedar oil and then soaked it in alkali until the flesh dissolved, leaving only the skin and bones. The majority of families—the poor and slaves—soaked the body in a solution for two months before interment.

In Renaissance Europe, a pioneering embalmer was an Italian surgeon named Pietro d'Argellata, who became renowned for his exquisite preservation of the corpse of Pope Alexander V, who died in 1410. The account of the deed appears in d'Argellata's treatise on surgery, *Chirurgia.*

The surgeon started by cutting open the pontiff's abdomen, using a straight line from the rib cage down to the pubis. He emptied the colon of feces, tied it up in two places, and removed all the intestines and viscera. Sponging the inside of the body with vinegar, d'Argellata next poured alcohol into the cavity, then sponged the abdominal walls dry. He filled the body with the balms of his day: "aloes, caballainies, succatrinol, acaciae, nucis xuperuni, Galluae, and muscatae." For additional stuffing, he inserted cotton in the abdomen. Sewing up the body, as he put it, "like a furrier," d'Argellata then powdered the pope's esophagus and larynx, and filled them with cotton and balsam. More cotton went into the mouth, nostrils, and ears, and as a garnish, he placed a solution of egg whites into the pope's anus. Finally, the surgeon wrapped the pontiff's body in waxed linen that had been soaked in turpentine, dressed the high priest in his papal cloak, and washed the corpse's face with salted rosewater. Thanks to this extraordinary procedure, it became possible to lay Pope Alexander V on a bier for eight days of public viewing before his entombment.

The next breakthrough in preservation took place in the 1700s, with the invention of injection embalming. After anatomists confirmed the circulation of the blood, the idea of pushing a caustic solution into the veins, and thus throughout the body, began to gain adherents. Credit for the origination of the process goes to Scottish physician William Hunter, who not only devised his method in the eighteenth century but also wrote it down. Hunter injected oil of turpentine and camphorated spirits of wine in the arteries, and for good measure, used camphor to pack the body cavity.

In the colonial period, the Americans fell behind Europe in the art of embalming. Thanks to the do-it-yourself brio of the early settlers, millions went into the grave untreated. In the mid-1800s, however, the United States regained the initiative, and the catalyst was the Civil War. A great death machine that devoured more than half a million men in the 1860s, the Civil War had the little remarked side effect of popularizing the use of injection embalming.

The hinge figure in this development was Thomas Holmes. In histories of funeral directing, Holmes bears the distinction of being known as "the father of American embalming." Born in New York in 1817, the son of a rich businessman, Thomas Holmes worked as a physician during the late 1840s and early 1850s, eventually securing the position of examiner for the coroner of New York County. He began to experiment with preservation processes, using murder victims as his study material. A few years later, when the Civil War broke out, Holmes chanced to find himself in Washington, D.C., just as thousands of new Union soldiers were flowing into the city. In an act that he believed to be altruistic, Holmes distributed pamphlets to the recruits that offered to embalm them at no charge if they happened to be killed. Holmes realized that for a man in his position, the war meant an enormous commercial opening. To stimulate business, Holmes exhibited the bodies of embalmed corpses around Washington. He was arrested for the stunt, jailed, and charged with creating a public nuisance. Bailing himself out (for the stiff price of three hundred dollars), he somehow got the job of embalming one of the first Union casualties of the war, Colonel Elmer E. Ellsworth. A former assistant to President Abraham Lincoln, Ellsworth had been shot in May 1861 in Alexandria, Virginia. After an injection embalming, Ellsworth's body was laid out for viewing in the East Room of the White House. The newspapers fell on the story, and Holmes's reputation was made.

As the war grew in scale, Thomas Holmes and several hired assistants did a fast trade, embalming thousands of Northern soldiers at a fixed fee per corpse. A "Holmes process" was developed, using the femoral artery as the point of injection, as well as a "Holmes injector," a device for inserting the fluid, which he patented in 1861. War deaths quickly outstripped Holmes's one-room operation. Within a year, embalmers cropped up in the ranks of every regiment, some with training from Holmes, many doing the job using manuals. Holmes covered all aspects of the booming market, selling his own embalming fluid, a secret formula he called Inominata, for three dollars a gallon.

Similar operations appeared on the Confederate side. In Nashville, a man named W. R. Cornelius began the war as a cabinetmaker and undertaker. According to his own account, Cornelius embalmed, buried, or shipped to their homes some thirty-three thousand soldiers.

For the families of dead soldiers, the advantage of embalming was that the remains of their loved ones could be preserved for shipment home. Of course, many thousands of soldiers went untreated and were buried where they died. But a family considered itself lucky when the regimental embalmer singled out their son for treatment, and the body came home by rail. Money could smooth the way, and prosperous officers and their families revolutionized the funeral business with their cash.

Captain Harleston practiced the same embalmer's art that had taken shape during the Civil War. The Mickey and Harleston firm used two instruments that had served morticians for decades: a force pump and something called a trocar. The force pump propelled the fluid into the body, and a trocar was a foot-long hollow needle, shaped like a crescent, that allowed the injection of preservatives into the torso.

To start, the Captain undressed the client's body, washed it, and then placed it on the "cooling board," a high gurney or table. By the time the corpse arrived, rigor mortis had set in, and it was difficult to manipulate the limbs. The Captain swabbed the body with either Durfee or Champion embalming fluid, and stopped up the orifices. He attached the force pump to a tube and looked for an artery—usually the femoral artery in the thigh worked best. The first step was to pump out the remaining blood, but the blood had often hardened and was difficult to drain. When the Captain had withdrawn as much blood as could be removed, he reversed the pump's action from suction to propulsion, and loaded the embalming fluid. He added a solvent to the liquid: this helped to dissolve blood still coagulated in the arteries, reducing its viscosity and making room for the preservative. Family lore reports that the most difficult embalming cases involved drowning victims. If the body had been in the water for two or three days, it was bloated, and the pump produced no circulation at all. With these corpses, the Captain merely injected fluid all over the body, using a syringe. But usually the pump worked, and when the arteries were full, the incision was sealed.

Fortunately, preservation no longer required evisceration. To conserve the body cavity, the Captain inserted the trocar in the navel, gently pumped fluid into the torso, and then sewed up the wound. The trocar was next inserted in the corner of the eye and forced into the brain cavity, which was filled with fluid. Withdrawing the instrument sometimes caused damage to the eye, but the eyelid usually covered the abrasion. If the body was a man, the Captain then shaved the face (hair continued to grow after death). He closed the eyes, using "eye-caps" beneath the lids to hold them shut, then shut the mouth, sewing the lips together. An assistant helped to dress the body, and as a final touch, the Captain might have applied liquid flesh tint, darkened or lightened as needed. He dusted the face with powder, placed a pillow under the head, and he was done.

* * *

Captain Harleston and Hannah Mickey worked well together, the brother making most of the decisions, the sister keeping majority interest. But things began to unravel as business improved. The trouble started with Hannah's family. Hannah Mickey had five children: Rosalie, Ellen, Marian, Edward, and Richard. The two sons, "Eddie" and "Dick," born in 1883 and 1889, respectively, had finished high school after their mother went into business with their uncle. When Eddie and Dick left home for college, they assumed they would come back to a career in the burial business. Eddie Mickey went to study in Georgia at Atlanta University; his brother, Dick, followed suit.

Eddie Mickey was scheduled to finish at Atlanta in 1905. That year, Hannah persuaded Captain Harleston to open a larger office that would accommodate her two sons. The two of them bought property on Calhoun Street, a wide boulevard that bisected the town like a belt across its middle. This new location consisted of a two-story wooden house, rather worn, but nonetheless larger than previous quarters on Meeting Street. They also bought a second building across the street, to be used for storage, and an empty lot as an investment.

A family member who knew him describes Eddie Mickey as a dandy. Short, with a puffy face and curious expression always on view, Eddie dressed carefully and was refined and soft-spoken. He is also said to have been a hypochondriac. Eddie often complained about his many ailments and could sometimes be found in bed predicting his own death. Fresh from college and sure of himself, Eddie Mickey showed up to work. Proud of his new high position in business, Eddie developed a fondness for driving mourners in "the landau," an open, stately funeral carriage drawn by horses. For each burial, Eddie would put on a top hat and cape and ride off with his chin raised, a supreme liveryman. No less a stylist, Captain Harleston evidently disliked his nephew's affectations, which got in the way of the Captain's own brand of officiousness.

Almost immediately after Eddie arrived, the trouble began in the office. Tradition says that Captain Harleston, then in his early fifties, could not adjust. He fought with his nephew, bossing him and making peremptory decisions. Eddie, just twenty-three, appealed to his mother, which only made things worse.

According to one family member, part of the problem was that Eddie Mickey was gay. Charleston society in general, and the black middle class in particular, did not like men who loved other men, and Eddie was forced to remain closeted. Not only did the Captain scoff at Eddie's pretensions, he cringed at the thought of what his nephew did with his friends. About 1910, however, to everyone's surprise, Eddie married. The woman was a suitable mate named Harriett McClennan, the daughter of a prosperous pharmacist. But it soon became gossip that the marriage was never consummated. Within a year, Eddie's wife had left him and spread word about why the marriage failed. This pushed Eddie's sexual orientation into the open.

The Harlestons shuddered with embarrassment—made only worse as the family partnership continued to deteriorate. When Dick Mickey graduated from college and joined the business, tensions rose again. In 1912, the Captain and Hannah realized they had made a bad decision. The Mickey brothers wanted to run their own firm, and the Captain wanted a personal fief, free from the Mickey boys' influence.

In May 1913, the Mickey and Harleston firm was dissolved in a burst of acrimony. A record of the transaction shows what happened to the company's assets. First, the firm's rolling stock was divided: "seven horses, one hearse, two undertaker's wagons, a buggy and surrey, three carriages, and an old buggy." Then the Captain bought his sister's half of two of the three properties: the storage site (at number 118 Calhoun Street) and a lot with a cottage (at number 121). Hannah and her sons kept the main office building. With accusations flying, the lucrative family business came to an end. Eddie Mickey, wounded, vowed never again to let his family be upstaged by the Harlestons, and the Mickey firm continued to limp ahead. Meanwhile, Captain Harleston, imperious and thin-skinned, decided he would set up his own funeral business, far from meddling relatives.

Chapter Five

IN June 1862, a hundred miles outside Charleston in remote Barnwell County, two slave farmhands had a son, whom they named Daniel Joseph Jenkins. It was a year into the Civil War, and marriage among slaves was not legally recognized. Neither were surnames, so the boy's parents—who were named Jack Kirkland and Nancy Jenkins—may have chosen separate last names after the Civil War, and given the child his mother's name, Jenkins, when he was a toddler.

Daniel Jenkins was just two years old when freedom came, and he grew up in the hardscrabble poverty of sharecrop farming. According to one of his later friends, Jenkins's formal schooling amounted to less than a year, and he spent the rest of his time digging in the fields with his parents. He married at nineteen, to a sixteen-year-old girl named Lena James, and about 1882, he and his wife moved briefly to Arkansas, where they worked a small holding of land. But the couple came back to South Carolina, living twenty miles north of Charleston, before finally moving to the city.

A vigorous and imposing man, Daniel Jenkins stood six and a half feet tall and had a powerful physique. He owned a mesmerizing voice and was handsome and composed. In the late 1880s, Jenkins became a wood hauler, while he and Lena lived modestly in rented rooms in Charleston. But something set Jenkins apart from the multitudes of black laborers who sweated, as he did, for a white man's pittance. When he spoke, he projected an aura of confidence and self-possession, and people fell silent, as Jenkins's eyes penetrated his listeners with a level gaze. He was an enthusiastic Christian, and as a member of the New Tabernacle Fourth Baptist Church, which stood on Charleston's Palmetto Street, he had an abiding faith that he was touched for a higher purpose. When the congregation at Fourth Baptist made Jenkins one of its deacons, his path began to widen.

What happened next is a bit of legend. On a cold December day in 1891, Jenkins, twenty-nine, went to the railroad yard to retrieve a consignment of wood. While loading the timber, he came upon four boys, aged about five to ten, who had sought to escape the chill winds of the afternoon by taking shelter in a railroad car. The black children were huddled together, speechless and shivering, in an abject state of neglect. Asking the boys why their parents had let them go out in such harsh weather, Jenkins learned that they had been abandoned by their several families and were without guardians. Jenkins brought the children home to Lena, even though there were already several natural children of Daniel and Mrs. Jenkins.

This unusual act of kindness might have remained the anonymous gesture of a single couple on the sea of poverty and broken lives that was the aftermath of slavery in the South, but the timber hauler took matters further.

At Sunday services at Fourth Baptist, Daniel Jenkins, who with his height made an excellent impression at the pulpit, appealed to the congregation on behalf of the homeless boys, and suggested the founding of an association that would help them and other parentless waifs. Everyone knew that unwanted children were being made homeless by the hundreds each year, and that they were suffering terribly.

Seven months later, on July 21, 1892, Jenkins received a charter from the state of South Carolina to operate a new philanthropy, to be known as the Orphan Aid Society. He began seeking out more wayward youths, and he and his wife accepted other children sent to them. The following October, the church elevated him from deacon to pastor of the congregation, and the Reverend Mr. Jenkins soon began to devote his considerable energies to the creation of a new home for black orphans.

The fledgling institution called the Jenkins Orphanage initially opened at 660 King Street, in Charleston, in a building that was little more than a shed. With the continual intake of children, Rev. Daniel and Lena Jenkins quickly found themselves guardians of several dozen foundlings. Appeals to African American churches could not satisfy the need for food and clothing, and the city government, observing the principle of whites first in charity, paid no attention. The Reverend Mr. Jenkins realized he needed a means of raising money that could provide the orphanage with a steady income.

Fortunately, Jenkins knew of at least one model for creating revenue. Twenty years after slavery, a famous black choral group from Nashville, the Fisk Jubilee Singers, sometimes passed through the South on fund-raising tours. Fisk, a black college founded after the Civil War, had been facing financial problems when a musician on the faculty struck on a fund-raising idea—a traveling choir that would perform for contributions. The result was the Fisk Jubilee Singers, a choral group that sang black religious music, especially Negro spirituals, in exchange for donations to the school. Appealing to audiences in churches and theaters throughout the country, the Fisk Singers raised large sums and ultimately brought the college back from the brink of bankruptcy.

Although he was not a musician, the Reverend Mr. Jenkins had probably heard about the Fisk Singers. He may have had them in mind when, sometime in 1893 or 1894, he hired a local black instrumentalist, P. M. "Hatsie" Logan, to begin musical instruction for the most promising of the orphans. Appealing to the public for their unwanted band instruments, Jenkins put battered trumpets and punctured drums in the hands of the orphan boys and hoped for the best. When the resulting "orchestra" played well enough not to embarrass the Baptist preacher, Jenkins led a contingent out onto the streets of Charleston for their first performance. The band hurried through a short, disorganized repertoire, after which the minister stepped forward to ask the bewildered crowd for donations.

Money flowed freely, and Jenkins was able to return to his office with enough to buy food for the following week. The Jenkins Orphanage Band was born.

In the first two years of Jenkins's charity, 360 boys and girls streamed into the preacher's care. Local newspapers dubbed him "the Orphanage Man." Having found a fund-raising tool in his children's brass band, Jenkins now began to look for a more permanent home for the orphans whom he called his "black lambs."

At the time, the city of Charleston was famous for its care of discarded children. The first municipal orphanage in the country had been founded within the city limits: the Charleston Orphan House, which dated from 1790. But an orphan in Charleston was far better off being white than black. Before the Civil War, nearly all black orphans had been enslaved. Yet after freedom, the public care of orphans was never extended to colored children. In 1855, the Charleston Orphan House was enlarged, and a magnificent Italianate building was put up for it. Five stories tall, with two long wings and a columned portico, the building stood at 160 Calhoun Street and covered nearly a city block. But while the Orphan House received generous funding from the city council and housed more than two hundred white children, segregation kept black children out. "Colored waifs" were left to fend for themselves on the street, until the advent of the Jenkins Orphanage.

In 1892, a large building stood empty in the midst of the "penal district," on the western edge of the city. The penal district had its epicenter at the intersection of Franklin and Magazine Streets, on whose southeast corner stood the old city jail. Next door to the jail, at 20 Franklin Street, stood the former Marine Hospital, a stately two-story Gothic building with pointed arches over the windows and some ten rooms. Built in 1833, the Marine Hospital had been designed by Robert Mills, an American architect whose work would later include the Washington Monument. It had two wings off the back that had served as hospital wards.

Making the rounds to city leaders, Jenkins approached the owner of the empty hospital, the South Carolina Medical College. His orphan project, he said, was feeding and training children that might otherwise fall into mischief. Without a group home, the children might actually commit crimes against white people. Responding to Jenkins's appeal, the board of the medical college agreed to release the Marine Hospital.

The Marine Hospital had initially cared for sick merchant seamen. Later, during the Civil War, it had functioned as a military clinic, and after the fighting the building had been converted into a temporary school for black children. The new home of the Jenkins Orphanage was in a neighborhood of wooden row houses and shacks with peeling paint. Its facade looked onto Franklin Street, which was paved, but in most places, dirt streets led here and there into cramped, narrow yards. The neighborhood was home to hundreds of black families, most of them working, and nearly all of them poor.

As a home for children, the Gothic building was an ambiguous gift, because it stood next door to the city jail. With its turrets, walls of stone, and a crenellated parapet, the jail projected a medieval, glowering aura a block in each direction. At night, neighbors could hear the shouts of prisoners through the jail's barred but open windows. During the day, new inmates arrived in large iron cages pulled by horses.

But the Reverend Mr. Jenkins decided to disregard the view of the jail. He made minor repairs to the old hospital and moved the boys and girls into their new home. Jenkins put his own office on the first floor, and the dormitories for the children in the long rooms of the old hospital wards, where dozens of beds could be arranged in a line. The music program, the engine of fund-raising, went into a practice room in the basement. Through the stairwell near his desk, the Reverend Mr. Jenkins could hear the brass band at its daily rehearsals, like the satisfying sound of coins dropping into a collection plate.

Meanwhile, about a mile away, on a distinctly better block, the sons and daughters of Captain Harleston and Louisa Moultrie were growing up. The Captain had five children who lived into adulthood: Katherine, Moultrie, Edwin, Eloise, and Robert. (Five more had died in infancy: a pair of twin boys, two more sons, and a daughter.) The eldest, Katherine (or Kitty), was born in January

1878 and raised on Laurel Street. When she was a young woman, Kitty made a surprising decision: she would go to work for the Jenkins Orphanage on Franklin Street. Kitty's father could not have been pleased that one of his daughters wanted to work for "the Orphanage Man." Many of the orphans were rough, and some had criminal records. For Kitty even to set foot in the orphanage meant a step down for her family. But for the Reverend Mr. Jenkins, to align himself with the prosperous and influential Harlestons was a diplomatic feat. When the smart young woman appeared, he definitely wanted to add her to his staff.

After Kitty, the next sibling in line was John Moultrie Harleston. In memory of his mother, this child of the Captain went by his middle name. Moultrie was tall, but he was flat-faced, with a dull expression in his eyes. Independent to the point of defiance, Moultrie's personality caused the Captain no end of grief. He refused to be bossed, and consequently, he and his father did not get along. When Moultrie was a young man, the Captain fixed on the idea of bringing him into the family funeral business, concluding that Moultrie would have the job of caring for the horses, of which the Mickey and Harleston firm owned many. In 1901, Moultrie graduated from Lincoln University, in Pennsylvania, and from there, he was sent by his father to study animal medicine in Philadelphia, at the University of Pennsylvania. With his father's insistence that he become a large-animal veterinarian, Moultrie spent two years in veterinary school. But just as Moultrie applied himself to learn the diseases that afflicted horses, by a twist of fate, the automobile was growing in popularity. Soon there were motorized hearses, and although the Captain didn't yet own one, everyone knew it would only be a matter of time. Moultrie didn't finish his education. He dropped out of veterinary school, drifted back to Charleston, and became a layabout, dependent for his livelihood on handouts from his father.

The third and in many ways most memorable of the Captain's children was Edwin, a middle child. Edwin Augustus Harleston had the same name as his father, but the boy had none of the Captain's domineering style. Neither was he actually his father's namesake. Captain Harleston was Edwin *Gaillard,* and the son was

Edwin *Augustus,* after a family connection now lost in memory. The proximity of the two names would cause years of confusion in many departments, from mail to journalistic accounts, to taxes.

Edwin A. Harleston was born on March 14, 1882. When he was a boy, he acquired the nickname Teddy, which served to separate him from his father. When Teddy was a child, Captain Harleston was still operating his cargo business, which meant that as most of the black population sank further each year into poverty, Teddy wore new pants and attended a school where tuition was required. The young Teddy had two promising talents: he was athletic and he could draw. With his brother Robert and other neighborhood boys, Teddy played a great game of baseball. (Robert Harleston was the first baseman; Teddy pitched.) Later, Teddy took up football, a dangerous sport at the time because the players' pads were made of thin rubber, and the helmets resembled leather gloves.

By his teens, Teddy had become an increasingly sensitive boy, and he had taken up drawing and painting. Artwork soon took over his imagination, so much so that when he graduated from high school, in 1900, Teddy presented his school with a painting called *Lincoln and His Cabinet.* The painting has been lost, but its title reflects the prevalent view of nonwhite Southerners. As the author of the Emancipation Proclamation, President Lincoln was an unassailable hero.

As a teenager, Teddy Harleston was unusually fine looking, with a square jaw, high forehead, and searching brown eyes. He attended high school at the Avery Institute, the only secondary school in Charleston for nonwhites. One had to pass an entrance test to enroll, and all of the teachers were white. The restrictions on attendance can be seen in the size of Teddy's graduating class, which consisted of fewer than ten people. Teddy Harleston was valedictorian at Avery in 1900, and from there he moved to Georgia to enroll at Atlanta University. In Atlanta, Teddy once again had white teachers. Many white scholars and missionaries had come from the North to train new black leaders for the South. And Teddy would turn out to be one of them.

The Captain's youngest son, Robert Harleston, was born in 1889. He went to Atlanta University but dropped out in two years, after which Captain Harleston forced him to study embalming. Robert's destination was the Renouard Training School for Embalmers, in New York City. It can be assumed that Robert took no particular joy in this branch of his education, but his schooling would help keep the family in linen and china.

Of the five brothers and sisters, Eloise Chapman Harleston, the Captain's youngest and most tender daughter, seems to have been the most vulnerable. A fawnlike girl whom the family called Ella, she was born in November 1883, and like her siblings, she had grown up in the home of her grandmother Kate Wilson, on Laurel Street. As a girl, Ella went to good schools and a respectable church, but when she was just thirteen, her mother died. Ella was naturally stunned by the event. After a damaging period of grief and uncertainty, the girl's aunt, Sister Sue Singleton, moved in with the family and took over Ella's care.

Photographs from her adulthood show Ella as a woman with intelligent, sad eyes. Her face was a near perfect oval, and she wore her dark hair long and pulled back behind her head. Her mouth was sensuous, and her lips often awakened in a little smile. Though Sister Sue had trained her in respectable chastity, Ella's sexuality would one day simmer to the surface, with an unforeseen outcome regretted by the entire Harleston clan.

As ruler of a colony of waifs, the Reverend Mr. Jenkins quickly acquired the stature of a patriarch. Friends called him "Uncle Joseph." The orphans and staff referred to him as "Parson Jenkins," or just "the Parson."

In the 1890s, the Jenkins Orphanage grew rapidly, and Parson Jenkins looked for a model for his new institution. The minister discovered his ideal in the example of Booker T. Washington, founder of the Tuskegee Institute in Alabama, and champion of black self-reliance. An educator of the children of sharecroppers and former slaves, Washington advocated self-help among black families in lieu of a possibly futile wait for assistance from whites. Jenkins began to pattern himself after Washington, who regarded manual labor as the equal of book learning, and obedient citizenship as the answer for black people to the cumbersome legacy of slavery.

Parson Jenkins, like Washington, believed the orphans needed skills above all else, so they could support themselves once they left his care. He set his "black lambs" to work in an array of jobs. The children, ages five to fifteen, were channeled through light classes in reading and arithmetic but intensive training in a trade. Soon, in addition to its brass band, the orphanage had shops that taught shoemaking, sewing, and blacksmithing.

Few writings of Jenkins have survived, but in a letter to a newspaper editor, the minister once wrote: "To allow orphan and destitute children to sit up in long dresses on public charity with a flower garden around them, doing nothing, is not in my mind. I have worked from a boy all of my days. Hard work and plenty of corn bread and molasses produce healthy children."

Of the various departments, the jewel of the operation was the orphanage newspaper, the *Charleston Messenger*. In 1893, the Parson bought a printing press from a machine shop in New York called Damon and Peets. When the press started turning out advertisements and fliers for local businesses, it quickly paid for itself. The printing department, run by the Reverend Mr. Frierson, grew until it was equipped with two linotype machines, two job presses, and several cases of type. Then, in a stroke of ambition, the Parson decided to put out a weekly newspaper. A larger press was bought, and beginning in 1894, the broadsheet went forth.

The *Charleston Messenger* carried items about black life in the city and ran national news of interest to nonwhites. The city's white-owned papers, the *News and Courier* and the *Evening Post*, ignored news of importance to black people. (Usually, the only Negroes seen in the white papers were those charged with a crime.) The *Messenger*'s advertising created a fresh income stream, and the paper kept the orphanage in the minds of black readers.

By the mid-1890s, the roll book of the Jenkins Orphanage listed 536 names. About two hundred "inmates" lived in the old hospital building; the rest were day students at the orphanage school. A small staff consisting of Jenkins, his wife, eight teachers, and two porters supervised the unruly mob. A white doctor named Edward F. Parker helped the children by running a free clinic. Life at 20

Franklin Street was harsh, but for many children, it was better than the neglect from which they had escaped.

Not all the children were youths abandoned by their parents. Some, of course, were the unwanted harvest of sex—infants carried to Parson Jenkins by single mothers and surrendered to his care. Others were older, sometimes merely the difficult siblings in large families, whose parents could no longer handle them. Quite a few had already had run-ins with the law. As a result, the orphans were a raucous, sometimes wild, bunch, given to fighting and running away, petty thieving, and mischief.

In the fall of 1893, the first real crisis descended on Parson Jenkins's young institution when a hurricane tore up the roof. Repairs were made, but the cost was enormous, and before long, Jenkins faced the problem of his creditors. Two years later, still saddled with debt, Parson Jenkins devised a plan to increase his prospects for donations: he would take the children's band on the road.

In the summer of 1895, Parson Jenkins, eighteen orphans, and a bandmaster took a steamship up the East Coast, stopping in various ports en route to New York. (Their itinerary has since been lost, but it's likely they spent two or three days each in Wilmington, North Carolina; Norfolk, Virginia; Baltimore; and Philadelphia.) The money the band collected on the trip proved insufficient, however, and arriving in New York, Jenkins found he had barely enough to take the children back home. Desperate, Parson Jenkins decided to take a further, still more risky, gamble.

"Some good white friends met me while in a spirit of despondency," Jenkins later wrote, probably referring to white ministers he met in New York, "and advised me to go over to England, saying that I would get barrels of money. Nothing doubting, neither counting the cost, I leaped out without a dollar."

Packing up the band, Jenkins negotiated passage to London and set sail for Europe.

"I only had half enough to pay our way, but the captain took us over anyway, expecting to make money on the ship," Jenkins remembered. "But in less than half hour's time after getting on the ship, we became seasick and remained so until the day before we landed."

Reaching London in early September 1895, Jenkins began parading his band on the streets. By this time, the band played more proficiently. The boys raced through fast marches, eased into simple sonatas, and tossed off popular melodies. But Parson Jenkins may not have known how staid England could be with its traditions; in music, the British marching band represented a heritage that was especially dour. The Jenkins band was garish, rhythmic, and was backed by a pounding bass drum as it weaved its way through the crowded streets. At the head of the march was the "bandmaster," a boy who was perhaps six or seven years old, who carried a baton in his hand. The idea of a tiny bandleader was a comic stunt that Jenkins had concocted, and it worked beautifully, often sending the crowd into convulsions of laughter. As the little figure in front danced and waved his baton, his "conducting" drew hoots and applause.

After several numbers, Parson Jenkins stopped the music, stepped forward, and appealed for money. Shillings and pence surfaced from hundreds of pockets.

According to several accounts, during one of the band's outings, Jenkins saw on the street an older and distinguished white man he recognized, named Augustine Smythe. A former major in the Confederate army who had become a prominent lawyer in Charleston, Smythe, then in his sixties, was in London on a European tour with his wife and three children. By accident, the family had stumbled across the band. Although for racial reasons Parson Jenkins and the attorney would have avoided each other in Charleston, in far-off London the Smythe family was happy to hear the orphan band, which reminded them of the reckless sounds they missed from back home.

As the Smythe family approached, racial etiquette required that Parson Jenkins smile and speak submissively to them. But even so, cheerful greetings were exchanged, and Augustine Smythe probably made a donation to the band. Then he and his family walked off to continue their sightseeing.

No sooner had the group made their good-byes than Jenkins faced a new surprise. As he wrote later, the band's "strange appearance created so much excitement and monopolized the thoroughfares" to such an extent that police officers began to materialize. The dumbfounded bobbies, not accustomed to black children who danced while playing the trombone, stepped in to put an end to the noise and issued the Parson a subpoena to appear before a magistrate.

"On the next morning," Jenkins wrote, "every newspaper published that we were stranded [in London]."

Augustine Smythe, back at his hotel, read about the incident in one of the dailies. Tradition in the Smythe family states that the well-heeled American wrote a letter to the authorities vouching for Jenkins, then loaned the minister money to keep his creditors at bay until the court date.

On September 8, 1895, Parson Jenkins arrived in the magistrate's courtroom, followed by the dozen or so orphan musicians, none of them older than fourteen. The charge was exploitation of child labor. Jenkins addressed the court, explaining that he had brought the children to London in order to raise money for his orphanage. He pointed out that the philanthropic institution could be found four thousand miles away, in his hometown of Charleston. The magistrate replied that under no circumstances could performers cause an obstruction in the public streets; otherwise, the police would interfere. He made clear that by law, no child under the age of eleven could be forced to perform in the streets for money, this being in conflict with child labor standards. The magistrate also advised Parson Jenkins to apply to the American consulate for money to return home. He then gave Jenkins a pound note.

The next day, the *London Daily Telegraph* published an editorial summarizing Jenkins's predicament: "Much may be done, no doubt, to raise money for an orphanage," said the newspaper. "But to let loose a brass band of thirteen Negro children upon an urban population suffering from nerves is likely to create almost as many orphans as it would relieve."

In a last attempt to retrieve success from disaster, Jenkins turned to a refuge he knew well: the Baptist church. The pastor of the

Metropolitan Tabernacle, a large Baptist congregation in London, agreed to let Jenkins make an appeal about his orphans at an upcoming Sunday service. (By coincidence, perhaps, twenty years earlier, the Tabernacle had been the scene of a fund-raising concert with the Fisk Jubilee Singers.) Climbing into the pulpit, the Parson poured out his lament for his black lambs and described the debt that awaited them back home. The congregation, helped in its catharsis by the orphan boys sitting in the front pew, took up an immediate collection, and in a matter of minutes presented the Americans with the equivalent of two hundred dollars in pound notes. Repeating his supplication at other churches in London, Bromley, and Kent, the Parson collected enough money to take the band home.

Back in America, Jenkins paid off the old debt, basked in his success, and celebrated the good fortune that had begun to come his way.

In 1902, Ella Harleston, the youngest daughter of the Captain, was a gentle and awkward eighteen-year-old, and a recent graduate of the Avery Institute, the private school for ambitious Negroes. Sometime that year, Ella walked over to the large white building at 20 Franklin Street, where the painted sign next to the door read JENKINS ORPHANAGE. At the time, Ella's sister, Kitty, was working for Parson Jenkins, and Kitty had introduced her younger sibling to the Orphanage Man in hopes he could offer her a job. A nimble speller, Ella was good with a typewriter, and she was well spoken. The director of the Jenkins Orphanage was not an educated man. He relied on better-schooled women to conduct his correspondence, and she was hired as a secretary.

By this time, Parson Jenkins was an eminent figure, known throughout the South, at least among colored people. African American newspapers called the Parson "a savior to the abandoned children of the race." White politicians praised him for taking care of black waifs (without costing white taxpayers anything). He had a right-hand man at the orphanage, his cousin Paul G. Daniel, who helped run the business side. Since the success in England, Jenkins and Paul Daniel had been enterprising. The Jenkins Orphanage had become the sole importer in the United States for the Collins brand of English band instruments. And after nearly a decade of appeals, focused on the lucrative brass band, Jenkins was personally comfortable. His clothes were new, and his house included one or two servants. He even wore a uniform of sorts: formal trousers with a silk stripe down the leg, a vest over a starched dress shirt, and a black formal jacket, with long tails down to the knees. The costume was intended to show the seriousness of his work and his personal high standing in the city.

For the first couple of years on the job, nothing unusual ruffled Ella's life. She was a better writer than her sister, Kitty, so she typed the Reverend Mr. Jenkins's dictation and corrected his many errors of speech and spelling. And at night she went home to Laurel Street, where she had a bedroom on the second floor. But in her third year as a clerk, things did not go so well. The Reverend Mr. Jenkins's attitude toward her seemed to change. He began to spend more time with her than he did with the other assistants, and he often asked her to stay late to finish extra work.

Despite his personal power and public good standing, Parson Jenkins's home life was suffering. His wife, Lena, had been ill for several years. She had contracted tuberculosis, which weakened her and often kept her confined. Jenkins and Lena were raising several children, and there was a risk that they, too, would become ill. In 1905, Ella Harleston was twenty-one, and curious. She admired the Reverend Mr. Jenkins and was attentive to his needs. On one occasion, when her employer kissed her, she did not refuse him. Later, when the minister actually forced himself on her, she responded. Within a few months, Ella was having sex with Jenkins.

Ella Harleston had come of age in the late 1800s. It would be decades before birth control became available to married women, let alone unmarried ones. Uttering the phrase "birth control" was enough to send many Southerners into apoplexy. To have children outside of marriage was not too uncommon among working people and poor farmhands. But if middle-class women chose to have nonmarital sex, they faced the worst social punishment.

In the spring of 1906, Ella must have been terrified when she discovered she was pregnant. To have a child without a husband represented a catastrophe. If it became known, the unwed mother would be ostracized and shame would haunt her family, who would never live down the humiliation. The Reverend Mr. Jenkins understood the social danger of Ella's pregnancy, but even worse, from his point of view, was the danger to his considerable reputation. Jenkins had seven children by his wife. Even if he wanted a divorce, he couldn't have one because divorce was against South Carolina law. Although Jenkins was sleeping with one of his employees, he had not stopped having sex with his wife. A little while before Ella became pregnant, Jenkins also had a daughter with Lena, whom they named Mildred.

The usual outcome of situations like this was that Ella, the young woman "in trouble," would be sequestered. After delivery, the baby would be given to a relative or handed over to an orphanage. In this case, neither solution worked. Ella could not be sequestered without informing her father, who would be cruel beyond measure. If Lena Jenkins found out, her health might suffer a bad turn, and the only place for a colored foundling in the city was the Jenkins Orphanage.

Ella Harleston's situation called for a dramatic solution. She and the Parson decided they could tell no one, and instead they devised an unusual plan: they would find a place to hide their baby. No matter where people looked, no one would be able to discover it. Ella's lover was a powerful man with considerable resources, and it was probably his idea to conceal the baby, because he alone knew where this could be done.

About September 1906, Parson Jenkins and Ella sailed for London. They probably left from New York during a trip of the Jenkins band up the East Coast. On the journey, Ella posed as her lover's young secretary, which in fact she was. But by now, gestation was advanced, and her belly was large.

On this occasion, the band seems to have stayed in America while Ella and Parson Jenkins went to England alone. When the couple arrived in London, the minister immediately took Ella to a small town in Lancashire called Wigan, a two-hour train ride from the capital. An industrial mill settlement, Wigan was nearly as far from Charleston as one could get and still inhabit an English-speaking community. It's not clear what connection Parson Jenkins had to Wigan. (Since the episode has been covered by years of shame, certain details of the story are lost.) He may have known a Baptist minister there from an earlier trip, or perhaps he was referred to the town by a London acquaintance from years before.

Parson Jenkins and Ella stayed in Wigan until November, when a girl was born to the young mother. A midwife named Alice Layland, who lived at 20 Lower Saint Stephen Street in Wigan, assisted in the delivery. The parents named the child Olive Ashton Harleston. By custom, a child of unwed parents took the name of its mother.

Olive Harleston was a tiny thing, and the only black child in a hundred miles. During a week of recovery, Ella and the Parson probably marveled at the baby in her crib, smiled secret, knowing smiles at each other, and counted their blessings. But within a few days, the couple had left England without their baby.

Ella and her lover could not face the public scorn at home, so they abandoned their child and made their way back to Charleston. The family of Alice Layland, the midwife, took Olive Harleston into her home. Parson Jenkins and his secretary returned to their former life together, answering letters, entertaining the interested approval of strangers, counting the money from the band. It was as though nothing unusual had happened. But a day did not go by when Ella, with her intelligent, sad eyes, did not look through the window at the hundred "black lambs" and think of some other child. She did not know when, or even if, she would ever see her baby. Ella wore her dark hair long and pulled back, and her lips no longer awakened in a little smile.

Part III

~

Eyes Sadder Than the Grave

Chapter Six

EDWIN A. "Teddy" Harleston was born on March 14, 1882. He is said to have been a painfully shy boy, who did not assert himself over others. Compared with the lives of the masses of black Southerners, his was safe and secure. Teddy attended high school at the Avery Institute, the only secondary school in Charleston for blacks. (Public schools ended instruction for black children with eighth grade.) It was in school that Teddy realized he could draw, a discovery that culminated in his painting *Lincoln and His Cabinet*. But even if he liked rendering figures, it probably never crossed his mind that he might work in art. In a climate where lynching defined the limits of black life, it would have seemed absurd for a colored person to consider becoming an artist.

In the early 1900s, nonwhite Americans were lucky to get enough schooling to read. The black population measured about nine million, among whom fewer than two thousand were enrolled in college. But Teddy belonged to the tiny elite. After graduating from high school in 1900, Teddy moved to Georgia, where he enrolled in Atlanta University. Atlanta was a thirty-five-year-old black institution whose curriculum focused on classics and the sciences. Its mission was to produce teachers and as many white-collar Negroes as possible. But in the early 1900s, Atlanta University consisted of a few buildings out in some fields west of the city. The school motto told much of the story: "I'll find a way or make one."

When Teddy first arrived in Atlanta, at age eighteen, he had his photograph taken, and in the picture, he looks pudgy, wary, and unsure of himself. Even young men with his advantages didn't travel much away from home. The school had no art department, which meant that Teddy's desire to draw fell dormant. Joining the freshman class, Teddy settled down to study Latin and the ancients.

According to a transcript, he spent his first semesters on Greek, algebra, Latin, history, and elocution. The elocution requirement was an attempt to fight the plantation dialects that some students brought with them from home. But Teddy's diction, drilled into him by punctilious elders, was already precise and correct, so he dispensed with the speech requirement and turned to a heavy diet of early literature. He worked his way through separate courses on Ovid, Pliny, Horace, and Aeschylus.

After a year or so, Teddy began to shake off his shyness. He found he had a talent for acting and tried out for plays put on by the drama department. It appears he had roles in Aeschylus, or perhaps in some adaptation of Homer, because at some point he took on the nickname "Teddyseus." In 1904, Teddy played the lead in a production called *The Shadow,* in which his character was a Native American named Nam-bok. The play's cast consisted entirely of "Indian" characters, and the action took place in a small village on the northwest Pacific coast.

He also joined both the football and baseball teams, becoming captain in football and chief pitcher in baseball. A faded photograph shows the football team, about 1903, with Teddy decked out in the skimpy leather padding of the period, holding the football, which according to the design of that time was plump and round.

Course requirements seemed endless. In a practical mode, Teddy took trigonometry, chemistry, and physics. Over in social sciences, he enrolled in history, sociology, political economy, and pedagogy. He proved to be an average student, with grades that ranged from a low of 50 in modern history (spring 1902) to a high of 85 in sociology (fall 1903). After four years, his overall average came in at 69.8—low, but not too far from normal in those days.

Part of the trouble with his performance was the amount of time Teddy spent on women. A photograph taken after a year or two in Atlanta shows him to have bloomed into a beautiful young man. His cheeks had become alluringly hollow, and his eyes gleamed with confidence. Women responded to him. Several letters addressed to Teddy have been handed down, each one written by a different college girlfriend. The letters alternate between gush and primness, but even their variety throws light on his loves.

One letter, from a woman named Inez Canty, said:

> [I] loved you and always have since the morning I walked into my Elocution class and saw you standing by the organ, for how could I ever cease to love the only man who has tried to show me the good that there is in living? I am sorry that you were so ill Xmas—if only I could have been there to wait on you! Ah! Ted, if I take what you say as a test of real love, then you are the only one who ever loved me. "The saddest words: It might have been." But there is no more danger now, for it has been many months since another's lips have touched mine.

In February 1903, a woman who signed only "CSM" wrote:

> My dear,
>
> I would rather stay in [the dormitory] and keep my afternoon work in order that I might keep my many pleasures. Greatest of all of these is mealtime, for then I am near you, even if it is a case of "so near and yet so far." I . . . miss seeing the "dear boy" three times a day.

A girlfriend who had left Atlanta to study music at Oberlin College in Ohio signed her letter "Chihuahua":

> I think of you and wonder, do you ever entertain a thought for a little maiden who said she would go with you? Of course you do. I really wish you were here. There are a good many nice fellows here, but the girls in my house are crazy over your picture.
>
> Yours always, Chihuahua.

And a fourth woman, who called herself "Little Bit," wrote this from Louisville, Kentucky:

> As it was written to me that you were sick, I am anxious about you. Being so far away from you, I don't want my Teddy to be sick. You must be sure and send me one of your pictures, Teddy, as I am just pining to see something that resembles you. This will make two letters you owe me now.
>
> Yours always, Little Bit.

It's difficult to know whether Teddy slept with his girlfriends. Although his sister Ella had relented and taken the Reverend Mr. Jenkins to bed, pressure remained high on college students to leave sex alone. Still, one letter hints at more. When middle-class girls became pregnant, and their parents knew it, the family often sent them away, sometimes to a convent, where the baby could be delivered out of sight. Another destination might be a home for unwed mothers, where the child was ultimately taken away for adoption. "CSM," who had written about missing her boyfriend "three times a day," sent Teddy this enigmatic note:

> Teddy,
>
> But it is all over, and I must go to the sure-enough convent for nearly three long months. It will be hard to reconcile myself to this change, but I guess I'll get used to it after a time, and feel strange when I come back again. One good thing. I won't have to keep quiet hour on Sunday when I am in the Home.

Whether "Teddyseus" had a child he never knew is an unanswerable question.

The start of the 1900s brought with it the hardening of apartheid. After the Supreme Court decision in *Plessy v. Ferguson,* restraints on unfairness had disappeared. A campaign against black citizens now spread in all directions—social, economic, and political—as states exercised their federal right to deride and despise the colored masses.

A key strategy was to starve black Southerners of education. After 1890, legislatures throughout the South started removing money from state budgets for "colored" schools and reapportioning it as funding for white ones. In South Carolina, the state education department adjusted its budget until, in a typical year, twelve dollars was allocated for each white pupil and one dollar for each black. A building boom ensued that resulted in hundreds of new schools for whites, while blacks continued to attend shanties. Students of color were also expelled from traditionally white colleges across the South. At universities, black degree candidates were no longer allowed in graduate schools of any kind.

In the general poverty of the South, white citizens struggled over scarce resources with blacks; white politicians and businesspeople wanted to put an end to black upward mobility before it had a chance to begin. In 1900, nine out of ten Southern blacks still lived in rural districts, where they worked as tenant farmers or pickers of cotton. The few colored families that had slipped the net now found jobs closed, and credit and business opportunities blocked. A call went up for "industrial education," meaning training for black workers in carpentry, sewing, brickwork, or nursing. Black leaders, such as Booker T. Washington of Tuskegee, cooperated in the movement. The model of a successful black businessperson became the dressmaker or the barber, not the store owner or landlord.

The disappearance of the vote had already occurred. Once black balloting ended, the political parties filled up with shrill bigots. In South Carolina, Coleman Blease, governor after 1910, became famous for his endorsement of lynching, which he called a "necessary and good" means of social control.

Though few whites actually participated in lynching, the majority took seriously the social separation of the races. In Baltimore, ordinances forbade people from moving onto city blocks where the other race (white or black) counted in the majority. In Virginia, city governments made maps of neighborhoods and labeled them either "B" or "W"; anyone who moved across the race line was charged with a misdemeanor. Black people were forced to use the rear door when entering businesses and the houses where they worked. This was also the period when the infamous symbols of apartheid, "colored" bathrooms and drinking fountains, became ubiquitous south of Washington, D.C. Separate railroad cars appeared, as did colored seating sections in theaters.

The creation of the Jim Crow order pushed black people into a cave and removed them from white consciousness. By 1910, when the process was largely finished, there was almost no point of contact between blacks and whites outside the workplace.

For a time, Southern blacks rebelled against the new system, which politicians described as the "settlement" between the races. In the first decade of the 1900s, blacks in twenty-five cities (from Richmond, Virginia, to Jackson, Mississippi, to San Antonio, Texas, though not Charleston) organized boycotts against trolley companies with segregated cars. But none of the policies were changed.

With the toughening of segregation, the Harlestons and their class lost some of their footing. White Americans no longer saw a distinction between blacks and "mulattoes." A person of color might have noticeable white ancestry, but whether she was college trained or an illiterate field hand no longer mattered. Previously, the colored elite had separated themselves from the black masses. Now they were forced into society with them.

A popular postcard summed up the change in atmosphere. It showed a young black girl, smiling broadly, and gave the caption "Just a nigger—Charleston, South Carolina."

At Atlanta, Teddy continued to draw and sometimes tried to paint. Though he had little formal training—there may have been some when he was a boy—his pictures were surprisingly good. He probably showed them to his professors and asked their opinion. But art was a luxury that almost no colored people could afford in their lives, either as consumers or producers.

Teddy won his best grades in sociology and political economy, under a young professor named W. E. B. Du Bois. Born in Massachusetts, Du Bois was a mixed-race New Englander very much in a correct Harleston mold. He had a bachelor's degree from Fisk University in Nashville and a Ph.D. from Harvard, and had done graduate work at the University of Berlin. When he met the teenage Teddy Harleston, Du Bois, thirty-three, was a worldly but little-known intellectual. He had written a dissertation about the slave trade and had published a sociology book called *The Philadelphia Negro*.

Du Bois apparently saw something in Teddy. The young professor, who was married and had a daughter, formed a friendship with his student. Du Bois reminded Teddy that there was a world outside the Jim Crow South. He told Teddy that he was privileged and drilled him in the idea of service and duty "to the race." It was up to him to decide what he would do with the advantages he had.

Teddy's friends at college were not always of the altruistic sort. His best companion was an aspiring business student named Truman K. Gibson. Gibson, who went by the name Gib, was a son of a prosperous family in Georgia, and he mainly wanted to make money. Another of Teddy's friends, Julius Westmoreland, shared Teddy's love for art, but like Gib, he, too, had a business career in mind.

After graduating, both Gib and Julius applied to Harvard Business School. When they were accepted and left Georgia, Teddy felt he had been abandoned.

Captain Harleston wanted his son to come home to Charleston, where he could help run the family business. Teddy was much better off than his father, who had never gone to college, but the last thing he wanted was to go home to join the Captain's world. Instead, after he graduated in 1904, Teddy stayed in Georgia, seemingly aimless. A family member says that "Teddyseus" had a wide streak of indecision; he procrastinated until things could no longer be put off. To support himself after graduation, Teddy worked as a waiter in a country club. That experience, bowing and scraping to well-heeled customers, many of whom called him "boy," could not have been a happy one.

From what he did next, it appears Teddy had fallen under the influence of his teacher Du Bois, as well as his friends Gib and Julius. It may have been in conversation with Du Bois that Teddy decided he would try to get out of the South; or it may have been his friends who were just off to Harvard. By chance, Du Bois was also a

Harvard alumnus. Whatever the reason, in the spring of 1905, Teddy sent his transcript up to Massachusetts. The idea was to stay in school and to join his friends Gib and Julius.

Teddy applied to the Harvard art department. The record doesn't reveal whether Du Bois recommended him, but he was pushing ahead to study painting. From a practical standpoint, this was a ludicrous idea. As opportunity vanished around the South, it was urgent for Teddy to acquire a profession that might keep him in a starched collar and out of overalls.

After just two weeks, an answer came from an admissions official named George W. Cram:

> Harvard University
> May 11, 1905
> Dear Mr. Harleston:
> I have examined your application of April 26th and I wish to inform you that on presenting evidence that you have received the degree of A.B. from Atlanta University, and have completed creditably the courses named in your application, you will be admitted next fall to the Junior Class of Harvard College.
> As a candidate for your degree, the Committee will require you to pass with grades averaging C to B, five approved courses each year.
> Very truly yours,
> George W. Cram

The letter of acceptance, however curt, must have lifted Teddy's spirits. He could get out of the South and stay away from the funeral business. But there was a stipulation: he would have to enroll as an undergraduate again. Having finished four years of college, he had little desire for more.

Turning over his options, Teddy, twenty-four, made the impractical choice. Ignoring his father's appeals to come home, in the summer of 1906, he packed up and moved to Boston. With the help of Gib, he found an apartment at 23 Harwich Street. According to a letter he wrote about the incident, Teddy then visited the Harvard art department. He was reluctant to enroll, because his pride was at stake. By requiring him to register as an undergraduate, Harvard sent the clear message that his "Negro college" wasn't good enough. "I was advised to enter the Museum School," he wrote, describing his meeting with Harvard faculty, "because Harvard's work in Fine Arts was mainly theoretical and historical."

That fall, Teddy decided against the Harvard offer. In September, he went to Boston's Museum of Fine Arts. He had taken the advice of Harvard's teachers and would enroll in the museum's art school.

The Museum of Fine Arts occupied a redbrick building on Copley Square, a few blocks from the Charles River. To the east lay the neighborhood of Back Bay, home to a solid nineteenth-century upper crust. To the west stood Brookline, a new suburb. Founded in 1870, the museum represented the genteel aspirations of Boston's largely Protestant, mainly Episcopalian, elite. For a generation, its patrons had been building one of the most important art collections in North America.

While growing up in Charleston, Teddy had rarely seen the inside of a museum. His hometown had just one institution, the Charleston Museum, though another, the Gibbes Art Gallery, opened the year he went to Massachusetts. Both institutions barred nonwhite visitors at various times. When the Charleston Museum finally allowed nonwhite patrons, the "colored hours" would be restricted to Saturdays between 3:00 and 6:00 P.M.

At Boston's Museum of Fine Arts, the art school occupied the third floor. Established in 1877, the school looked to European movements in art and translated them for American students. Most of the instructors had trained in France. They regarded themselves as messengers of modern taste, which naturally they saw as an import.

The School of the Museum of Fine Arts (SMFA) admitted Teddy immediately. And luckily by this time, Captain Harleston's resistance had worn down, and he was offering Teddy grudging support. At SMFA, Teddy was the only black student in a new class of eighty-five students. According to an annual bulletin, tuition was $125 per year, and the school had 235 pupils.

When he stopped to think about it, Teddy must have been thrilled. He was doing the most unlikely, most pleasurable thing he could imagine. He had arrived at a safe haven. His friends Gib and Julius, enrolled at the Harvard Business School, lived a short (unsegregated) trolley ride away.

But not everything felt like success, and Teddy must have noticed he was surrounded by a strange peer group. A photograph of Teddy with some of his schoolmates shows that, not surprisingly, many of the other art students were both white and rich. In the picture, the men are uniformly well dressed and look exuberant, as though they had just won a squash match. The women are equally precise with their clothes, although more subdued. Teddy noticed that the school was especially rich in proper white girls, who were often dilettantes, sent to the school by their parents to round out their education. By the standards of black Southerners, Teddy's life experience had been soft; but he was far from the class rank of his fellow students, some of whom were merely languid heirs and heiresses.

Teddy may have felt exhilarated finally to be in art school, but he was probably lonely as well. Not only was he the only black person in his class, but he was a Southerner. According to the school's records, most of his fellow students came from Massachusetts and Connecticut. In the minds of many New Englanders, the South was a suspicious, even barbaric, place.

The shock of having access to a white institution was great enough. But Teddy's surprise must have increased when he showed up for one of his first required courses: nude drawing. Every week, the drawing instructor, a painter named Frank Benson, brought a new white woman into the studio, where she nonchalantly undressed. The models were often Italian immigrants, working-class women who needed money, and who left immediately after class; but that hardly mattered. In some parts of the South, a black man could be killed for staring at a white woman on the street. (According to the prevailing etiquette, a Negro was compelled to cast down his eyes if he met the gaze of a white woman.) Now Teddy was attending a class that required him to gaze.

One of Teddy's sketchbooks from the period survives. It contains a small graphite nude of a young woman seen from behind, her body curving at the hip to form an S shape. Both Teddy's rendering and the model's pose borrow from nudes by French painter Jean-Auguste-Dominique Ingres.

When Teddy Harleston enrolled in art school, there was virtually no such thing as a black painter. He became one of perhaps only ten serious "Negro art students" in the United States. The best-known African American artist of the day, practically the only one, was Henry Ossawa Tanner. Teddy's letters and notebooks show that he thought about Tanner and took the model of his success to heart.

Born in Pittsburgh in 1859, the son of a Methodist bishop, Tanner had studied at the Pennsylvania Academy of the Fine Arts in the 1880s. He painted portraits of black people for several years, but the market for pictures with black faces was tiny, so he turned to religious subjects. In his late twenties, Tanner moved to Paris, where he produced the first of a series of famous paintings whose narrative content and polished execution appealed to middle-class French taste. Beginning with *Daniel in the Lions' Den* and *The Resurrection of Lazarus,* Tanner moved speedily through biblical themes. In a few years, Tanner's fame was exported back to America, where his paintings found their way into galleries and museums.

Throughout his most productive years, Tanner lived in Europe. In addition to his Paris apartment on Boulevard Saint-Jacques, Tanner had a rural studio in the small Picard village of Trepied, in eastern France. It could not have been lost on Teddy Harleston that Tanner had chosen France and rejected his homeland. Tanner stayed in Europe most of his life, possibly because he knew he could not gain the same level of acceptance in America.

All of Teddy's teachers in art school were white. He studied anatomical drawing with a draftsman named Philip Hale, and painting with an artist named Edmund Tarbell. Tarbell served as director of the school. Born in 1862 in West Groton, Massachusetts, Tarbell had studied in Boston at the School of the Museum of Fine Arts and afterward in Paris. In 1889, at age twenty-seven, he started teaching. The following year, Tarbell won a prize at the National Academy of Design, a conservative atelier and arbiter of discreet taste in New York.

So that they might absorb the proper values, students were required to take an "antique drawing" course from faculty member William Paxton. Paxton sent his class downstairs from the school on the third floor into the museum galleries to copy Old Masters. A notebook containing some of Teddy's studies shows that he penciled details of paintings—a torso by Rembrandt van Rijn, a leg by Jacques-Louis David. Imitating artwork he could not even see in the South, Teddy honed his technique.

By the early 1900s, the French domination of American art was complete. All of Teddy's teachers had studied in Paris. Frank Benson, who taught figure drawing, was an American Impressionist. Born and raised in Salem, Massachusetts, Benson trained at the Académie Julian in France. When Teddy enrolled, Benson was forty-three and had already been on the school faculty seventeen years. Benson seems to have been a pretentious man, because his pupils were known to call him *"Cher Maître"* (Dear Master). Perhaps this was because they revered him, or he may have forced them to use false honorifics.

Frank Benson was a well-known painter who had frequent shows and who became wealthy selling his art. Like French artists of the period, Benson cultivated a painting technique that concentrated on the representation of light and the capture of a fleeting moment in time, figurative values he taught in the studio. As a teacher, Benson pushed his European aesthetic on his students, especially the style of the Impressionists. But as an artist, Benson was a flatterer. He specialized in painting rich people at leisure. His *In the Orchard* (1891), a large canvas almost six feet square, shows five young people, finely dressed and relaxed, sitting in wood furniture in an apple orchard. *Mother and Child in Boat* (1892) depicts a poised mother with a baby in her lap, sitting in a rowboat surrounded by glittering water. Under the influence of Benson, Teddy and the other students at SMFA were encouraged to depict handsome, prosperous Americans, taking their relaxation on manicured lawns, or holding parasols, or seated in rooms with expensive decor.

Under such guidance, Teddy soldiered on, taking daily notes of his teachers' comments. One notebook that survives shows various recipes he used for color, from a time when artists still mixed their own paints:

Mr. Paxton's recipe	ochre	permanent blue
	chrome	black
	vermilion	white
Mr. Tarbell's recipe	rose madden	zinc yellow
	cobalt blue	black
	white	

Just as he had been in Atlanta, Teddy was a striving but uneven student. The nude studies on canvas that Teddy painted were not as good as his pencil drawings. In one, he captures the entire body of the model, but the space she occupies seems flat. He could represent the figure but had trouble with perspective. By the end of his first year, Teddy was runner-up for a prize in drawing, but he had failed William Paxton's course in anatomy. His anatomy sketchbook, full of not-quite-good drawings of skinless human bodies, tells the tale. In the drawings, the muscles are red, the tendons blue, and labels are added to both. The final result looks like weak medical illustration. Pulled out of the old South and thrust into a pseudoaristocratic New England milieu, Teddy may have felt like an anatomy specimen himself.

As though to underline Teddy's feeling of being an outsider, during his second year, a student from his hometown enrolled at the school. Her name was Louisa Stoney. A proper young woman from a family that also included the mayor of Charleston, Thomas Stoney, Louisa Stoney seems to have come to Boston to burnish her upbringing as a lady. At home, Mayor Stoney was friendly with Captain Harleston. But it's not likely that in Boston, apart from occasional greetings, Louisa Stoney and Teddy Harleston socialized. The two students knew that a colored man and a white woman who spoke to each other, even a thousand miles from home, ran the risk of disastrous gossip.

No letters from Teddy's girlfriends in Boston survive. For companionship, he had Gib and Julius, but he also spent time with a black couple, William and Ada Hinton, who had moved to Massachusetts from Georgia. William Hinton was a research physician who had graduated from Harvard Medical School, then gotten a job there. Teddy had known Ada at college. The Hintons lived in the town of Canton, Massachusetts, a short train ride from Boston, on a small farm, and Teddy often spent weekends with them. He felt comfortable in their company. They were cultivated, and they had nothing to do with the all-white art world.

Captain Harleston did not like his son's chosen vocation, but he tried to support it. He sent money, though not enough to cover the bills. Kitty Harleston, Teddy's oldest sister, then single and living on a teacher's salary, also sent infrequent money orders; these were gone within a month. Teddy gradually slipped from a respectable standard of living down into an artist's poverty. Family tradition says that he did all sorts of things to earn money. He worked as a waiter at Harvard's Colonial Club and gave art lessons. He worked as a crewman on Canadian steamships that ran out of Gloucester, Massachusetts. As an art student and a ship's hand, Teddy began to affect a bohemian lifestyle. He became increasingly poor and more self-reliant. He spent his money on art materials and dressed in used clothes.

He did get occasional art commissions, but they could not have paid much. A letter from 1912 documents one of Teddy's earliest portraits. In April, F. B. Sanborn of Concord, Massachusetts, wrote the art student to suggest the artist meet him at 98 Chestnut Street to begin sittings for Sanborn's portrait. "Suppose we begin at 2:30 p.m.," said the client.

A photograph shows Teddy's new, somewhat tough art-school persona. In 1910, a painting student named Philip Adams asked Teddy to sit for a painting. The result was a portrait called *The Sailor*. The painting attracted the attention of the *Boston Post,* which sent a photographer to shoot Teddy next to his own likeness. In the picture, Teddy wears a rough open shirt and slouches in his chair, his hair tossed. He looks not only like a sailor but also like a first-rate rumpled student.

* * *

It took Teddy six years to make his way through art school, because he sometimes dropped out for a semester, returning when he could again pay the fees. But by 1912, he had accumulated an impressive group of paintings. They resembled somewhat the work of his instructors. But the people in Teddy's work looked plainer, less fancy than the New England leisure set. And in each picture, there was something in the eyes.

He had become a specialist in eyes. Teddy Harleston's paintings, even the student work, are as sensitive about the depiction of the gaze as John Singer Sargent's. A self-portrait he executed shows the product of all those years in school. In the painting, Teddy is young and fit, but he stares at the viewer with experienced eyes. His eyes are no longer shy, but worldly and resigned. By leaving home, Teddy had given himself veteran eyes, eyes that were sadder than the grave.

Chapter Seven

IT was Thursday, March 4, 1909, and in Washington, D.C., the inaugural parade of President-elect William Howard Taft moved slowly through the streets. There had been a blizzard the previous night. Snow and slush filled the wide swath of Pennsylvania Avenue, and Taft's procession crept along as best it could. The crowd waved, but the big Ohio Republican (the president-elect weighed more than three hundred pounds) was not the only spectacle on view. A marching band from Charleston had also been asked to parade. The rattle of drums echoed between the buildings, and a child waving a baton danced ahead of a group of young black trumpet and trombone players. As Taft rolled past, two children could be seen carrying a banner painted with the words "Jenkins Orphanage Band."

William Howard Taft, fifty-one, a graduate of Yale and the son of a distinguished Ohio judge, was not known for his joy around black music. He had come to the presidency via his cabinet post as secretary of war under President Theodore Roosevelt, who tapped him as his successor in the Republican Party. But five years earlier, Taft, as governor of the Philippines, had been present at the opening ceremonies of the Louisiana Purchase Exposition, a trade fair in St. Louis, Missouri. The Jenkins band had performed for the dignitaries. It seems Taft liked what he heard, and he later approved their invitation to his inauguration.

There was another reason the president-elect wanted the Jenkins band nearby. Taft's Republican Party was the party of Abraham Lincoln. Though Southern blacks had lost the vote to Jim Crow, most blacks in the North, out of loyalty to Lincoln, still voted Republican. At a time when the Republican Party no longer pushed black candidates for any office, word of mouth would get around to Taft's colored constituency about the inclusion of the Jenkins band. It would be a cheap payback for their support.

The inaugural ceremony took place in the Senate chamber, inside the Capitol. When it ended, Taft emerged to deliver his speech. He began by acknowledging that many white Southerners, Democrats all, disliked him and held resentments dating from the Civil War nearly fifty years earlier. Then he offered a somewhat controversial opinion. The Fifteenth Amendment, which gave the vote to black Americans, should be enforced. Black people in the South should have voting rights.

"While the Fifteenth Amendment has not been generally observed in the past, it ought to be observed," said Taft. "The Negroes are now Americans. Their ancestors came here years ago against their will, and this is their only country, and their only flag. They have shown themselves anxious to live for it and to die for it. Encountering the race feeling against them, subjected at times to cruel injustice growing out of it, they may well have our profound sympathy and aid in the struggle they are making. We are charged with the sacred duty of making their path as smooth and easy as we can." Taft's words represented more payback.

Later that night, at the Pension Building, tuxedoed crowds gathered for the inaugural ball. Though the Jenkins band had been entertaining, the boys were probably sent away from the door. Whatever President Taft's views of his "sacred duty," loud black children would not have been welcome in a room full of Republicans.

Ella Harleston was twenty-seven years old, to Parson Jenkins's forty-eight. According to a brochure put out by the Jenkins Orphanage, by 1910, Ella worked as a teacher, supervising the orphan girls. Her lover's wife, Lena, was still alive, though barely, and many of Lena's own children with the Parson were grown.

Four years had passed since Ella and the Parson left their daughter Olive with a stranger in Wigan, England. Olive Harleston barely knew her parents, and under the care of Alice Layland, the midwife, she had become Olive Layland, the nurse's mysterious black daughter.

In addition to being forced to give up her child, Ella had been made to suffer back home. The birth had not remained secret, and the Harleston family was the subject of much snickering gossip in mixed-race circles. Captain Harleston disliked the embarrassment, which caused him to carry a vendetta against Jenkins and to treat his daughter with scorn. Teddy Harleston also ignored his sister, punishing her with silence. The frigid judgments of her own family and the Harlestons' feud with her lover left Ella desolate and withdrawn.

By contrast, Parson Jenkins paid very little for his wayward libido. His public reputation as "the Orphanage Man" surged upward every year. "Among the most influential Negroes in the state today is Rev. D. J. Jenkins, the President of the Colored Orphanage," one newspaper said. "The successful institution which he controls is a splendid monument to his industry. Mr. Jenkins has proven himself a public benefactor. His mind bristles with practical ideas."

The standoff between the Harlestons and the Parson lasted for years. In September 1912, earth between the families finally moved. That month, Parson Jenkins's wife, Lena, died from tuberculosis. He worried about the damage his reputation might suffer from marrying so soon after his wife's death, but not so much that he would abandon Ella, whom he adored. The couple waited a short time after Lena's burial, then, in the custom of the day, the Reverend Mr. Jenkins approached Captain Harleston to ask permission to marry his daughter. The Captain became furious. It was bad enough that Ella had been seduced, but Jenkins came from the poorest and darkest sector of society. The marriage would not only advertise his daughter's sex life; worse, it would mean a loss of status.

The family story goes that Captain Harleston was in a choir rehearsal at his church when, across town, the elopement took place. One of the Captain's brothers ran into the sanctuary and yelled, "Ella is running off with Reverend Jenkins!" Another man shouted, "Wait a minute! I'll get my gun!"

At the New Tabernacle Fourth Baptist Church on Palmetto Street, Ella was in the midst of a hurried wedding to the Reverend Mr. Jenkins. Captain Harleston ran toward the church with his armed friends. The service ended, and the couple got word they were being pursued. The Parson and his bride hurried to make their getaway. They ran for the riverboat docks and boarded a ferry going inland, toward the middle of the state. Captain Harleston arrived at the wharf just behind them, waving a pistol, as the boat pulled out from its slip.

And this was how Ella Harleston began married life. A few days later, from his unplanned honeymoon, the Reverend Mr. Jenkins sent the Captain a letter.

> October 4, 1912
>
> Dear Sir,
>
> I am inspired to write you this letter, because I regret equally as much as you the occasion that happened between us. The letter that I received from your daughter, Kate, and your message to your daughter [Ella] over the phone Monday morning ordering her home for good unnerved me. . . . Thus I became throughout the day uncontrollable. Your not listening to my explanation after I mentioned the license was the climax upon which I married your daughter.
>
> I know this is painful to you, as much as it is to her and myself; but it was a short-cut to avoid longer dissatisfaction. I therefore ask that you overlook the past and let us look forward to a brighter future. Both you and I are public men, and to some extent our reputations are at stake, and we cannot afford for others of our connection, for whom we are living, to hang their heads in shame. I promise you to make your daughter happy, and so long as I live, she shall not want. . . . My reason for leaving the city was that I was angry; but I am over it.
>
> Respectfully yours,
>
> D. J. Jenkins

Ella's language was impeccable where her new husband's was not, and she may have written this letter to her father. In the climate of cold fury, it was an attempt at a thaw.

Two weeks later, Ella and her husband returned to Charleston. They moved into the Jenkins house a couple of doors from the orphanage. There is no record of the Captain's response to the Parson's letter, but the undertaker did eventually swallow his anger. In future dealings with his son-in-law, he would be distant, but he would leave the gun at home.

The Reverend Mr. Jenkins's marriage to Ella turned out to be a good commercial decision. After the wedding, Mrs. Ella Jenkins began to run the orphanage as her husband's business partner. Jenkins operated in the public sphere, keeping up the conversation with whites, raising money, and recruiting support; Ella ran the office, handling the correspondence and keeping the books. Cool, literate, and sweet, she became the lubricant for the rough parts of the machine, and she knew how to count the income. A story goes that on the nights before meetings of the orphanage advisory board, which consisted of only white men, Ella often would stay up till dawn to make sure their accounts were in exact order.

Life at the orphanage began with the intake of new children. After several years of accepting only inmates whose parents were both dead, Jenkins began to take the offspring of single parents, or "half-orphans." He charged the parent a monthly fee, but it was rarely collected. Boys and girls usually entered by one of two means. Most often, surviving family sent a one-page application requesting a place for a child. Ella answered it, turning down children when they were too old, or when too many siblings already lived under the Jenkins roof. New inmates also came from the police. When police patrols picked up juvenile offenders, their first call went to the boy's family; the second call was to the Jenkins Orphanage.

One November, in a typical case, the Charleston police department delivered three boys to Parson Jenkins. An officer accompanied them and presented this brief profile:

Name:	Benj. Seals
Age:	13 yrs.
Address:	lives with aunt
Mother:	Dead
Father:	Not known
Note:	Bad character—solitary confinement requested

Name:	John Mouzon
Age:	15 yrs.
Address:	2½ Rose Lane
Mother:	Vinnie Bradley (dead)
Father:	Not known
Note:	Resides with sister

Name:	Sherman Hosey
Age:	6 yrs.
Mother:	
Father:	
Address:	(former) 61 Bay Street
Note:	Mother arrested, child brought to orphanage by police officer to be taken care of until other arrangements could be made.

When a fresh inmate arrived in his office, escorted by Ella, Jenkins subjected him or her to "the treatment." This was a hazing action, the better to control the child later. Seated silently at his desk, the minister would study the child, who stood beside Ella. If it was a boy, he was either cocky (a young criminal in the making) or sullen (abandoned by his mother). After a minute, Jenkins got up, walked slowly across the room, and put his hands entirely around the boy's head. The Parson had large hands that could swallow a sizable skull. He would feel the shape of the cranium, like a phrenologist, and look off in the distance in contemplation. Continuing the examination, he peered into the boy's mouth and ears and stared at his hands. Finally, Jenkins sat down, the theatrics having impressed the child with his mysterious authority. A minute later came the verdict: "Put him in the print shop." Or,

"Send him to the farm." Or (the prize assignment), "Put him in music." The new inmate was whisked off and assigned a cot in one of the dormitories.

Though the number of children at the Jenkins Orphanage changed from year to year, the rolls hovered around two hundred. Boys, who were always the majority, slept in the rear of the building at 20 Franklin Street. Girls lived nearby, on Beaufain Street, in separate housing. By 1910, the whole population was supervised by a staff of some thirty teachers, "matrons," administrators, kitchen workers, and porters.

The daily regime began at 7:00 A.M., with a chapel service. Breakfast followed. A heavy, cheerful woman named Mrs. L. J. Walker supervised the kitchen and dining room, where the children ate at long wooden tables. The Parson, describing the orphanage diet, once wrote, "When I was a boy, I was put to work and given six quarts of corn meal, two pounds of meat, and a quart of molasses a week. I am persuaded that plenty of cornbread and molasses is the best food in the world for children. It keeps down illness. However crude it may seem, it is most profitable to any race or nation."

The school for the children extended from first through seventh grades. (High school was neither provided nor expected: Charleston wouldn't open its first public high school for black students until 1929.) Classes were supposed to last several months a year, but teachers spent endless hours merely gaining control over the students.

For most of its inmates, the Jenkins Orphanage was neither an academy nor a music conservatory but a trade school. Parson Jenkins was trying to create an "industrial education" for the black masses. The orphanage had a tailoring shop and a shoemaking department, both for boys. According to a pamphlet, "The girls are taught cooking, dressmaking, hat making, general housekeeping, and how to be good maids and nurses." It might have been industrious, but the labor was preindustrial: clothes made by hand, washing scrubbed on a board.

Moneymaking operations grew every year. The Parson opened a bakery called the Poor Boy Bread Company, whose ovens soon produced a thousand loaves a day, which were sold throughout the city.

A farm was acquired outside town. About 1900, a Brooklyn clergyman named Joseph Wild had given the orphanage a hundred-acre piece of land at Ladson, South Carolina, twenty miles north of Charleston. The farm was named the Jenkins Industrial Reformatory, and it became a kind of boot camp. Boys who were too defiant to work at a shoemaker's bench went into the fields at Ladson, and the farm soon provided much of the food for the hundreds of inmates, from sweet potatoes to corn, to the occasional hog.

Often, the children with the worst jobs ran away, only to be retrieved by the police. A letter to the Reverend Mr. Jenkins from Clarence Lunz, a probate judge in Charleston County, tells the story of three fugitives. "Arthur Washington, Frank Graham, and Willie Heyward," wrote Lunz, "who were brought before me this morning for sleeping in a vacant house and for stealing, have informed me that they recently escaped from your home. Therefore I am returning you them by Detective McNeil for your care."

Daily life in the orphanage had a hard edge. Parson Jenkins himself doled out beatings, and physical punishment could come at any time. "He was all right to me," a former orphan recalled, "but if anyone got out of line, then Jenkins would whip their ass." The children also roughed each other up. In music classes, an older boy, holding a drumstick, would stand behind a younger one during practice. If the beginner missed a note, the tutor would swipe him with the drumstick. A band member remembered how a music teacher named Amos Gilliard would arrive at rehearsal and put a brick on the table. Gilliard would then shout, "I'll throw this at the first boy who makes a mistake!"

The back-and-forth of punishment and hell-raising continued around the clock. In one case, which was not too rare, police entered the children's dormitory to arrest a boy named Ford Perry. When Perry's guardian, an aunt in the town of Spartanburg, wrote to ask for details, Ella wrote back calmly. The police had taken him, along with his clothing trunk, she said. They had informed the Parson that Perry had both "a pistol and stolen jewelry." Furthermore, said Ella, Perry had "fitted a key to the matron's room and used the phone, running up a large bill." The aunt in Spartanburg wrote back to apologize.

A good view of the daily clash can be seen in a two-page memoir written by a former inmate. The rough English of the boy's story gives an idea of the schooling the orphanage offered its children:

> Ka ling a ling! Ka ling a ling!
>
> With "Big David" coming through the domitories ringing a big handbell, in about five minuets the boys woold hafter get dressed and get ready to go to chapel, for about in five mineutes after "Big David" got through ringing his bell, Rev Jenkins came along with a long cow whip. And anydody that where still in bed ol Rev would sting them with tha cowhide a they would break a record getting to chapel. Benny Lance was a hard sleeper and every morning "Ol Rev" would hafter lift him out of bed with that cowhide whip, and Benny would almost touch the sciling he'ed jump up trying to find his clouths, runing and spiting over every thing. When everybody got in chapel including the girls—the girls lived on Befame Street so I guess they had to get up by four thiry to get to chapel by six: they had to walk a block and a half to get to chapel on #20 Franklin St. When everybody got there they called roll to find out who were missing. Every night a few of the boys would jump over the fence next to the dog pound. The county jail was right next door to the orphage. The boys would go out and burgreised and steal everthing in sight. And when they came back and got in bed, some of the boys pretending to sleep was watching them all the while. And as soon as they went to sleep, they would get up and steal whatever they had stolen. They called it "rolling them." In those days, the light skin Negro woudent associate with the dark skin Negro. The light skin Negros had a church on the way where we had to march to our church, so the kids ran them out of there thoughing bricks in there on them. They had to move.

The mainspring of the fund-raising machinery was the Jenkins band. The "band" was really two or three marching groups, whose names referred to their players' ability. "Jenkins #1" got the best assignments and traveled to the most interesting cities—New York, Sarasota, Boston. "Jenkins #2" was next best and got to Washington, D.C., Baltimore, and Jacksonville. Brass bands staffed with boys made up half the music program. A series of girls' choirs also appeared ("Choir #1," "Choir #2," etc.). Each sang a repertoire of religious music, hymns, and spirituals.

In the summer, two of the bands and one or two choirs traveled north up the East Coast; in the winter, they went south into Georgia and Florida. The bands moved as self-contained communities, each act with a manager, a cook, and a valet to handle the uniforms. (The band often got new uniforms on the summer stopover in New York, from a clothier called Brown & King Company; red was the frequent color choice, with matching hats.) Traveling ahead of the band to the next town, the manager checked with three powers in every locale—the mayor, the police chief, and the chamber of commerce—to make sure the band could play on the streets. A typical permit for the children to play was issued in Yonkers, New York, and consisted of a letter of introduction from the mayor:

> William J. Wallins
> Mayor
> Permission is hereby granted to Rev. D. J. Jenkins, to have the Jenkins Orphan Band play on the public streets in the City of Yonkers at such places as are indicated and set apart by the Captain of the Precinct.

During the early years, the orphans traveled in the dingy nonwhite seating section of the Clyde Line, a steamship company whose passenger boats served Charleston and a dozen other port cities. Later, Jenkins bought buses and trucks to haul the children and save on fares. The Parson deputized young men as drivers, who took the groups on the road for two or three months at a time.

One former driver was a nephew of Ella's, Harleston Fleming, the son of her sister, Kitty. Harleston Fleming recalls that when he was a junior in high school, the Parson asked him to drive one of the bands up to New York.

"That particular summer," Fleming remembers, "Jenkins had a girls band. I was assigned to them, and they rode in the back of a truck. There was a canopy over it, with about ten girls in the band, and a matron back there with them. These were teenage girls, who played the same instruments as the boys, but who wore dresses instead of uniforms. The band had a cornet, trombone, baritone horn, and the smallest tuba you could get, along with a drummer. When we arrived in a town, I wouldn't have to drive them to their performances. They would just walk two blocks this direction or that, and start another show. We didn't confine ourselves to the black neighborhoods. In fact, the black neighborhoods had the least money, so they veered toward the white. We seldom stayed in one place more than two days, and when the collections began to lessen, we would move to the next town. The band would play every day of the week, except Sunday, when we would find a church to attend.

"Rev. Jenkins didn't travel with us. He moved from one unit to another unit, following the choirs and bands in a separate car. When we got to New York, I became Rev. Jenkins's driver, and left the truck to someone else. Jenkins didn't believe in anything else but Lincolns. In his eyes, those cars had a reputation for durability. He needed a durable car, because he would come out of wherever we were staying, and say, 'Okay, we're going to New Jersey today.' Or, 'Today it's Virginia.' While I drove, he slept in the back seat of the car. Things went that way all summer, and he wore me out."

In the Jim Crow South, no hotels would accept black guests, so the children had to hunt for beds. Churches usually provided, and the bands camped out on parish hall floors, or parishioners put up band members in their homes in groups of four or five.

The boys' bands brought in the most money and were the stars of the tours. One account book from the early 1900s shows that two bands together took in some nine thousand dollars in six months. Collections were made on the street or from the stage. When he traveled with the groups, the Reverend Mr. Jenkins kept an eye on the proceeds and sent the money home by wire to Ella, who deposited it in the bank. But when he was away from the band, the minister trusted a band manager with the cash. The boys did their best to skim, picking up coins tossed at their feet. Sometimes their chaperones also touched the till. Family lore reports the Parson had a saying, often repeated: "The first count is the best count." He meant that when he could personally oversee the collection, there always seemed to be more than if someone else had passed it on.

In addition to street corners, the bands played at Baptist churches, such as the Tremont Temple in Boston and the Abyssinian Baptist Church on 138th Street in New York. (Parson Jenkins became friendly with the minister at Abyssinian, Rev. Adam Clayton Powell, and the concert became an annual event, on Labor Day night.) The band also had gigs in white hotels. In Florida, they became regulars at the Ponce de Leon in St. Augustine, the Royal Palms and the Breakers in Miami, and the Royal Poinciana in Palm Beach. The more the bands hammed up their act for white audiences, dancing and goofing off, the bigger the tips.

When the boys arrived back home in Charleston, their ritual was to stop the bus two or three blocks from the orphanage, so the band could march back in triumph, playing at the maximum volume. On these occasions (and whenever the band paraded locally), they would invariably raise a crowd of followers—a hundred or more children, most of them black, who trailed behind. The bands developed annual routines: every Christmas Eve night they would serenade prisoners at the Charleston city jail, next door to the orphanage, then march off to play under the windows of two or three prominent white families. For local children, being associated with the Jenkins band meant practically being famous. Numbers of children actually wanted their parents to put them into the orphanage under the Parson's care, so they could become musicians and travel.

In black neighborhoods, band members were heroes, but to many whites they were like rambunctious pets. White newspapers took a dim view and coined a name for the Parson's act: "the Pickaninny Band." The word "pickaninny" was a condescending way of referring to black children. A pickaninny was a white person's idea of a black boy—a dirty, ignorant, and mischievous urchin. Parson Jenkins no doubt knew this, but he nevertheless allowed promoters to advertise "the Pickaninny Band," knowing that it would boost donations.

The Jenkins bands came out of three traditions in American music. The first was the minstrel show, which dated from the 1840s. Minstrel shows were troupes of white musicians and comics who performed in blackface (dark wax or shoe polish), pretending to be

Negroes, and toured a tent-show circuit in the South. For fifty years, until about 1900, minstrel acts were the most popular entertainment for whites in small Southern towns.

In a typical minstrel show, the curtain went up to a big musical number like "There'll Be a Hot Time in the Old Town Tonight." An interlude would follow in which a comic might play a semieducated Jim Dandy, and launch into a tirade of fake preaching, or a lecture with malapropisms and invented Latin. A middle section of the show was the "olio," songs that went back and forth between individual blackfaced musicians. There was fast banjo music and high-stepping dance, followed by ballads and mawkish songs. The finale was often a medley that the whole troupe performed, ending with a chorus and line dance. The minstrel act rested on the twin pretenses that the performers were both black and foolish. After the Civil War, black performers began to appear on white stages for the first time, playing the same "darkey" roles perfected by white comics and wearing blackface over black skin.

One of the well-known minstrel acts was *The Creole Show,* which opened in Boston in 1890. It took the name "Creole" because it put dark black men onstage with light-skinned women. The women were thinly dressed, which added a sexual element missing from other shows. The finale of the show consisted of a dance called the cakewalk. A cakewalk was a parody of formal dances; it mocked the quadrilles that ended the ballroom parties of the white elites. In a cakewalk, couples in fancy dress would promenade around the dance floor to syncopated music, high-kicking and strutting in a wild and funny way. There were two explanations for the name *cakewalk.* In the first, it was said that slaves used to dance in front of the master's house at Christmas, and the couple with the best steps won the master's prize, a cake. In the other explanation, black roadhouse saloons were said to hold competitions for dessert sweets, and the best walkers would "take the cake." After *The Creole Show* opened, every black minstrel act had the dance, and cakewalk contests appeared all over the East Coast.

By the time of the founding of the Jenkins Orphanage, it was practically impossible for black performers to please any white audience unless they included these high-stepping dances and put-on stupidity. Parson Jenkins knew this and did his best to provide the entertainment that brought in the most money.

During the first decade of the 1900s, one of the singing groups at the orphanage was known as the Springfield Minstrels. Although their sheet music hasn't survived, the Springfield Minstrels probably sang a genre of popular music known as coon songs. A coon song was a tune with mocking lyrics about black people. A typical example was the popular 1896 ditty "All Coons Look Alike to Me." A black comic named Ernest Hogan wrote the song and performed it while wearing blackface.

In the early 1900s in New York, Broadway theaters flourished, and pressure increased to allow black performers who had previously been restricted to tents and small auditoriums to come onto the big stage. The first Broadway show that was written, produced, and managed by nonwhites was *A Trip to Coontown,* a musical comedy that opened in 1898. Later that same year came *Clorindy—the Origin of the Cake-Walk,* produced by musician Will Marion Cook and poet Paul Laurence Dunbar. Although the orphanage musicians were children and spent most of the year in a provincial Southern city, their teachers were adults who kept abreast of changing tastes. The genre of the "Negro revue," with jiving, dancing coons and mugging black comedians, brought minstrel acts onto the Broadway stage. When the Jenkins bands began their tours, it was part of what was expected.

A second tradition that provided a platform for the Jenkins bands was ragtime. In the second half of the 1890s, a new black music, the "rag," broke into white taste. Black piano players in roadhouse joints and on the minstrel circuit had developed the syncopated form. Ragtime took its name from the way the pianist would "rag," or corrupt, the traditional four-four beat. The sound of a rag was heard when the pianist played the bass or octave on one and three in the bar, and midrange chords on two and four. The result was a pounding and walking noise, a kind of "ragged time." Before it was named as a new style, the early inflections of ragtime could also be heard in the syncopations of marching bands with their back-and-forth, call-and-response play between instruments. But when Scott Joplin, a black pianist from Missouri, published his *Collected Rags* in 1899, the style had transcended the military bands. Joplin's "Maple Leaf Rag" became wildly popular and launched a ten-year craze for rag music in mainstream American taste.

Ragtime, cakewalks, and coon songs all made their way into the Jenkins repertoire. A program from one of their concerts shows the mix. In a typical show, Jenkins band #1 started the set with "Razzazza Mazzazza," a 1906 piece by the American songwriter Arthur Pryor. Pryor was a former trombonist with John Philip Sousa's band, who had left Sousa to make his own career. "Razzazza Mazzazza" was a rag in march tempo, about 120 beats per minute, with a dancing interplay of parts, perfect for horns. A bit later in the show came Pryor's "Frozen Bill Rag," from 1909, whose syncopated melody resembled the more famous song "Put Another Nickel in the Nickelodeon." The band would fill out the set with slower rags from composer F. H. Losey, with song titles like "Sliding Jim" and "The Gold West." These songs had a good barrelhouse sound and were less jittery than the Pryor pieces. The band's sets usually closed out with a number that gave the boys room to dance and make the crowd laugh at them. For this part of their show they often played a cakewalk, brisk and loud, called "Coon Band Contest."

The third source for the early sound of the Jenkins bands was vaudeville. Minstrel shows aged, ragtime grew stale, and the center of gravity for popular music moved elsewhere. A vaudeville show was a live variety act with two hours of nonstop entertainment, combining music, comedy, dance, juggling, magic, clowning, and mime. Vaudeville was the most popular form of entertainment in America between the 1890s and the 1930s, when the introduction of sound film killed it off. The name of the form came from an adaptation of the French phrase *voix de ville*, or "voice of the city," meaning urbane or worldly. Most towns had a vaudeville theater, and troupes would travel from one city to the next throughout the year. Where minstrel acts were regional, vaudeville went out from New York to the provinces and to smaller cities. Middle-class audiences bought the tickets, with blacks sitting in a colored section at the above rear (a mezzanine known as the "crow's nest").

Vaudeville was divided into chains of theaters, or circuits. The circuits hired the same group of acts from the same booking agents, such as William Morris and Keith-Albee, and theaters carried a brand name, like Orpheum, Paramount, or Hammerstein. Vaudeville had inherited much of its coon-song preferences from minstrel shows, so that when black performers appeared, they had to wear blackface and make fun of "Negro" dialect. But very occasionally, a black performer could cut through the racist haze and be heard.

Year after year, the Reverend Mr. Jenkins and his bands traveled up and down the coast, on the fringes of vaudeville, playing marching band standards with a rag beat and goofing their way through coon stunts. The pickaninny act paid the bills, but finally, a breakthrough came that took the band to a new level. An agent in New York who handled Broadway bookings (his name has been lost) hired one of the Jenkins bands away from its usual venues to do a new show on Broadway.

The act was to be a musical version of *Uncle Tom's Cabin*, the novel by Harriet Beecher Stowe. Published in 1852, *Uncle Tom's Cabin* was the largest-selling book of the nineteenth century. With its themes of slavery, faith, and love, the story appealed to white readers to sympathize with the African American characters' plight as slaves. After 1900, the book was adapted in theatrical and, later, film versions. Thomas Edison, using his new movie machine, the kinetoscope, made the first film, a fourteen-minute version of the novel.

The Jenkins *Uncle Tom* bore some of the minstrel influence. In Stowe's book, the title character is a strong, hardworking family man; onstage, however, he was transformed into a humble, shuffling old Negro. Another character, Topsy, a wayward slave girl, is a tragic figure who, after being bought and sold as a baby, doesn't know her parents. In the 1913 *Uncle Tom*, Topsy became a comic figure—a dirty, ignorant, roguish pixie—a pickaninny.

In New York, the Reverend Mr. Jenkins probably winced his way through rehearsals but allowed his kids to go on with the show. The lineup of performers included few boys older than fourteen. A ten-year-old orphan named John Garlington was the bandmaster, the prancing child who led the group with his baton. (The real bandleader was a twentyish cornet player named Alonzo Mills, who had grown up in the band.) Another orphan, Augustine "Gus" Aitken, twelve years old, played cornet. Bill Benford, eleven, was on tuba; Stephen Wright, in his early teens, played drums; and a boy named Joseph Watson was on trombone.

The producers of the show believed the only way to keep white people in their seats was for blacks to play dancing, jiving coons, so the Jenkins band obliged. When the show closed, the Parson pocketed the money, and the band headed home. But a few months later, a surprising thing happened. During the run of *Uncle Tom*, a vaudeville agent named Jules Hurtig had evidently seen the band. In February 1914, Hurtig wrote Jenkins and proposed a business venture. Hurtig said he had been in touch with an entrepreneur in London, a certain gentleman who was putting together an international fair. This fair, which was as big and serious as could be, would be called the Anglo-American Exposition. The fair was intended to celebrate a century of British-American "peace and progress" since the War of 1812 and the defeat of Napoleon, and to praise the British Empire. It was to be held between May and October 1914, at Shepherd's Bush, four miles west of the center of London. Hurtig wanted Jenkins to bring a band to London.

On April 16, 1914, the agent wrote Jenkins from his office at 1545 Broadway, in New York, with details:

> Dear Sir,
>
> I just sent you a wire advising you not to make any engagements for the Pickaninny Band until you receive this letter. If we can make satisfactory arrangements and the terms reasonable, the Band will be used only for Concert purposes for the Exposition patrons. They will no doubt be honored by having the privilege of appearing before the Royal Family. This distinction will give them such prestige which will make them a great attraction in America.
>
> I must know at once what you expect to receive per month for their services. We pay all transportation, also their Board and Lodging. We will also provide for them new uniforms at our expense.
>
> Yours very truly,
> Jules Hurtig

The Parson took the idea to his wife. Ella Jenkins must have been stunned. In nearly two years of marriage, she had had no more children. She had not seen her daughter, Olive, who was now seven years old, since the girl's birth. To Ella, this trip to London was an opportunity for her to see her only child.

Jenkins agreed and sent his terms. The agent wrote back with confirmation, plus a counteroffer: "I can send contracts for ten weeks engagement to open in London May 14th, and pay you one thousand dollars for the ten weeks, and all transportation and board. —Jules Hurtig."

The ragtime and coon act had finally paid off. The Jenkins band was heading for England. For Parson Jenkins, it would be his third trip. But whereas on the first visit, in 1895, he had marched the band through the London streets, this time they would play a serious venue on a real stage. There would be no jiving.

On May 2, Ella, the Parson, and a group of eighteen boys set off on the first leg of the trip, bound for New York.

This particular band included a new face, someone who had become the most skilled musician of any in the program. He was proficient with several instruments and well versed in European music. His name was Edmund Thornton Jenkins, and he was, in fact, the twenty-year-old son of Parson Jenkins. Edmund's father had asked him to help with the band, because in England they would play for a real audience. It was a bit of favoritism that the Parson would live to regret.

Three days after leaving Charleston, the Jenkins party sailed for London.

Chapter Eight

THE letter to Teddy from his friend T. K. Gibson, who was writing in 1912, contained a ridiculous offer. Gib had graduated from Harvard Business School and had gone to work for the Atlanta Mutual Insurance Company, a black-owned firm on whose stationery he wrote:

> I am mighty anxious to get up a calendar for Atlanta Mutual Insurance for next year (1913). I want you to get up a sketch of something which would show up well in one or two colors. I will see that you are paid $25. Here is my general idea: A ship representing the Atlanta Mutual Insurance Assn.—with *real colored faces*—plowing through a sea of sickness, death & accidents, and of racial inexperience and non-cooperation, to a harbor of health & happiness. Maybe you could also draw a crowd of happy Negro policy-holders flocking to this Atlanta Mutual ship. 50,000 will be printed—they'll surely help you. —Gib

It was not the kind of commission Teddy, the bona fide portrait painter, was waiting for. But after six years in Boston as an art student, he had run out of money. Gib and other friends from college (not to mention old girlfriends) had gone on to jobs and nice berths in the black bourgeoisie. Teddy had been forced to move several times to reduce his rent, because of his lack of income. "Delighted to know you will attempt the job," Gib wrote, after Teddy agreed to do the illustration.

Though the drawing of the insurance ship with happy Negro passengers has not survived, another letter shows Teddy, at the end of his tether, had begun to cadge people for loans. Mattie Marsh, his former high school teacher in Charleston, a white woman, wrote her old student in Boston: "Here, after all these years, I am glad, oh-so-glad you wrote to me, whether the money is ever paid or not."

Teddy had taken all the classes and whiled away all the time he could. He was delaying as long as possible his return to the South. After many seasons in the relative fluidity of New England—whose white people, if they didn't embrace him, at least acknowledged his talent—Teddy feared that going home would be certain entombment.

But Captain Harleston had lost patience with his son. He had been tolerant and had even paid some art school tuition. As Teddy turned twenty-nine, and finally thirty, the Captain began to insist he come down from up North.

In 1913, the feud between the Harlestons and the Mickey family reached a crescendo. That May, in a poisonous dispute, Hannah Mickey and her brother had dissolved the Mickey and Harleston funeral business. In the aftermath, the somewhat diminished Mickey Funeral Home was born, leaving Hannah in partnership with her sons, Eddie and Dick.

Captain Harleston waited a few months. All he had from the breakup was a pair of housing lots with tumbledown buildings on Calhoun Street. But he had no intention of letting his sister outmaneuver him in business. The Captain was determined to own the largest "Negro firm" in Charleston. To get it, he settled on a strategy: he would put up a new building and would start a separate funeral operation. Construction began in 1913 at 121 Calhoun Street, four doors down from the Mickey business. It was mockingly close, at a site the Captain had gotten from his unfortunate sister. The Harleston Funeral Home went up in six months.

It was a wooden building about the dimensions of a tennis court and three stories tall. Most structures in Charleston sagged from age, but this one was tight and sturdy, with pretty details like leaded glass. On the first floor, the front room was decorated as a chapel to be used for funeral services. There was a reception area where a body was laid out in its casket, often with the lid raised, and the family could come to view the face of the deceased. A showroom for caskets occupied the middle room, and there was the Captain's private office in a small room off the coffin display. (He had a liking for ferns and hung several near the window.) Behind the office was the Captain's bedroom, and an embalming room was built in the rear of the building, with a door that opened onto the backyard. The location of the embalming operation was a concession to the neighbors: bodies could arrive via a path to the rear of the building and be unloaded away from view.

On the second floor of his new headquarters, the Captain had his builder design a large ballroom running the full width of the building and nearly the whole length. The room was to be rented out for meetings and dinners. (In a glossy brochure, this was known as "Harleston Hall.") The income this produced would act as insurance against a slow season, when deaths did not occur frequently enough.

Finally, on the top floor, there were two apartments.

The local white paper, the *Charleston News and Courier,* carried an item about the new Harleston Funeral Home, "the largest building constructed by a colored resident" in the history of the city. It became an immediate landmark. Everyone who walked by got his or her bearings at its front door. Down the street, the Mickey family tried to look busy, but their cramped quarters seemed nearly embarrassing in the shadow of their relative's fresh-painted monument.

Captain Harleston's three sons and two daughters were all grown. Ella, a great disappointment, was married to Jenkins. Kitty, the eldest, had moved to St. Augustine, Florida, where she had married an undertaker named Maithlun Fleming. (Kitty's marriage delighted the Captain: it represented a possible business expansion.) Moultrie, the dropout from veterinary school in Pennsylvania, had nothing to do with the funeral home; he ran a small wood yard in the nearby neighborhood of Ansonborough. Teddy's younger brother, Robert, worked for his father intermittently.

None but Teddy was really available to help with the new firm. Overwhelmed by his new venture, the Captain increased his demands on Teddy to come home. When hectoring didn't work, he decided to use real estate as an enticement. He knew Teddy was having trouble paying his rent: in Charleston, he would have an apartment on the third floor, rent free, with an extra room to use as a painting studio. To prove his good intentions, the Captain had a skylight cut in the roof of one of the rooms to bring in the best light.

Teddy was tiring of being penniless, and finally he relented. In 1913, at age thirty-one, he said good-bye to his friends in Boston and took the steamboat to Charleston. On the trip home, he contemplated his absurd future: he would work in the burial business, "the dismal trade," and paint in his spare time.

After seven years in New England, when he arrived in the South, Teddy felt the impact of a cultural shock. New "colored" doors had been cut in the dirty alleys next to all the buildings, through which he now had to walk. In some places, it was forbidden for whites and blacks to use the same stairway at the same time. Colored and white workers no longer worked in the same room together. Even the city directory was segregated, with "white" and "colored" pages, the latter of which were literally dyed brown and therefore hard to read.

At the time Teddy had gone to college, in 1900, simple social separation of the races had sufficed to keep white supremacists happy. When he returned, police-enforced apartheid was the norm. It was depressing enough to find rear seating on the trolleys and separate bathrooms in stores. He had never faced these in Boston, but he expected them in Charleston. But something was different: all of everyday life seemed designed to make him feel small.

There were strange, insidious differences. In Boston, people who read books in public could be seen carrying editions of Emily Dickinson or Henry James. In Charleston, whites who were reading seemed to prefer racial fiction. The books of Thomas Dixon, a Baptist minister and novelist, set the tone. Dixon's best-selling historical novels were everywhere (two in particular, the ones with the awful titles: *The Clansman: An Historical Romance of the Ku Klux Klan* and *The Leopard's Spots: A Romance of the White Man's*

Burden). Teddy knew that Dixon's stories glorified the rise of the Klan and encouraged vigilante terror against black people.

And there were things not insidious. As the months passed, Teddy noticed, like the ticking of a clock, the regular reports of lynchings. They came from every corner of the South: two one week, four the next, one in a "good" week.

Teddy was pessimistic about the deal he'd made with his father. He knew he couldn't make a living as a painter but suspected he was entering a trap. Settling into his new third-floor apartment, he found he had a roommate, his brother Moultrie. (The Captain had rented a second apartment to paying tenants.) Lazy and a little boorish, Moultrie quickly proved himself to be a bad companion, borrowing money and rarely paying it back. On weekends, Moultrie sometimes left town without explanation and didn't return for days.

There was another disillusionment. Teddy observed that his father had acquired both a new mistress and two new children. It seems Mamie Blanchard, the Captain's dutiful concubine, whose house he had once outfitted with electricity, had outlasted her usefulness. Now there was a younger woman. The new girlfriend had a similar name: Mamie Randall. The Captain was fifty-nine. At thirty-three, Mamie Randall was two years younger than Teddy's sister Kitty.

It seems the Captain met the new Mamie when he first moved the funeral business to Calhoun Street. She was born Mary E. Bacot in 1880. In the early 1900s, Mamie was married and living with her husband, Joseph Randall, a musician and a waiter, in a cottage at 121 Calhoun Street. When the Captain bought the lot, Mamie became his tenant, and it was possibly on one of his rent-collecting visits that she became his lover.

Mamie Randall was glamorous compared with the previous mistress. Long and slender, she had an unusual mole on her left cheek. She worked at home as a dressmaker, was personable, and had a beautiful smile. Mamie had been married for a dozen years and she and her husband had six children when she decided to start seeing the Captain on the side. Captain Harleston had a daughter with Mamie, Alethia, in 1909, and another daughter, Catherine, in

1911. Mamie continued to live with her husband throughout the affair. But according to a city directory, in 1913, when the Captain apparently tore down the Randalls' house to begin building the funeral home, Joseph Randall moved out and was never heard from again. The Captain installed Mamie and her children in a new cottage a block away, at 98 Calhoun Street. To Teddy's disappointment, Mamie was a frequent visitor, and he now had fresh sisters a good thirty years younger than he was.

The big street called Calhoun ran for about two miles from its westernmost beginnings at the Ashley River to its eastern termination at the Cooper River. The funeral home stood at the eastern end of Calhoun Street, close to the Cooper. Many of the old mixed-race families were known to people this neighborhood, though darker Negroes surrounded these lighter clusters like circles around a bull's-eye. Calhoun Street had taken its name from John C. Calhoun, a South Carolina politician and writer who decades before the Civil War had created the doctrine of states' rights and secession. It was an irony not lost on the city's black bourgeoisie that they lived on a boulevard named for a man who wanted to keep black people in slavery.

Teddy had left home before the Captain moved the business to Calhoun Street. His "new" neighborhood was not entirely colored. At number 115 stood the home of the Lewisohn family, who were Jewish immigrants from Germany. One of their number, Ludwig Lewisohn, later edited the *New Palestine* magazine and was an associate editor of *The Nation* before becoming a novelist. Two of Lewisohn's novels, *The Broken Snare* and *The Case of Mr. Crump,* had to be published in France, because American publishers thought they contained too much sex. Down the street at 160 Calhoun, the site of the Charleston Orphan House, more whites could be found. The Charleston Orphan House was one of the grandest and most ornate buildings in the city, and it was full to the bursting with white waifs.

Next door to the funeral home stood Zion Presbyterian Church, a stone building skirted with an iron fence. Across the street was Emmanuel African Methodist Episcopal Church, a big, dominating building with a tall steeple. Emmanuel held a thousand mourners, Zion Presbyterian two thousand. The two large black congregations frequently filled the Captain's calendar.

A few blocks east of the Harleston Funeral Home lay the banks of the Cooper River. From that point, going south for half a mile, stood the docks of Charleston harbor. This neighborhood, which lay behind the Harleston place, was the old seamen's quarter. Officially, it was called Ansonborough, but most people knew it as "the Borough." The Borough was black, poor, and falling down, a jumble of wooden cottages and old mansions that had been broken into tenements. Churches stood next to drinking houses, which leaned against brothels. At the north end of the Borough, toward Calhoun, the area remained black, but with each block, the skin of the residents got lighter, and household income went up.

Calhoun Street, home of the lightest and richest Negroes, stayed busy. At noon there were sidewalk philosophers playing checkers and deliverymen making rounds. In the afternoon came the street peddlers on their circuit. The city was full of these traveling salespeople—vegetable and fish men and women who pushed big carts down the rutted lanes between the old houses, selling oysters or shrimp, sometimes merely sharpening knives. They came rattling up the block, singing a song about their goods. One of the old lyrics told of the best fish money could buy, at least that particular day:

Shark steak!
Don't need no gravy!
Put 'em in the pot,
Make they own gravy!

Fighting disappointment, Teddy set about to learn the funeral business, but it was clear right away that he hated it. During the day, when Captain Harleston disappeared to the backyard to see his mistress, or went out on an errand, Teddy had to mind the office. Someone always needed to be on hand, in case the phone rang with news of a death. This part of the routine was like a dungeon to the artist's mind. The office smelled of bay rum, a kind of cologne Captain Harleston used. Teddy sat among the ferns, his eyes focused on the distance. When finally the phone did ring, Teddy used a neutral and correct way of answering: he spoke the number (a mere four digits: 2022) with a slight rise in his voice, like a question. "Two zero two two?"

To lighten his load, Teddy coined nicknames for things. The funeral home he called "the Firm." Its workers were "the Force." The Force included Teddy, his brother Robert, their father, and a middle-aged porter named Moses, who handled the horses and deliveries.

According to a family member, one morning Teddy was preparing to go downstairs from his apartment to the Firm and the daily vigil with the phone. He was shaving in the bath, when a funeral products salesman showed up, unannounced. The salesman, who was white, was promoting a new embalming fluid. Teddy, flustered, rushed down to greet him, shaving soap still on his chin. Then something telling occurred. In the middle of his pitch, the salesman began to call Teddy by his first name, "Edwin." The code of Jim Crow allowed white people to call nonwhites by their first names, whereas colored people had to use formal terms of address, "Mr." and "Mrs.," when speaking to whites. The salesman did this repeatedly, ignoring Teddy's appeal to be addressed as "Mr. Harleston." Teddy was furious. When the salesman finished his product description, Teddy thanked him, then turned and marched, cursing, back into the house. As he returned to shaving, he grew so angry that he began to cut himself on the face. He pulled down the razor on every curse, and the blood flowed down his cheeks.

The Harlestons were part of a tiny society, with only a few colored families who shared their special caste. Their peers could be listed by family name: DeCosta, Dereef, Ellison, Felder, Holloway, Mickey, Weston, a dozen or two others. All of these families were mixed race, and most had been free before the Civil War, like the family of Teddy's mother, Louisa Moultrie.

Not far from the funeral home, there lived a family named Forrest. The Forrests were of the "right" class and counted among their seven children four unmarried sisters. The girls' father, Augustus Forrest, had come from a family that was free before the Civil War. He had been an accountant, although he had suffered a stroke and was confined, speechless, to a wheelchair. Their mother, the former Elvira Moorer, was the daughter of a Confederate general and his slave mistress. She had taken on all duties of the house when her husband was incapacitated.

Of the Forrest women, the middle two, Marie, twenty, and Elise, twenty-two, together began to bring an end to Teddy's despair.

By 1913, Teddy's brother Robert had been seeing Marie Forrest for two years. Marie was slender and small, with an angelic expression permanently etched on her face. She was attending Howard University in Washington, D.C. The story goes that one day, Robert asked Teddy to carry a present to his girlfriend. Teddy obliged and took the gift to the Forrest home at 97 Morris Street. Marie was out of the house, but Elise happened to be home, and she answered the door.

Elise Beatrice Forrest was petite, at five feet one, with big brown eyes and straight brown hair. At twenty-two, she was nearly ten years younger than Teddy. The two stood in the doorway, facing each other. Teddy was poised. In contrast to Marie, her more studious sister, Elise was an impetuous young woman. Elise would later say that Teddy behaved in a shy and polite way; that he had given her the package, tipped his hat, and swooped away, leaving her standing on the step. But he had left an impression. He was fine-looking, with smooth skin and penetrating eyes. His cheekbones were high, and he had a straight nose over a strong jaw.

She had not gone to college and had taken a job in a hat store where a cousin ran the shop; Elise was an assistant milliner and was learning the trade. And it was she who made the first move. The next time Robert came to the Forrest home to see Marie, Elise sent him back with a note for Teddy. Elise invited the artist to a party and told him to "let Robbie know" his answer. Teddy said all right, he would go—and so began their affair.

The language that the black bourgeoisie used to correspond with one another was tightly formal, at least before the first kiss. For one of their first dates, Elise sent a note inviting Teddy to a club dinner at the home of a Mrs. Beaubain, "if it is convenient and agreeable to you. Should I not hear from you before noon of Tuesday, I will look for you at 9 or 9:30 p.m. Sincerely, E.B.F. [Elise B. Forrest]."

At the time, colored couples could not go out in public for the kind of public entertainment whites took for granted. The parks were off-limits to Negroes, and no restaurants served blacks. Even though motion pictures, a recent dazzling invention, had been playing in theaters for six or seven years, movie houses did not yet admit colored people. But Teddy and Elise adapted. They both attended Plymouth Congregational Church, which meant they saw each other on Sundays, after services. They had tea, or took walks. (Teddy had resolved never to use the trolley system, because black people had to board and sit in the rear. He walked most places, which took him twice or three times as long. It was a small protest, but Elise had noticed it.)

Elise found Teddy beguiling. He was handsome, older, and worldly. A family member who knew Elise reports that she was a virgin. Teddy was not. Teddy showed Elise his paintings and talked to her about art history. For his part, Teddy thought Elise was fresh and funny. She played ingenue to his discerning ways.

In January 1915, Elise sent a letter to her boyfriend that gives a sense of her innocence. She had written it at midnight, in front of a fire in her parents' house. Teddy was home on Calhoun Street, sick in bed with the flu. In the letter, Elise said that earlier in the evening she had attended a church fair and costume party—put on by Plymouth Congregational Church—called the "Mother Goose Bazaar."

> I have just come in from the bazaar, am ready for bed . . . am in my favorite seat, on the rug before the fire writing you. Shall I tell you about the bazaar first? Got home at 7, and dressed for the affair, wore [my sister] Marie's lace over blue, with the blue pleating at the bottom. Had my hair fixed in the latest. And gee! You should have seen my face. [I] burnt a cork and made up my face, "perfectly adorably." Haven't washed it off yet.

Colored families had attended the event in costume. Elise, like Teddy, was light-skinned. She had put on ashes from the cork—a cheap makeup—to look like a darker Negro, which in Elise's world was a kind of costume.

> Are you any better dearest? Can't you get up tomorrow, I wonder? When you do, please phone me. Remember I'm wanting you badly . . . With love and a kiss,
>
> Your L.

Teddy might have taken advantage of Elise and seduced her. But after his sister Ella had gone off and slept with Parson Jenkins, he had begun to treat women with greater respect. Teddy wanted something more from Elise than fawning attention. He tried to cultivate in her an interest in art. On one occasion, it seems Teddy gave Elise a gift of a camera, in an effort to enlarge her visual sense. The camera itself appears in a family photograph. It was a "No. 3 Brownie Camera," one of the first mass-produced cameras manufactured by the Eastman Kodak Company. The No. 3 Brownie sold for four dollars and was among the most popular apparatuses of its day. Under the influence of her boyfriend, Elise began to take photographs.

Among the black middle class, for a woman to work as a shop girl in a hat store was all right, though not something to boast about. The most respectable job for a young woman was as a teacher. At the time, however, it was illegal for colored women to teach in the public schools of Charleston. By law, all jobs in the city's "colored" schools went to white instructors. Women of color were forced to accept lower pay and to teach only in the one-room schoolhouses far from town—the shacks and shanties of the state's feeble rural education program.

To please her family, and perhaps to raise her standing with Teddy, in November 1915 Elise took a teaching job. She was assigned by the state to a village called Fowler, seventy miles north of Charleston, where she was to run the grammar school. Situated ten miles from the nearest train stop, the settlement of Fowler had only a couple of hundred people, and nearly all of them were destitute farm families who lived barely above subsistence. At the time, public education did not mean routine medical examinations, free textbooks, and compulsory attendance. Elise's schoolhouse was an unequipped and windowless shack.

For Elise, a relatively pampered city girl, the move to Fowler was a shock. She rented a room in a decrepit cabin occupied by a minister and his wife. (Elise later recalled that at night, before getting into bed, she had to pull back the sheet and sweep the bugs off.) "My school house is very poor," she wrote Teddy curtly. "I am undecided about remaining here." To judge from her letters, Elise had fallen in love with her boyfriend, and she was lonely in her new job. From her cabin in the woods, Elise wrote Teddy:

> My dear Edwin Augustus,
>
> This is Thanksgiving Day, and there are lots of things for which I am thankful. You are chieftest among them all. I am thankful to have met you; I am thankful to have had the opportunity of knowing you so well as I do.
>
> You asked me not to think hard of you. I never shall. Rest assured on that score. You are a man, and human. I do not see you with a halo about your head. Neither do I look upon you as a beast, but somewhere between the two I have placed you, the One Man, the embodiment of men, the Alpha and Omega of my too short life.

Few of Teddy's letters to Elise from these years have survived, but from the frequency of her mail, it seems he stayed in touch when she was away. And though Elise was unaccustomed to the lives of poor sharecroppers, she appears to have taken an interest in a few people and tried to entertain herself.

> I shall learn to ride horseback before I return. Ben will teach me. You don't know Ben, but he is as fine a specimen of young American Negro you will find anywhere. Black, with a wide mouth, white teeth set apart, flat nose, large nostrils, and soft dark eyes. And best of all, a large heart.

When Elise went off to teach, she had taken her Brownie camera with her, perhaps at Teddy's urging. As weeks passed, she craved two things from her boyfriend: news from home and Kodak film.

> You are exceedingly kind, too kind, so kind. A kiss for the best boy alive. Now then—send me Jenkins's paper [the *Charleston*

Messenger]. Then, too, the films, dear boy, how about them? I shall show Ben how to use the camera so he can snap me on the horse.

With the camera in hand, Elise sent Teddy photographs of herself to keep his interest alive. Some of them may have been more seductive than Teddy had expected. "Oh you saucy one!" Teddy wrote. "I had your letter with the two pictures and my! Oh my, some blessing."

In the summer of 1916, Elise came home from teaching in the countryside. She had realized she had little taste for one-room shacks and went back to the hat shop. Elise remained preoccupied with Teddy. She dated him, doted on him, and smothered him with attention, but Teddy wasn't sure he wanted so much of her. Charleston already felt suffocating, and he didn't know whether he could endure it. He avoided committing himself to his girlfriend, just in case he had to leave the South.

Elise cast about for ways to lure Teddy and make his love grow. She bombarded him with letters, even when they were in town together. The letters are sometimes thin and grope for things to say. One piece of mail was again written from bed, this time from beneath a mosquito net:

My dear Edwin,

The moon is gorgeous! Sunday night I sat on my back porch and just wished. There is a mosquito in my net and I shall kill him as soon as I am through this letter. Now I'm going after that mosquito.

Elise tried jealousy to get Teddy's attention. One night, after a party she had attended, she wrote her boyfriend a list of names of the men she had danced with.

Ted, none of these men are going to take your place unless they are allowed to call evenings, and calling evenings means getting better acquainted, and affectionately so. There are several reasons why I do not wish to become too closely attached to any of them. Firstly and mainly, none of them appeal to me. Secondly, for three years I have been using one brand of XXXX, and don't want any substitute.

Neither mosquitoes nor dance partners held Teddy's imagination, but Elise's camera seemed to attract him. Teddy, isolated, forced into the family business, wanted someone of like artistic mind, and Elise had the Kodak. At some point, he began to talk to her about art—not merely about art history, but actually about somehow sharing his work with her. At first, the idea was a fantasy. Teddy would be a painter, and Elise would make photographs. Whether any of this was possible, it took his mind away from the Firm.

Business was good at the Harleston Funeral Home, and the Captain doled out the profits. Yet the aging patriarch (the Captain was now sixty) kept tight control. He refused to put Teddy on a salary, preferring to keep him dependent. As fees rolled in, Teddy discarded his old clothes and acquired a wardrobe. He and his father lived well. They ate off of good china, drank from stemmed glasses, and wiped their mouths from napkins kept in silver rings. In good years, there were servants, and there was an added benefit from the Firm. The funeral home had a livery service—carriages that took mourners to the cemetery—and during a service, the livery drivers wore capes and black hats. But when no burial was taking place, which was most of the time, the drivers were available to take the Harlestons to appointments. Teddy disliked using these chauffeurs, but the Captain relished the display. It made him a prince, and Calhoun Street was his principality.

The Harlestons were an exception, not only to black society but also to white. Everywhere else, their city was crumbling. The plantation economy that had long supported Charleston had finally failed, leaving no wheel of industry to drive tax collections. South Carolina had seen economic growth, but it was mainly in the construction of textile mills in the northwest section of the state. These factory jobs, by a ratio of ten to one, had gone to white workers. Eighty percent of employed black men in the state were peasants without property—either sharecroppers or agriculture workers paid by the day. In Charleston, despite its black masses, there were fewer than a hundred black-owned businesses. Most of these were barbers, one-room groceries, or tailors.

In the funeral line, Teddy may have helped his father, but he hated it. Captain Harleston pressured Teddy to take on more responsibility, but the son could not commit himself to being a mortician. He turned to Elise for solace:

> I presume I may be useful [around the funeral home], although I grow so weary of the purposeless periods there, and when you are not in town it is awful. These accumulating years of loneliness are telling on me.

In particular, the Captain thought the embalming operation needed attention. He was tired of it and wanted Teddy to do more. The state of South Carolina required certification of embalmers, and so far Teddy had been operating without a license. He would have to get his credentials in order, and to do this, he needed to take a training course.

Since the burial business had become a profession, in the late 1800s, several schools had opened to cater to morticians. The best known of them was in New York. The last thing Teddy wanted from life was a season in New York to study embalming, but his father had him in his grasp. In the fall of 1916, Teddy agreed to the Captain's terms. He would go into training at the Renouard School for Embalmers in Manhattan.

"I'm glad you have made your decision and are complying with your father's wishes," Elise wrote Teddy, lending her support.

Elise began to beg Teddy to let her come with him. In New York, he would be out of sight, and among other women, while she would be marooned in the South. Elise wrote furiously to friends of friends, trying to find a place where she could stay near her boyfriend. Within a few months, she had found a teaching job, this time on Long Island.

When his girlfriend announced that she would be coming to New York, Teddy put up no opposition. The two of them would move their affair up North. In the fall of 1916, the couple took the Clyde Line steamer for New York. In her bag, Elise had packed the Brownie camera.

Chapter Nine

WHEN they arrived in London, in May 1914, the Reverend Daniel and Mrs. Ella Jenkins were bewildered. In America, the couple had grown accustomed to sarcastic treatment of the orphan band, but this time they had been led to expect something different. The Parson and his wife thought the children would be given their first respectful hearing among whites at the Anglo-American Exposition, a fair of all the nations. Instead, the Americans discovered their favorite pupils would have to perform on a stage wedged between freak shows and carnival rides.

In the early 1900s, international expositions were nearly as common as sporting events. Between 1850 and 1914, some fifty world's fairs were organized in the United States, France, and the United Kingdom alone. Most were glorified trade shows. They hauled out new consumer goods, added a few dazzling inventions of science, and threw the mixture against a nationalist backdrop. At the London event, the center of the fair occupied the so-called White City buildings, which were museumlike pavilions built specially for the purpose. But away from them, planners had built an amusement park called Merryland, to entertain the more simpleminded of England's hordes. It was here, off of a tawdry midway strip, that the Jenkins band had been asked to play.

Parson Jenkins and Ella swallowed their disappointment and settled in for an eight-week English "vacation." Their contract ran for two months, until mid-July.

Despite her frustration, Ella did not want her family at home to forget that she, alone among the Harlestons, had managed a trip abroad (her second, including the British holiday when she gave birth to Olive). She bought a postcard of Trafalgar Square and sent it to her brother Teddy. "Am expecting to go to Paris," she boasted. "Trust all are well, Love to all."

The Parson's son Edmund was a lover of serious music, and he had been enticed to England with the promise that he would not have to play the band's usual coon songs. A violinist by training, Edmund Jenkins had brought a portfolio of opera and overtures, with a few popular songs thrown in as his only concession to the old minstrel legacy.

Edmund Thornton Jenkins was born in Charleston on April 9, 1894. He grew up in the odd half-world his father had created, moving between the comfort of the Jenkins house and the destitution of the orphanage. In the family parlor, Edmund took private music lessons from a white teacher named Mr. Dorsey. But his father wanted to give him a taste of the family business, so when the Jenkins band traveled, the boy went along.

His friends called him "Jenks." At five feet six inches, Jenks was a full foot shorter than his father. Under rigid training, he became proficient on clarinet, piano, and violin. He was the Parson's most dexterous son and went quickly to the head of the class, helping to teach the young orphans the basics of music. Jenks grew up to become a handsome, frail "Negro aristocrat," and he seemed to carry himself with a touch of arrogance. Photographs of him as a young man show an overdressed, aloof dandy with a proud countenance. His preferred costume was a tight black coat and a scowl. Jenks grew to hate his father's insistence that he lead the delinquent children on marches in the street, because he believed the orphan boys were beneath him. They were also tougher than he was and probably made fun of him because he was a small violinist with precious airs.

In 1910, at age sixteen, Jenks went to Atlanta to finish his education, enrolling at Morehouse College, in the music program. In Atlanta, Jenks studied under a violin master named Kemper Herrold, who refined the student's technique and introduced him to the European repertoire. On weekends, Jenks played the organ as well as the cornet at the city's First Congregational Church.

Four years later, at the end of his senior year, the young dandy received a telegram from his father, urgently calling him home to Charleston. The Jenkins band had been invited to England, and he would be needed to play with them. For Jenks, the trip meant traveling again with a roomful of noisy urchins. But this time, because it was London, he would be able to listen to some real European music. And he would make sure the boys played something other than ragtime.

On the printed program of attractions, they were billed as the "American Piccaninny Band," a misspelling of the mocking nickname. There were three stages at the fairgrounds where music could be heard: the "Court of Honor," the "Elite Gardens," and the bandstand in "Merryland." Whereas the Court of Honor hosted the band of the First Battalion, Lincolnshire Regiment, and the Elite Gardens welcomed another stately military band from His Majesty's Twentieth Hussars, Jenks and the boys took their place near the sideshow entertainment. Not far from their stage were the carnival rides, with names like "Caves of Mirth," "Texas Twister," and "Revolving Flip-Flap." The "101 Ranch" attraction, the "Indian Reservation," and the "Kentucky Dance Pavilion" provided more American flavor. And it was difficult to overlook the "Human Marvels" exhibit—in all likelihood, a freak show with midgets, the deformed, and strange human specimens.

Jenks played clarinet in the band, which did five shows a day. Their schedule began at 11:45 A.M. with an hour-long concert, and they resumed playing at 2:45 for a full two hours. At 6:00 P.M. came a ninety-minute performance; finally, they played a half-hour concluding concert late at night, wrapping up at 11:00. The twelve-hour workday was exhausting. On the Reverend Mr. Jenkins's first trip to England, in 1895, a magistrate scolded him for overworking the children. This time, the managers of the fair insisted on drilling the boys like coal miners.

The audiences who preferred His Majesty's Twentieth Hussars were prosperous and sedate, but the crowds in Merryland were generally loud and poor. They might have wanted to see an "Uncle Tom" act from the American South, but when the band played, the sound that emerged from the black boys stunned the scullery maids and dockworkers milling about in front of the stage.

Thanks to Jenks's classical bent, the band's show was a fantastic mix of the concert hall, the street, and vaudeville. To start, the boys played a waltz from *Il Trovatore* (The Troubadour), the 1853 opera by Giuseppe Verdi. (Set in fifteenth-century Spain, *Il Trovatore* tells the story of the troubadour Manrico and his nemesis from a rival clan, the Count di Luna; the two men are antagonists who love the same woman and who turn out to be brothers.) After the Verdi, the band segued into a piece from Offenbach's *Les Contes d'Hoffmann* (The Tales of Hoffmann). Jacques Offenbach was a German-born composer who lived most of his life in Paris, where he composed frolicsome operas with Romantic themes. *Les Contes d'Hoffmann,* which premiered in 1881, was his final work, a joining of frivolity and nightmare whose narrative action unfolds in Venice. The Jenkins band played the barcarole, or gondolier's song, from the opera.

Jenks, the precocious lover of serious composers, leaned heavily on his conservatory tastes. The barcarole of Offenbach was a lovers' duet in waltz time, in which two performers sing to each other about a night of sex. Although there is no evidence that vocalists accompanied the band, the final verses of the piece give its flavor:

Burning zephyrs,
embrace us with your caresses!
Burning zephyrs,
give us your caresses!

Lovely night, oh night of love,
smile upon our joys!
Night much sweeter than the day,
oh beautiful night of love!

When the serious pieces proved too much, the band launched into a few rousing marches. They struck up the patriotic "National Emblem" march by American composer Eugene E. Bagley. Then came a stirring version of "Stars and Stripes Forever," the John Philip Sousa standard.

In midconcert, the band took another unexpected turn when the orphans pulled a sheaf of popular songs from their folios. In the second decade of the 1900s, one of the most popular songwriters in America was Irving Berlin, and Jenks had brought along a few of his pieces. Berlin, a Russian immigrant to the United States, had published his first song only in 1907, but by 1914 he was an established songwriter, with a musical on Broadway called *Watch Your Step*. The band played Berlin tunes with a Southern theme, such as "Down in Chattanooga" and "When the Midnight Choo Choo Leaves for Alabam'."

Day in and day out, the band flaunted its talent. They became the talk of the fair. Picture postcards went on sale showing the "Famous Piccaninny Band" in their military-style uniforms. But despite their abilities, the boys were never allowed to forget that they were Negro children from America. The postcards showed the band's ten-year-old urchin "conductor" pretending to lead the act. Pressured by the fair's promoters, the band always closed a set with a racial stunt, to make sure they left the crowds laughing. The finale was the old cakewalk "Coon Band Contest," perfect for a jazzy step.

On June 25, two elderly sisters came together to the White City. Both were monarchs. Queen Alexandra, the sixty-nine-year-old widow of Britain's King Edward VII, arrived with her sister Empress Marie, the sixty-seven-year-old Dowager Empress of Russia and widow of Czar Alexander III. On this particular day, the Jenkins band was allowed to play at the Court of Honor. Nothing is known about how the sister-queens enjoyed the unusual boys from the American South, but the women seem to have sent home good word of mouth, because a short time later, George V, the reigning British king, let it be known that he would like to visit the fair and would expect to hear the band.

A flurry of panic invaded the fair's front office. The Reverend Mr. Jenkins was hurriedly asked whether he would extend his contract an additional three months, until October. The Parson agreed, and a further fee was negotiated, with the proviso that this time he would be paid in gold. Word continued to spread about the strange, funny, skilled black boys in uniform. An invitation came from the Bristol International Exhibition, a competing fair several hours away. The director of the Bristol show wrote the Parson, "Could you arrange for your 'Pickaninny Band' to play at this Exhibition for one week?" Jenkins declined, pleading the longer run in London.

On July 13, to Ella's and the Parson's astonishment, a letter arrived from the administration confirming the king's appearance. The pointed note had come from a fair manager named Gerald Kiralfy, who had one nervous question for the Reverend Mr. Jenkins:

> Dear Sir,
>
> I shall be glad to know whether your band can play "God Save the King," and if not will you please arrange for them to learn it well in readiness for the special performance which is to take place shortly.
>
> Yours faithfully,
> Gerald Kiralfy

Ella and the Parson were at the peak of their good fortune, and they knew it. In late July, Jenkins composed a letter to the governor of South Carolina, Coleman Blease, advertising his accomplishments. Blease, a handsome man with piercing eyes, his head topped by a mane of swirling white hair, was a fiery orator, popular among poor whites. Governor Blease poured hate on his state's black citizens at every speaking opportunity. He is said to have boasted that once, after a lynching, he received in the mail a severed finger of the black victim, which he then planted in the garden of the governor's mansion. On July 23, Parson Jenkins wrote Blease:

> Honorable Sir,
>
> It is sympathy and pity that I have for the little waifs and outcasts of my race that force me to write you these few lines in their behalf.

You are the headlight of South Carolina: to you the humble Negro must look for help and comfort, and I cannot help but believe that if you are called upon in the right spirit, you will do your part.

The salvation of the South between the white and the black man lies in the careful training of the little Negro boys and girls to become honest, upright, and industrious citizens. Open the door and turn out every juvenile criminal; send them to the county from which they came. Let them build reform schools. If you will do this, Sir, you will be called blessed by nations born and unborn.

Teaching the Negro to read, to write, and to work is not going to do the white man any harm. Nine of the Councilmen of London called on me yesterday and congratulated me on the work I am doing for my race. I feel much encouraged. If boys taken from the depths of the lowest dives can be taught and trained in such a manner as to gain the respect of the people of England, how much more can be done if the Governor and Lawmakers of South Carolina would simply co-operate with me?

Rev. D.J. Jenkins

There is no evidence that Governor Blease responded to Jenkins's appeal; most likely, he probably dismissed it.

Sometime during their "vacation," the Reverend Mr. Jenkins and Ella took a train to the town of Wigan, in Lancashire, to pay a visit to their daughter, Olive. (Jenks was left behind in London.) The parents had not seen their girl since she was a newborn, and Olive was now seven years old, quiet, and English in every way. She was a piano student and pupil at a church academy called the All Saints School. She lived with the Layland family, off a road known as Gideon Lane, and she was the only black child in her circle of white friends. Years later, one of her schoolmates, Laura E. Gilbert, would recall that Olive "used to play all the games in the street—top and whip, rounders, hockey, skipping rope." Laura Gilbert, who was white, said that she liked Olive so much that she once announced to her parents she would no longer drink plain milk; henceforth she would drink only chocolate milk, because it might help change her skin to Olive's color.

When the Parson and Ella arrived, the reunion must have been bitter. Another friend of Olive's remembered that "a well-to-do gentleman" had come with his young companion, "and as she wouldn't take the girl back with her," the couple left again without their daughter. Olive returned to school wondering which was her real family—the white Laylands or the black couple named Jenkins.

About the time Ella was visiting her daughter, far away, a thousand miles from England, events in Europe were turning in such a pattern as to bring an end to the Jenkins's luxurious holiday. Ella, a fairly provincial woman, had no reason to pay attention to European politics, and the happenings that would soon affect her began obscurely enough. On June 28, the Hapsburg heir to the throne of the Austro-Hungarian Empire, Archduke Franz Ferdinand, was paying a visit to Sarajevo in Bosnia-Herzegovina. Ferdinand's uncle, the emperor of Austria, had annexed the province and sent his nephew to placate local dissatisfaction. Archduke Ferdinand, fifty-one, was in the Bosnian capital with his wife, Sophie, the duchess of Hohenberg. As the royal motorcade stopped at a corner, an assassin named Gavrilo Princip approached the open-topped car, pulled out a pistol, and shot the archduke and the duchess to death.

Gavrilo Princip was an agent of the government of Serbia, which wanted Bosnia for itself. In the following weeks, the government of Austria-Hungary debated what reprisal to take against Serbia. Emperor Franz Josef took a hard line, planning an invasion; his ally, Kaiser Wilhelm II of Germany, agreed to help. Other monarchs in Europe, including George V, took sides. After four weeks of intrigue, on July 29, Austrian artillery opened fire on the Serbian capital, Belgrade.

At the most, the Reverend Mr. Jenkins and Ella followed these events in the papers, but they were concerned for Olive and otherwise preoccupied with their success at the fair. The couple had decided to try to take their minds off things by taking a wider European tour. On the day of the bombardment of Belgrade, Ella and the Parson were sailing across the English Channel, en route to Paris.

On August 1, Ella sent word to Teddy in America that she had finally gotten to France. "Am here on a visit," she wrote from Paris, on a postcard of the Eiffel Tower. Then, in an uneasy aside, she added, "Everybody is at fever heat over the war question."

By this time, Ella probably had heard that the German army was planning to invade France. The patron and protector of Serbia was Russia, and by treaty, Russia was allied with the French. In the twisted logic of the interlaced agreements, Germany's plan to assist its Austrian ally meant a campaign aimed at Paris.

On August 1, the cabinet of French President Raymond Poincaré authorized mobilization of the French army against Germany. When the news hit the papers, Ella, from her Paris hotel, wrote a frightened note to Teddy: "Will return to London Monday, leave for Liverpool Wednesday, and sail for New York Saturday."

Britain was now drawn into the dispute. A German invasion of France would have to pass through Belgium, which was neutral. The United Kingdom, bound by treaty to Belgium, issued an ultimatum to Kaiser Wilhelm that if his army occupied Belgium, Britain would be in a state of war with Germany. The kaiser refused to relent, and the day after Ella and the Parson reached London, Britain declared war.

In his haste to get out of the country, Parson Jenkins had bought tickets on a ship called the *Laconia*, a Cunard liner scheduled to leave for New York on August 8. But with the excitement and dread, the British government suspended civilian ship traffic: ocean liners might be needed to move soldiers. Out of work and stranded, with eighteen dependent boys, Ella, the Parson, and Jenks holed up for a long wait.

Fortunately, the band had collected its last fees in gold. According to family lore, there were a number of rich Americans in London at the start of World War I. They included Henry A. Duke, a member of the powerful Duke family of North Carolina. The Dukes controlled the American Tobacco Company and had made a fortune in cigarettes. When the banks closed and transactions in foreign currency were temporarily suspended, Henry Duke, marooned in London, was in trouble. For foreigners, gold was the only tender left, and word got around the American circles that a certain black man had a big store of gold. Several came to Jenkins for a loan, which he gladly extended. When Henry Duke came, however, the

Parson remembered the existence of the Duke Foundation, back home in the South. Jenkins told Duke that he would loan him whatever he needed to keep him comfortable and even buy his ticket home. Furthermore, he wouldn't ask Duke to pay him back. But the Duke Foundation should be aware of a particular orphanage in Charleston and consider a contribution to it. Duke agreed, and then borrowed a sum of bullion large enough for first-class passage. The next year, the Duke Foundation made its first of many annual gifts to the Jenkins Orphanage.

In September 1914, the Reverend Mr. Jenkins, Ella, and the band boarded the S.S. *Campania* at Liverpool and sailed to New York. The war would be conducted without them.

In the hurried evacuation, one musician was left behind. Edmund Jenkins, Ella's stepson, had decided to stay. It wasn't that Jenks had a special interest in the war. He had no intention of surrendering to the mass hysteria of the enlistment; furthermore, because he was American, this would not be expected. But after four months, the musical prodigy from the South had gotten a feeling for life outside the prisonhouse of Jim Crow. He saw open paths in England that he could have never dreamed existed in America. Jenks wanted to play in concert halls, for people of means, not on street corners for dirty workers and their laughing wives. As Europe's armies clashed, Jenks retreated into a tiny practice studio in London. He took several sheets of empty music paper from his portfolio and started to write.

Part IV

~

NIGGER RICH

Chapter Ten

ROBERT Othello Harleston, Teddy Harleston's younger brother, was thinner than his siblings and effeminate in appearance. He stood about five feet seven, looked slight from any angle, and had an unusual face: it had a perfect oval shape that resembled mask. Perhaps to compensate for his outward delicacy, Robert dressed himself in three-piece suits, slicked back his hair, and wore shirts that had been starched stiff.

Born in February 1889, he was the baby in the family, and everyone called him Bob. He grew up in Charleston and after graduating from high school went to Atlanta University, but he soon dropped out and moved home. His father then sent him to study undertaking in New York at the Renouard Training School for Embalmers. When he returned, Bob took the South Carolina licensing test for funeral directors; and at age twenty-one, with lackluster energy, he started work at the family funeral home.

In the summer of 1911, Bob began to see Marie Isabelle Forrest—and this was before Teddy Harleston began dating Marie's sister Elise. Marie, born in 1893 and eighteen at the start of the flirtation, was a beautiful young woman. Photographs made by Elise show Marie as alert and sensitive, and two paintings of her done by Teddy portray her as pallid and vulnerable. She had attended high school at the Avery Institute in Charleston and enrolled at Howard University in Washington, D.C. At that time, only about one in a hundred nonwhite women had the chance to go to college, and Marie was among the few.

Bob and Marie began seeing each other when Marie was home from Washington on a school break. She went back to Howard in the fall and, in October 1911, wrote a teasing letter to her new boyfriend:

> Dear Bob:
>
> I thought that you took an unusual length of time in answering the last letter that I wrote you before leaving home, so I was simply returning the compliment by putting into practice my old stunt, which is meanness. I am also very sorry to hear that your see-saw is at last broken. I only wish that it was for good. I have two pictures of you. One, in which you are standing with a gun in your hand, head down, was given to me. The other, in which you were looking at the camera, I stole. Do you mind?

Howard University was a staid place that patterned itself after the elite New England colleges. Its faculty expected solemnity from students and taught them to expect a life of struggle under the heel of Jim Crow. But by Marie's report, some Howard students spent much of their time playing pranks on one another. She wrote her boyfriend about one escapade:

> I will now proceed to give you a little news of the day. Friday evening the sophomores planned to have their annual banquet. The freshmen decided that there would be no banquet. When the time arrived, the freshmen stole several sophomores, tied them, put them in an automobile, carried some out into a suburb, and tied them to trees. Others they took to houses in the city, tied them in rooms, others they brought back to the college hill and tied them down in the science hall. The rest of the sophomores and the freshmen then held a hand-to-hand fight in Miner Hall. The freshmen broke the windowpanes in the kitchen and rushed in on the sophs who were in the dining room. Chairs were broken up and made clubs of—bowls,

glasses, and syrup jars just flew like wind back and forth. The deans had to be called. They succeeded in stopping the fight after several men on both sides were bruised.

Marie Forrest finished Howard in 1914, and in her graduation photo, she stands among about thirty students. Afterward she moved home to Charleston and took a job as a schoolteacher. Marie had managed to hold on to her boyfriend from a distance, and in the spring of 1915, when Bob was twenty-six and she was twenty-two, they were engaged. On June 9 of that year, they married in a nighttime service at Plymouth Congregational Church.

The couple moved into the apartment on the third floor of the funeral home across the hall from Bob's brothers, Teddy and Moultrie. Fifteen months later, Bob and Marie had their first child, Gussie Louise Harleston.

Years later, Gussie would tell her own story:

"I was born in Charleston, on September 28, 1916. My parents were Bob and Marie Harleston. My mother gave birth on the third floor of 121 Calhoun Street, next to the front window facing Calhoun nearest to the west side of the building. The midwife was a woman named Sarah Jones. In the South at that time, when a woman had a baby, she had to go upstairs for a month because she was considered unclean. Maybe that was just among black people, I don't know.

"I was christened Gussie Louise Harleston, but I always thought that name sounded coarse. My mother had a brother named Gus, and her father's name was Augustus Forrest, so I may have been given their name. I think Louise came from Louisa Moultrie, my father's mother's name. But however Gussie Louise came about, I didn't like it."

Two years into the marriage, the lives of Marie and Bob hardened. The war in Europe was on, and after long observation, the United States had entered it. Bob applied to the army and was sent to attend officers' training camp. At the time, there were almost no nonwhite officers, and the U.S. Army had built a separate training camp for colored officers outside of Des Moines, Iowa. The camp went up against much protest from the upper army echelon, most of whom opposed commissioning any black officers at all. (Bob's brother Teddy also wanted to fight in the war and campaigned for admission to the same training camp, but ultimately was refused.)

Marie was one of the few black college graduates in town, and although she was a new mother, she looked for a job. The ban on black teachers in Charleston did not apply to the Avery Institute, a private school for black students, so Marie was able to get work at Avery, her alma mater, where she taught "domestic science"—sewing, cooking, and decorating. She was one of the school's first nonwhite teachers: all of Avery's previous faculty had been white, and most had come from the North. A photograph made by Avery of "our first colored faculty" shows the faces of a dozen teachers, including Marie, who looks nearly white, and a little shy.

In the fall of 1916, Elise Forrest arrived in New York City and made her way to the town of Kings Park on Long Island. She had taken a job at the Howard Orphanage and Industrial School, forty miles east of Manhattan. Founded in Brooklyn after the Civil War as the Howard Colored Orphan Asylum, the institution had moved to Long Island in the early 1900s and took in only black youth. It stood near the north shore, on a five-hundred-acre piece of farmland. When Elise came to teach, the school was a few brick buildings among working fields, which were tilled by students during off hours. It was declining from lack of funds and would close its doors the following year.

Elise had left Charleston not so much for work as to be near her boyfriend. The little town of Kings Park was home to several asylums and hospitals, each full of staff and social workers, but Elise rarely left the compound of the Howard Orphanage and had few social contacts outside a handful of its teachers, whom she saw at meals. The orphanage was the only black facility in the local constellation of welfare institutions, and few white professionals wanted to have anything to do with their black counterparts. Elise taught her classes, then went back to her rooms and thought mainly of Teddy, who lived an hour's train ride to the west.

It appears that Elise's feeling for Teddy had grown to something near an obsession. "I am a little nervous tonight," she wrote him. "Started laughing at supper and—well I wasn't thinking of you just then so you can't say I had hysterics over you for I did not. I was unduly nervous and felt like crying and so there. But I'm going to bed soon and sleep it off."

After his own arrival in New York, Teddy had moved in with a friend in Brooklyn and enrolled at the Renouard Training School for Embalmers in lower Manhattan at 289 Fourth Avenue. During the week, Teddy spent his days at the morticians' school. But every two or three weekends, Elise would take a train into New York to visit him. On these visits, she took a room at the YWCA in Harlem. She and Teddy would spend the day at a museum, or walking in Central Park—simple pastimes made poignant by the reality that in their hometown they could neither use the museums nor sit on a park bench. At dusk, Teddy would bring Elise back to the YWCA.

The Renouard Training School for Embalmers took its name from Auguste Renouard, a physician and native of Louisiana. Born in New Orleans, Renouard was a twenty-two-year-old medical student when the Civil War broke out in 1861. He joined the Confederate army as a surgeon and became familiar with the embalming of soldiers. After the truce, Renouard moved to the frontier town of Denver, Colorado, where he ran an undertaking business. He specialized in sending the well-preserved bodies of western settlers back to their homes in the East on a slow stagecoach ride that could take weeks. But Renouard was soon dogged by rumors about how he obtained his supplies. It seems his more prosperous clients wanted manufactured caskets rather than handmade coffins; but since all of those caskets had to come from the East by stagecoach freight, there was often a shortage of good ones. Rather than lose business, Renouard decided that recently buried caskets might do just as well, so he began salvaging them from cemeteries. When he was discovered digging up graves, he quickly left Colorado before he could be prosecuted.

In 1878, Renouard had surfaced in Rochester, New York, where he published *The Undertaker's Manual: A Treatise of Useful and*

Reliable Information; Embracing Complete and Detailed Instructions for the Preservation of Bodies. The book led to his rehabilitation in burial circles, and he furthered his scholarly reputation with a series of articles on the proper demeanor of funeral directors for the trade journal *Casket and Sunnyside*. Eventually, Renouard moved to New York City and found a job as chief demonstrator at the United States College of Embalming. When the school began to fail, Renouard bought it from the proprietors and gave it his own name. Auguste Renouard died in 1912. When Teddy arrived, the school was in the hands of Renouard's two sons.

Teddy's letters contain nothing about his days in the laboratories of the school, where he evidently worked on corpses and learned the newer methods of preservation. In addition to the school, the Renouard family ran a company that sold an embalming fluid, purported to be less toxic than most. Teddy studied the new fluids, as well as the delicate process of pumping coagulated blood out of the body and streaming in preservatives.

On days off, Teddy seems to have taken himself to look at art. His letters show enough familiarity with the Metropolitan Museum of Art to suggest that he escaped there frequently. Teddy hated the Renouard School, but he was a diligent student, and when final examinations were held, he scored the highest of any pupil at the school.

Elise had brought her Brownie camera, and to pass the time, she took photographs. One of the earliest that survives shows the dining room at the YWCA. It's a snapshot more than a photograph, with little attention to framing and no singling-out of a subject—nevertheless, she saved it. In the photograph, well-dressed black customers sit at long communal tables, and among the diners sits the bass singer Paul Robeson. Between the years 1915 and 1919, Robeson was a student at Rutgers University in nearby New Jersey, and he took meals at the YWCA when he came into New York, because it was inexpensive and friendly.

Teddy's embalming course lasted just four months, and at the end of it, he was given a certificate. Accreditation requirements for embalmers varied from state to state; in South Carolina, one needed a diploma from embalming school and three months' apprenticeship. Teddy already had the apprenticeship, and by the end of January 1917, he was back in Charleston.

Elise stayed on at the Howard Orphanage another three months. During that time, so that Teddy wouldn't forget the smell of her body, she sent him a lock of her hair: "I had to cut my hair so am sending you a curl." She returned to Charleston at the end of spring, but, not wanting to teach at another school in the countryside, Elise went back to work in the hat store. Life returned to old rhythms, and Teddy and Elise again started talking about their future.

The giant clashes of World War I, more than two years along in Europe, distracted Americans in large numbers, but few black people paid much attention until 1917. When the United States entered the war, in April of that year, a wave of patriotism washed through the black South. For many months before the American declaration of war, the army, in its preparations, wanted nothing to do with black men. In South Carolina, colored recruits had simply been turned away at induction centers. But in May 1917, Congress passed the Selective Service Act, which allowed nonwhites "to register on equal terms with whites," and by June, blacks represented more than half the men enlisting in the state.

This eruption of the war into black society sabotaged Elise's careful plan to return to the safe courtship of Teddy. Just as she and her boyfriend had begun seeing each other again, Teddy and his brothers, Bob and Moultrie, put in for training as officers, hoping to become lieutenants in the new, separate companies of colored troops. For Elise, the events seemed to torpedo her plans for the future.

First Teddy took a fitness exam and received the classification of 1-A from the local draft board. He got in deeper when it became obvious the army had no idea where to look in black society for prospective officers. Teddy and a few other members of the colored elite volunteered their connections to help find suitable men. In their normal lives, Teddy and his family had almost no friendly dealings with whites: all interaction between the races was strictly business. But as the military threw its net into black society for the first time, Teddy advised an officer from the headquarters of the Southeastern Department of the Army, E. B. Garey, and made a rare white acquaintance. Teddy provided Garey with lists of names of black middle-class families and became Garey's confidant, so much so that when Teddy put together his own application for the officers' corps and solicited letters of recommendation, he included this curt but friendly note from Garey:

> To Whom It May Concern,
>
> Mr. Edwin Harleston has rendered to me very valuable service in compiling data in connection with the establishment of a Training Camp for colored citizens to be established at Des Moines, Iowa. It is my judgment that he is morally and mentally equipped to enter the Training Camp at Fort Des Moines. I believe that he is a double A man.
>
> Sincerely, E. B. Garey

Elise was devastated. For four years, she'd been trying to get Teddy to marry her, as his brother had married her sister. Now she contemplated whether her boyfriend would be killed. "And now comes the draft," she wrote Teddy in September.

> The winter we were in New York I had reason to believe we would by this time be "next door to heaven." Christmas coming makes two years we have surely lost. There are only a few months left, and you may be gone-forever. And I? I surely don't intend having you go to camp leaving me in Charleston. As near as I can figure it, Spring will find you in camp as they are planning to have this draft in France by June, according to the latest report. Ted, is this the way? Every few words I write I must stop to choke down the sobs and smile so the rest won't guess my thoughts. I had planned to go to the woods today to have a good cry all alone. What are you planning, Ted? Anything at all?

Moultrie and Bob went off to the officers' camp in Iowa, which opened in July 1917 with 1,250 trainees; but for some reason Teddy

In Charleston, South Carolina, in 1870, William Harleston, a white sixty-six-year-old former slave owner, and Kate Wilson, his forty-four-year-old black lover and former slave, chose this house at 28 Laurel Street as a home for Kate and the couple's eight mixed-race children. Because no pictures of either William or Kate survive, the house is the only surviving image of their relationship. The street name, number, and shape of the house have all changed: the same building is now 50 Ashe Street, and it is vacant and collapsing.

William Harleston and Kate Wilson's arrangement, though taboo, was common throughout America. Some years earlier a Virginia newspaper accused President Thomas Jefferson of having an enslaved black lover, Sally Hemings. A Massachusetts cartoonist named James Akin poked fun at Jefferson's reputation, calling him a "Philosophic Cock." Jefferson and Hemings are disguised as rooster and hen, and the small print reads, " 'Tis not a set of features or complexion or tincture of a skin that I admire."

The eight children of William and Kate, including their second son, Richard, born in 1845, were mulattoes, a word that derives from the Spanish mula, *meaning mule, a cross between a mare and a jackass. Richard Harleston became a housepainter in the city of Charleston and had this photograph taken about 1890, when he was forty-five years old.*

In 1865, Richard Harleston, then twenty years old, married nineteen-year-old Agnes Graham of Red Bank, South Carolina. It was a custom for multiracial men to "marry light, and never dark," and Agnes was as light-skinned as a woman could be without actually leaving colored society and passing for white. The couple settled in Agnes's hometown and had five daughters.

Edwin G. Harleston, Kate Wilson's fifth child, was nineteen years old when his father died and his inheritance was stolen by a white relative. Cast into poverty, Edwin worked as a cargo boatman and eventually a skipper of his own schooner (which gave him the lifelong nickname of "Captain Harleston"), before going into the lucrative business of undertaking, which became the instrument of his family's return to relative wealth. Afterward, Captain Harleston developed a fondness for formal clothes and stern photographs, as in this image from about 1920, when he was sixty-five years old.

Captain Harleston's wife, Louisa Moultrie, whom he called "Lassie," came from one of the few black families that had been free during the slave period. Free mulattoes represented the upper class of Afro-Americans; therefore, Captain Harleston "married up" at his wedding in 1877. This charcoal drawing dates from about 1890.

The Harleston family belonged to the "colored elite," a tiny, light-skinned class of influential nonwhites, many of them the children of white men and black women. About 1920, in Charleston, Edwin G. "Captain" Harleston (seated, center) *hired a photographer to make this portrait of his family in the stiff drawing-room style of the time. His wife, Louisa, was long dead, so Captain Harleston surrounded himself with his grown children* (left to right): *Eloise Harleston Jenkins, Robert Harleston, John Moultrie Harleston, Edwin A. Harleston, and Katherine Harleston Fleming.*

After his wife died, Captain Harleston found a lover, Mamie Randall, with whom he had a daughter named Alethia (or "Leetie," left, *photographed in the early 1920s), and another daughter, Catherine. Leetie stands in a doorway with her elder half sister, Eloise Harleston Jenkins* (right rear), *her half niece Gussie Harleston* (right front), *and her half nephew Harleston Fleming. Mamie is said to have gone into business as an abortionist, making a living by ending the pregnancies of white women who had had dalliances of their own.*

Until the early 1900s, most black families in the South buried family members without paid help from an undertaker, washing and dressing the body and making a coffin from pine planks. Many also kept up the West African death ritual of decorating fresh burial plots with "grave goods," consisting of mementos and keepsakes of the deceased, as in this 1912 photograph of a grave in South Carolina.

Seeing a rich market, Captain Harleston (left) *learned the burial business with his sister Hannah, who was the widow of an undertaker, before opening the Harleston Funeral Home at the turn of the century. This advertising card dates from about 1915, when Captain Harleston was briefly in partnership with his son Robert* (right).

Captain Harleston built new headquarters for the Harleston Funeral Home in 1913, at 121 Calhoun Street in Charleston. In this 1920s photograph, one of Captain Harleston's workers, a disabled black man named Moses (right, with walking stick) *talks with an unidentified man in front of the funeral office.*

A typical funeral cortege in the early 1900s included a hearse (far right), *carriages of mourners, and drivers in brass-buttoned livery. McAlister's Stables in Charleston, which doubled as a burial business, had black drivers (but no black clients), while the Harleston Funeral Home had similar liverymen (but no white customers).*

Edward C. "Eddie" Mickey (standing in front), *a nephew of Captain Harleston, ran the competing Mickey Funeral Home, which he had inherited from his mother, the Captain's sister Hannah. His extravagance led him to buy too many limousines and hearses (this collection of his cars dates from about 1925), which helped drive him into bankruptcy during the Depression, when the Mickey Funeral Home failed and Eddie Mickey moved to New York.*

King Street, the main commercial street through Charleston, around 1910. In the early 1900s, Charleston was poor and provincial, and many of its white citizens looked to the past. This photograph was taken during a reunion of Confederate veterans nearly fifty years after the Civil War.

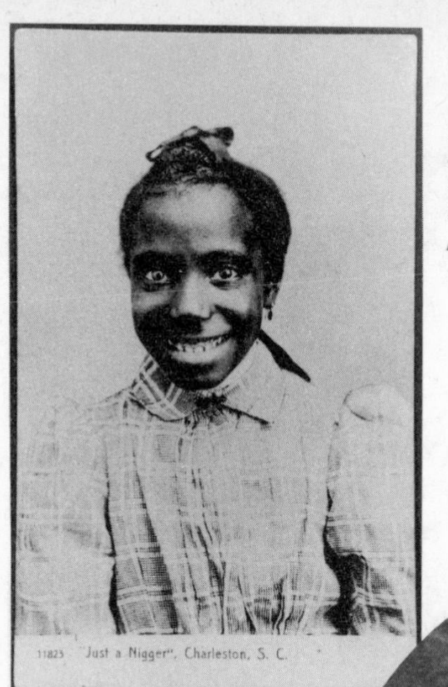

Charleston was 57 percent black in 1900, but people of color lived as aliens in their own hometown. "Jim Crow" laws meant unequal schools, separate seating, "colored" doorways—and images like this postcard were commonplace.

Reverend Daniel Joseph Jenkins (right, *with his first wife, Lena James) was a self-taught Baptist preacher who was born in slavery (and dark-skinned), but "married light" in the early 1880s, finding his wife, Lena James, from among the South's colored elite. Together they founded the Jenkins Orphanage, a home for abandoned black children in Charleston.*

Reverend Daniel Jenkins, photographed about 1900, believed himself a missionary who could help the throngs of "motherless and fatherless waifs thrown on the cold charity of the world."

The Jenkins Orphanage, founded in 1891, occupied a stately neo-Gothic building with pointed arches and slender columns, which had previously served as a hospital. By 1900, the orphanage housed some two hundred black children (who slept in dormitories made from the former sick wards) and gave crude schooling to hundreds more local boys and girls.

$10,000 Needed for Orphans and Destitute Children.

THE ORPHAN AID SOCIETY ASYLUM, 20 Franklin St., Charleston, S. C., was chartered by the state of South Carolina in 1892. The Jenkins Orphanage Institute and Greenwood Orphan Industrial Farm are all conducted under the auspices of the society which is run on faith—536 children and 9 teachers on roll in the Orphanage and School, 20 Franklin St. The farm consists of 100 acres of land, donated by Mr. J. Wild of New York City. The Orphanage is in great need of money, food and clothing to keep it up—$10,000 is needed to feed and clothe and to erect a building on the farm where the children and teachers may rest after a hard day's work. These are orphan boys out of the Asylum and are playing hard, trying to help raise the money to put up the building this winter. The President and Board of Managers are anxious to erect the building on the Industrial Farm, which is 17 miles from the Orphanage where the children can be taught daily in the art of farming and trades of all kinds.

We believe every man, woman and child who sees and hears these little orphan children read note at sight and play their brass band will help in some way. Some love one, perhaps, will give a Cottage to the poor orphans whose parents are lying in yonder cold icy grave, as you do not know how soon your son or daughter will be left behind in the cold charity of the world to be baffled about from a hand to mouth struggle. Help the orphans! And let help be the pass word to your friends! Tell them to help and send their names to the Rev. D. J. Jenkins, President and Founder, 20 Franklin St., Charleston, S. C. Send money by Post Order, registered letter or check, and boxes and barrels by steamer or freight. All packages send direct to 20 Franklin St., Charleston, S. C.

Members of the Board: Rev. D. J. Jenkins, President and Founder, J. H. Hagans, Vice-President, S. L. Green, Local Treasurer, T. E. Richardson, Secretary, Misses E. A. Thornton, Corresponding and Financial Secretary, M. E. Brown, E. A. Clark, Mrs. Emma Johnson and F. F. Fields.

Carolina Savings Bank Depository Advisory Board: Mr. Geo. W. Williams, President Carolina Savings Bank, Dr. F. L. Parker, Dean of Medical College, Mr. F. R. Frost, Attorney at Law, Mr. W. G. Harvey, Jr., Cashier of Enterprise Banking Co., Dr. Edward Parker, Secretary, Hon. John P. Ficken Mayor for Reference.

Subscribe for THE CHARLESTON MESSENGER, the leading weekly of the South, which is published in the interest of the Orphanage and acknowledges all donations sent to the Orphanage, also tells of it internal workings.

THIS CIRCULAR 5c. EACH.

To raise money, Reverend Jenkins solicited a few battered horns and hand-me-down uniforms and created the Jenkins Orphanage Band. Beginning around 1893, selected boys were given rough music lessons and then released onto the streets. As the music improved, the children (led by Reverend Jenkins) began annual tours up the East Coast to New York and Boston.

Edmund Thornton Jenkins, born in 1894, son of Reverend Jenkins and his wife, Lena, played clarinet and saxophone with the Jenkins Orphanage Band, but fled their chaotic rhythms for college, where he studied violin and composition. The effete Edmund eventually attended London's Royal Academy of Music and settled down as a composer and nightclub musician in Paris during the 1920s.

The Harleston and Jenkins families first came together in 1905, when Eloise "Ella" Harleston (at age twenty-two, the youngest daughter of Edwin "Captain" Harleston), worked as Reverend Daniel Jenkins's secretary. The minister, who was married and forty-three, seduced her, and Ella became pregnant—a crisis that culminated with the couple's plan to abandon their baby in England, far from prying eyes. This done, Ella eventually married her lover (after his first wife died) and lived a proper middle-class life with Reverend Jenkins for twenty-five years.

Olive Harleston Jenkins, born in 1906 in England, was the love child of Ella Harleston and Reverend Jenkins. Her parents abandoned her from birth until age fourteen, when the orphaned girl finally came "home" to America. But Ella was so scarred from having an illegitimate child, she never allowed Olive to call her "mother."

Edwin A. "Teddy" Harleston, the third child of Captain Harleston, captained his college football team, studied Latin and ancient literature, and befriended militant black intellectual W. E. B. Du Bois (his sociology teacher and a founder of the NAACP). The Harleston family hoped Teddy would become an exemplary member of the "black bourgeoisie," but Teddy wanted to be an artist.

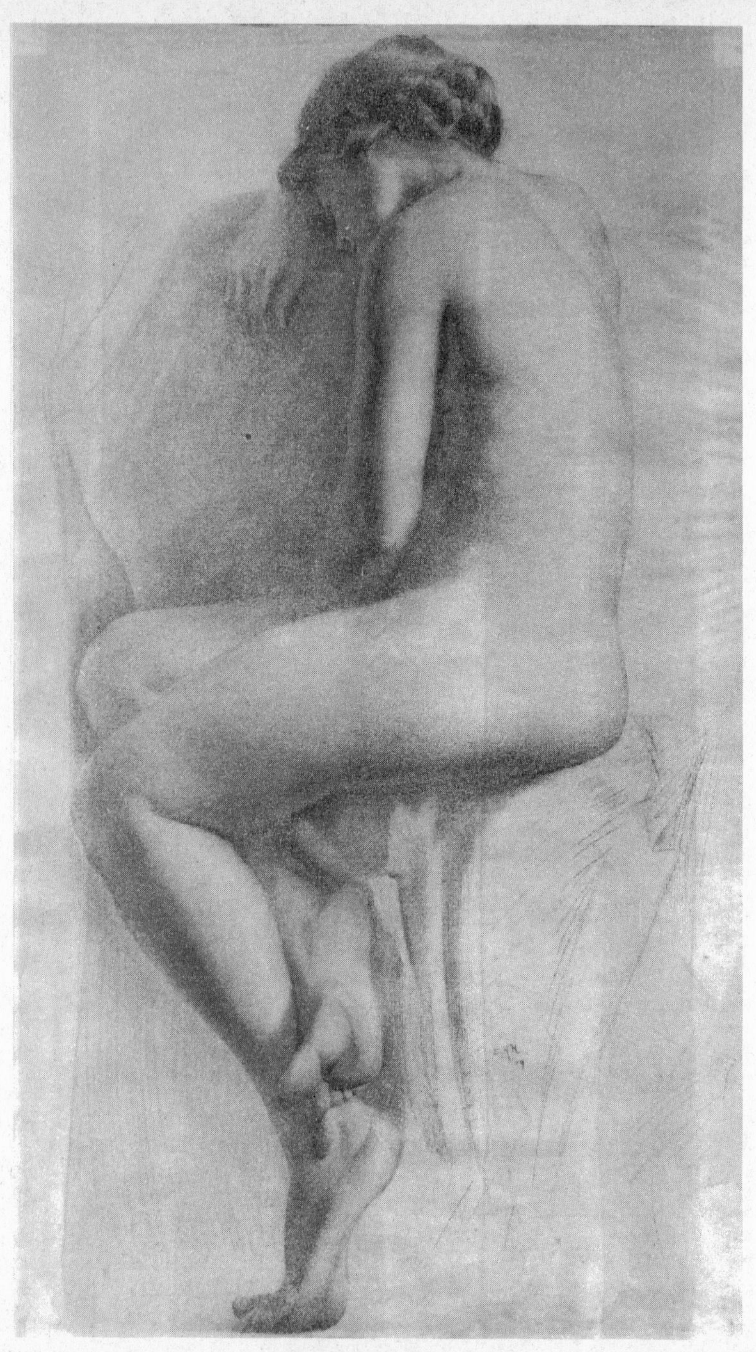

Nude Study, *Edwin A. Harleston, charcoal, 1909. After college, Teddy was accepted into the art department at Harvard but chose to attend Boston's School of the Museum of Fine Arts instead, where among dozens of studies he produced this charcoal in 1909.*

Teddy worked part-time on a Massachusetts ferryboat and lived as a frugal bohemian. In 1910, a fellow student named Philip Adams asked Teddy to sit for a painting called The Sailor *(on the easel), which showed Teddy in a sensuous pose, far from the straitlaced milieu of his middle-class Southern family.*

Self-Portrait, *Edwin A. Harleston, oil on canvas, 1916. Teddy's father chided him to give up the unreliable career of artist and come home. At age thirty-one and after twelve years away, Teddy relented, and moved in above the Harleston Funeral Home. He set up a painting studio and produced this sad, enigmatic self-portrait.*

The Renouard
Training School for Embalmers
289 Fourth Avenue
New York City

The Bearer........E. A. Harleston........
having qualified for admission and paid the tuition fee is entitled to all lectures and demonstrations for the year commencing
JAN 2 - 1917
....

This card is not transferable, and must be shown when requested.

No. 2806

A. Renouard,
C. A. Renouard,
Lecturers and Demonstrators

Moved by a sense of duty to his father, who had helped put him through college and art school, Teddy Harleston enrolled in a three-month course at the Renouard Training School for Embalmers in New York City.

From his previous life as a rough-edged artist, Teddy reinvented himself as a well-groomed businessman—a handsome, soft-spoken, trustworthy heir.

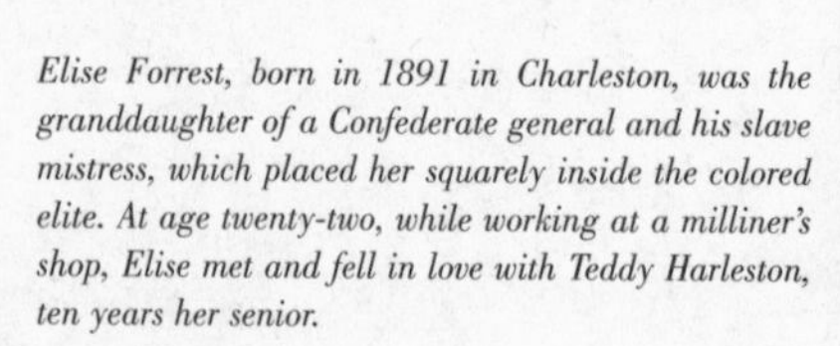

Elise Forrest, born in 1891 in Charleston, was the granddaughter of a Confederate general and his slave mistress, which placed her squarely inside the colored elite. At age twenty-two, while working at a milliner's shop, Elise met and fell in love with Teddy Harleston, ten years her senior.

Portrait of Elise Forrest, *Edwin A. Harleston, oil on canvas, ca. 1920. Teddy made this portrait of his girlfriend Elise and then wrote her in a letter that he often had "little chats" with her picture. They saw each other for seven years before they were married (and didn't sleep together until their honeymoon).*

In 1919, Elise (left) *moved to New York City and enrolled in photography school (at her boyfriend's expense); the following year, she was visited by* Edmund Jenkins (right), *who had arrived from Europe. The photograph was probably taken on the balcony outside Elise's rented rooms in a Brooklyn brownstone.*

Teddy Harleston with his two-year-old niece, Gussie Louise Harleston, about 1919, in the yard of the Harleston Funeral Home. During World War I, Teddy's younger brother Robert enlisted in the service but returned with a case of tuberculosis, which quickly spread to his wife. Physicians insisted the infected couple try to save their children, Gussie Harleston and her infant sister Sylvia, by giving them away. Teddy agreed to take Gussie, and Sylvia went to Teddy's sister Ella. When Gussie's mother was confined to a tuberculosis sanitarium in the mountains of North Carolina, she wrote Teddy and asked him to buy her daughter a pair of black shoes and white socks; and Teddy sent his sister-in-law this photograph as proof he had carried out her request.

Elise Forrest and Teddy Harleston married in September 1920. They agreed that they would launch a partnership in art and photography, and together they would raise Gussie Harleston. Elise took a photograph of Gussie, with a Kodak "Brownie" camera on her lap, and made the first business card for the family project.

did not receive orders to join them. He remained in Charleston, swept up in the patriotic glee, and waited. He began to make paintings with war subjects. One, *The Gas Attack,* shows three black soldiers charging up a hill during an assault. In it, two of the three figures wear gas masks, while the third has a mask that has slipped from his face. The enemy has launched a gas attack, and the man without the mask grabs his throat, choking. However, *The Gas Attack* is not one of Teddy's better paintings. At best, it's a good piece of illustration and fair propaganda for black soldiers who would fight.

Teddy received a draft notice on October 12, 1918, but again, for some reason, he was never called up. He turned his attention to portraits, painting black men in uniform—this time seated and from the waist up, in the traditional style. His painting *Doughboy* shows a young soldier looking nervously at the viewer.

When the Armistice came on November 11, 1918, Teddy was almost disappointed. His brothers had their uniforms and guns, though they had not made it to Europe. Some four hundred thousand black Americans had served in the war, about half of them in France. Teddy was left with his pictures of soldiers.

The war ended before Bob Harleston could be sent overseas, but according to family tradition, he returned from training camp in Iowa in a mysteriously weakened condition. He and Marie resumed their lives, and soon Marie became pregnant again. Their second child, Sylvia, was born July 6, 1919. Bob's health continued to decline, and he had disturbing symptoms—frequent cough, shortness of breath, and spitting blood. Finally a physician gave the diagnosis: it seems Bob had contracted tuberculosis while in training. Just as the shock of the finding had begun to subside, Marie came down with her own symptoms. The couple that had begun life so effortlessly now faced the possibility that they and their two children, Gussie and Sylvia, might not survive another year.

In the early 1900s, Charleston was a focal point of a tuberculosis epidemic, and black society was the hardest hit. But little was done to stop this disease called "consumption." It was thought to be a problem of climate and of fresh air, so treatment was largely environmental. In homes with tubercular patients, families would often build a screened-in addition that extended from the side of the house, so the patient could live outdoors. These rooms without walls were said to strengthen the lungs, and dozens of houses had them, especially in black sections of town. In reality, the outdoor rooms also appeared because families feared contamination from their own relatives. A designated sister or parent would pass food to the infected person through a hole in the screen.

Gussie Harleston would later recall:

"My earliest memory is of sitting in the seat of a horse-drawn buggy, about 1920, watching my mother as she was lifted by my father and carried along the sidewalk, up the steps, and into our house. My mother was only twenty-six, and the tuberculosis had drained her. She couldn't climb the stairs and had to be carried. My father resisted the disease better than she did, and lifted her from carriages when it had to be done. My father took my mother in his arms like a limp scarf, walked over the threshold, and started up the stairs.

"When I was two and a half, my sister, Sylvia, was born. When they found the tuberculosis, the physicians then told my mother not to have any more children. Her answer was that having children was the reason she married my father. Doctors had too much power then. My mother obeyed, and stopped getting pregnant. After that, my mother was forced to give up both of us."

Gussie was taken from her parents at the suggestion of physicians who said she would catch the disease. Her sister, Sylvia, was also taken away and sent to the home of Ella Jenkins. Ella, Sylvia's aunt, was anxious for a child of her own (Olive lived in England) and welcomed the baby. Family lore says that Ella loved Sylvia and treated her as if she were her own child.

As a way of helping with their treatment, Captain Harleston arranged to send Bob and Marie to a town called Summerville, twenty miles from Charleston. This nearby village was dry and full of pine trees whose sharp scent was thought to clear the lungs, and many tuberculosis patients had already moved there. The Captain set Bob up in business with a new Summerville branch of the Harleston Funeral Home. The office was in a building on a street called Black Jack, which took its name from a tree that was common to the area, the blackjack oak.

The late summer and fall were deadly for tubercular patients, with the moisture and heat helping the contagion to multiply. So during August and September, Bob and Marie moved to Asheville, North Carolina, a mountain town near the South Carolina border with recovery camps for consumptives. The climate in Asheville was cool and dry, even during much of the summer, and its hillsides were thick with allegedly healthful pines. In Asheville, Bob improved, but Marie lost weight and grew weaker. Meanwhile, a physician reported that Marie was "very much run down and frail [and] feels continually nervous."

From the cooler climate of Asheville, Marie wrote family members about her slow death. In one note to her brother-in-law Teddy, Marie describes wasting away: "Dear Mr. Teddy—Everybody says that I look good. I am ninety-one pounds now. Would give anything if I could reach one hundred." While his wife declined, Bob generally disregarded the advice of his doctors who had told him that he should limit his activity and eat a healthy diet. In one of Marie's letters, she scolds her husband, whom she calls "Mr. Bob":

> Mr. Bob lost half a pound but he has regained it along with a pound more. Now he is 138½. Looks fine, eats well, sleeps (over well). Sometimes he coughs, and the sound still makes me shudder. It is hollow and sounds very strange. Bob tried himself again today. He came down a hill then went back up bicycling. The physician told him to be careful of all strenuous exercise, as it might cause a hemorrhage.

Later in the same letter, Marie revealed how much she missed her daughter Gussie, who had just moved in with Teddy and Moultrie.

> How do you like Gus? Do you think she is still as ugly as she used to be? Has she sung and played the piano for you yet? Please take her one of her dolls. She used to cry for one every night. Say, Mr. Teddy, Elise said you wanted to give Gus something, but did not know just what, so I told her to tell you to give her a pair of shoes

> and two pairs of white socks. Mr. Moult wrote that he has bought her a pair. If his are white, then you buy black; or if he bought black, then you buy white.

Teddy did the errand and sent his sister-in-law a photograph to prove it. In the picture, he holds Gussie aloft for the camera, and on her dangling feet is a new pair of black shoes.

The baby Gussie Harleston had curly brown hair, which her elders combed out and dressed up with bows. She grew into a rather plump toddler under Teddy's supervision. "Gussie is getting fat," he wrote Elise. Several months later, he repeated, "Gus is doing fine, but I believe she is getting fat." Photographs from her childhood show that young Gussie had two common expressions: a smirk when she was being coddled and a scowl when she was not.

When she was grown, Gussie would remember:

"To be given up for adoption was the best that could have happened to me, because I was given to my father's brother Teddy to be raised. I don't know what my mother had seen in my father. Compared with Uncle Teddy, he wasn't much, but my mother just had to have her Bob Harleston. I loved Uncle Teddy. He was an artist, and he was glamorous, intelligent, and handsome. Women melted over him, and he hardly knew it. I moved in with him when I was just a little thing, and he shared an apartment with his brother Moultrie on the third floor of the funeral home, above the Firm. Uncle Teddy and Uncle Moultrie were two aging bachelors, and as a little girl, I was lucky to be the only woman in the house. I often wished Uncle Teddy had been my father, and I still think of him that way."

Teddy Harleston was thirty-seven when he became an adoptive father. Family members remember that his manner was peaceful, and that he was melancholic and soft-spoken. He entered rooms quietly, and he had a gentle laugh, if it could be drawn out of him. Teddy's brother Moultrie was less attentive as a surrogate father and, by most accounts, lazy. Although Captain Harleston tried to control him, Moultrie was his own man, who ignored the Captain's wish that he work in the funeral business. Similarly, Moultrie brushed aside the responsibilities of parenting Gussie and left most of them to Teddy. Gussie says:

"Uncle Teddy had a sad atmosphere around him. I won't say he was a disappointed man, but he lived under the rule of his father. My grandfather had forced him to join the family business, and Uncle Teddy hated it. But as a dutiful son, he had returned from years of living in the North, where his freedoms were great, to Charleston, where they were nil. When I arrived in Uncle Teddy's life, I became the focus of his world. He thought whatever I did was funny, and he completely spoiled me."

When Elise was out of town, Teddy wrote her to report on the new baby: "Gussie is with us, and believe me her two uncles here have a great time taking care of her. She seems more lovable than ever, and is as perk as a beetle." Teddy doted on his niece by dressing her in pricey clothes and enrolling her in piano lessons. "I got her shirts, and now she has another sweater and several more bloomer dresses, and a new buffalo hat, and everything," he wrote. Teddy washed Gussie's clothes himself, and when he was in a hurry, put them in the oven to dry. In a story told by Ella Jenkins, one afternoon when she was shopping Ella bumped into a breathless Teddy on the street. Teddy wheeled around and said, "I don't have time to talk. I've burned up all the baby's clothes and underwear in the oven, and I need to buy more."

Gussie was happy when she was the center of attention. With Teddy as her audience, she learned to perform for adults, and before long, Teddy sounded like an actress's proud parent.

"Today the Sunday school had a program," he wrote Elise, "and Oct 23rd, for the first time in her eventful life, little Miss Gussie Louise Harleston ascended the platform, made her stiffly graceful bow and recited a whole verse! Tomorrow I must get her a present. She is learning script letters and can read 'it is a cat,' so if you are away don't be surprised to return and find her taking dictation on a typewriter."

Gussie remembers, "I was young enough to withstand the change of being taken from my parents, and the Harleston house was full of attention. Captain Harleston was playful with me. He would bend down and say, 'How you does?' And I would correct him: 'Grandpa, you don't say *How you does?*, but *How do you do?*' He would say, 'Right, now how you does, girl?' Captain Harleston smelled like bay rum, which was the scent of his cologne. And the whole office smelled of it, too. He would take me on walks around the neighborhood. Everybody hailed him, because he was a big muckety-muck, and he smiled because he was with his little grandbaby."

Captain Harleston and Teddy competed to see who could give Gussie the most comforts. One Christmas, Teddy told Elise: "As usual, Santa Claus forgot me, but I found pleasure in making a few gifts to as many folks. Gussie was delighted with her many gifts, including a bicycle."

From the Captain, Gussie got other enticements. "The Captain had a chauffeur whenever he wanted," she remembers, "and he would have his driver take me to school. He also owned a pile of real estate, rental property. Once a month he would take me with him to collect the rents. The tenants were simple people. They would greet him, 'Af'noon, Mr. Harleston! Is that your little grandbaby? Isn't she somepin'?'"

Orphaned, but with parents who were still living, Gussie had a contradictory childhood. "There's an expression among black people, 'nigger rich,'" she says. "It means you have more than other colored people, and you inflate the amount further by talking a good game. But compared to rich white people, you don't have a lot. In fact, you don't have much at all. That's what we Harlestons were, nigger rich."

Chapter Eleven

AFTER years of neglecting his artwork, Teddy began to spend long days in his Charleston studio. He started by painting his family one at a time. Teddy invited the Harlestons to sit for him, making portraits of his older sister, Katherine Harleston Fleming, and his brother Moultrie. He painted Captain Harleston, and then he did a portrait of his sister-in-law, Marie.

Marie seems to have come in for some sittings, but it was an hour's travel to Charleston from her home in Summerville, and she was weak, besides. As an experiment, Teddy propped a photograph of Marie on his easel and used it for his model. (Elise had taken the picture.) In her painting, which sharply resembles the photograph, Marie is wearing a lace blouse, buttoned up to the top, with her hair pulled back from her face. She looks like a cool, prosperous lady as might have been executed by Mary Cassatt or Frank Benson, Teddy's teacher in Boston. Marie's expression, however, is richer and more human than any Benson gave his subjects. It's possible to see at what cost she has lived.

After Marie, the Harlestons lined up for more portraits. Teddy painted his aunts, his other brother, and his cousins. But he wouldn't let one family member sit for him: his sister Ella. Teddy was still angry with Ella over her affair with Parson Jenkins. It was a double standard—the Captain and Mamie Randall could be understood, but not Ella. While paintings of all the Harlestons survive, there is none of Ella Jenkins.

Making pictures of his family, Teddy felt released. He saw that his talent was intact, and his ambition began to return. In a letter to Elise, Teddy explained that he was trying to represent "our group," meaning colored people, in all their complexity. "[I want to paint] our varied lives and types with the classic technique and the truth, not caricatures," he wrote. "To do the dignified portrait and take the picturesque composition of arrangements or scenes, showing the thousand and one interests of our group."

The "caricatures" Teddy worried about were probably the prevailing images of colored people. Though black Americans almost never appeared in fine art or movies, images of Negroes were ubiquitous in everyday life. Pharmacies sold "Darkey" brand toothpaste. Groceries stocked canned goods with brand names like "Coontown Yams," whose labels showed grinning black children with big lips. There was an ocean of knickknacks and kitchen products—housewares, lighting fixtures, toys—made in or with the image of Negroes. White households in many parts of the country had "mammy" cookie jars, in the shape of a big black woman wearing a bandanna, or "Sambo" ashtrays, showing a thin, acquiescent servant seeming to catch the ash. "Buckwheat" dolls were made for children to knock around, and sculptures of black jockeys appeared on lawns in the new suburbs.

The mammy images could have been part of the reason Teddy had stopped painting. It wasn't easy to make art when the world was full of so many spoofs of the very subjects he wanted to represent. White artists could, and often did, see their work as upholding the tradition of Joshua Reynolds and Jacques-Louis David, whose dignified pictures of aristocrats ennobled both the painter and the painted. But if Teddy went to the grocery, he might stumble upon a mound of walnuts that the grocer had arranged next to a sign that read "Nigger Toes."

The photographs Elise had taken in New York weren't very good, but Teddy was intrigued by them, and he began to talk with her about how she might do better work. Elise suggested the next step: what if she went to school to study photography? Relieved that the danger of the war had passed, Elise pushed the plan. Teddy wrote: "I like the idea of your going to the school of photography. If you be good I think I can arrange it."

Teddy and Elise took the idea one step further. After he left art school and joined his father's funeral business, Teddy's morale had been sapped, and he felt he hadn't been painting enough. But what if Elise became a photographer, and the two of them went into business together? Could they set up an art studio in Charleston? Would Teddy be able to paint, and Elise make portraits? It was a fantasy, but the couple liked the sound of it. Elise took to the notion: she had learned from her boyfriend's discussions of art and she wanted to wade in more deeply. Teddy encouraged her because such a scheme not only would promise to give him an artistic partner but also might stimulate his own productivity. If they got married and then opened a "Harleston studio," they might have a chance at happiness.

The idea of establishing an art studio, at least for a nonwhite couple in out-of-the-way South Carolina, was radical, at best. There were simply no models for it. In New York, the well-known artist-couple of photographer Alfred Stieglitz and painter Georgia O'Keeffe had only just set up shop in 1918. News about them would probably not have reached Teddy's ears in Charleston. That a colored twosome in the South would collaborate in art was ludicrous to anyone who heard mention of it. Elise must have discussed the idea of marrying Teddy and sharing his studio with members of her family, because one of her sisters wrote her skeptically, "If you think you could keep house and engage in your work and stand the strain of marital relations you are imagining more than you could ever realize. Of course, no one ever takes advice, so it's useless to try to advise."

But the more Teddy and Elise talked, the more their plan began to address practical issues. For Teddy, the trouble with doing paintings was that sittings took time. With his obligations to the funeral home, he didn't have the free days and weeks needed for a client to sit. His portrait sessions could be interrupted the moment a new client walked in the door, or a fresh body appeared from the morgue. If Elise were a photographer, she could help solve the problem. A client could sit for several initial drawings. Elise would then photograph the sitter in his or her formal pose, and Teddy could finish the painting from the picture. It would be an unequal, even sexist, division of labor, but the bigger scheme was so unusual that it hardly mattered.

By the second decade of the 1900s, only a handful of black photographers had managed to have real careers. Among the earliest were three black brothers named Goodridge, who worked in Michigan in the 1880s. Glenalvin, Wallace, and William Goodridge were all born in York, Pennsylvania, to a wealthy father who had helped move runaway slaves on the Underground Railroad. When the Confederate army approached York in 1864, the family fled for its safety, moving to Saginaw, Michigan. In Saginaw, the Goodridge brothers became portrait and landscape photographers, using the old ambrotype process, and operated a studio that stayed in business until 1922.

When Elise began thinking about going to photography school, there was at least one true portrait artist she could look to, a black Georgian named Cornelius M. Battey. Born in 1873 in the town of Augusta, C. M. Battey had moved to New York in the early 1900s, where he became a photographer. Battey was one of three partners in the Battey and Warren Studio, at 509 Eighth Avenue, in Manhattan. But in 1917 he was hired by Booker T. Washington to head a new photography department at Tuskegee Institute, and from his Alabama studio, Battey produced a long stream of magnificent portraits of black Americans of his day.

In Charleston, the earliest black photographer to open a business was probably Arthur MacBeth, who had a studio from the late 1800s until 1910, when he moved to Baltimore. The second and possibly the only other black photographer was Michael Francis Blake, a graduate of Avery Institute who opened shop in 1912 and kept at work for more than twenty years. Teddy probably knew both of them, and Elise at least knew of Blake. Teddy's letters show that he was acquainted with C. M. Battey, the Tuskegee photographer, and talked about him with Elise.

There were other, more remote models: Arthur Bedou, a pioneering black photographer in New Orleans, and Paul Poole, who had a studio in Atlanta. Even so, it was one thing for these few black men to have pushed ahead and established themselves, but it was quite another for a black woman to attempt a career in photography. Nonwhite women in photography worked in such obscurity that it's unlikely Elise would have heard of any of them. In 1904, the directory of the National Negro Business League in Washington, D.C., listed Mary E. Flenoy as a photographer and member of the league. Flenoy, who was originally from Danville, Illinois, evidently operated a small studio in Washington, the only black woman to do so. In 1910 in New York City, Jennie Louise Welcome, using the name Mme. E. Toussaint Welcome, ran an ad in a magazine for black subscribers that listed her as the proprietor of a studio. "Mme. Welcome" was born in 1885 as Jennie Louise Van Der Zee, in Lenox, Massachusetts. Her parents, John and Susan Van Der Zee, had been maid and butler to General Ulysses Grant. Her brother, photographer James Van Der Zee, who famously documented Harlem life in the 1920s, would later eclipse Jennie Welcome's business success.

There is no evidence that Elise was aware of more than one or two black photographers of her day—and yet she persevered. Elise and Teddy inquired about photography schools and found one of them that concentrated on portraits, the Emile Brunel School of Photography, on Broadway at Thirty-second Street, in New York City. The two contacted the Brunel School and asked about fees. The idea was that Elise would study photography in New York, and when she finished, she would be able to work with Teddy, documenting his paintings and making the studio portraits from which he could work. As they waited for word from New York, Elise prepared to leap into the unknown.

Five months after the Armistice, in April 1919, the *Mercury*, a U.S. Navy ship carrying soldiers returning from France, arrived in Charleston harbor. Some of the soldiers on the troop transport were South Carolina residents, while others merely found themselves coming home via the South. The *Charleston News and Courier* published an invitation to citizens to greet the ship at the dock, but when Teddy and other family members showed up to do just that, the police turned them away. This was apparently a whites-only gathering, as though only whites could celebrate the victory in Europe. Teddy sent a letter to the newspaper in protest:

> I do not know, Mr. Editor, who was responsible for the order, but it certainly caused a number of people an unnecessary and unwarranted humiliation. In future announcements state when possible whether it is the "public" that is invited or "whites only." Perhaps, I do not need to tell you, Mr. Editor, that we know how to stay away.

That would seem to have been the end of it, but for a malign feeling among the white navy crew and some of the soldiers aboard the *Mercury*. Ominously, other military homecomings had been followed by incidents of racial violence. Some servicemen had apparently never gotten over the political deal that had allowed black men to enlist, and white-led attacks on black neighborhoods ensued. In Charleston, the "bluejackets" aboard the *Mercury* quickly recognized that they were in a black town, and at least some of them were unhappy about it.

On May 10, 1919, one of the sailors from the ship shot and killed a black civilian under mysterious circumstances. There is little to suggest the victim provoked the attack. In most cases of theft or assault committed by a nonwhite, the newspaper accounts were very clear about who was to blame, but in this instance, reports about the shooting were vague and did not accuse the black victim or offer the justification of self-defense. When the police arrived to make an arrest, the officers were menaced by a mob of sailors from aboard the ship. In the next several hours, a riot ensued in which gangs of sailors stampeded through the streets, seizing and randomly beating black locals. In the end, two black men, James Talbert and Isaac Doctor, were killed; seventeen others were injured.

The Charleston riot was one of twenty-five other race riots across the country that year, many of them involving white soldiers returning from the war. In Chicago, twenty-three blacks and fifteen whites were killed; and in Washington, D.C., six blacks were killed. After the war, violence against black Americans surged, and the number of lynchings increased by a third. In 1918, fifty-eight black Americans were hung or burned to death, and in 1919, seventy-seven were killed. Eleven of the seventy-seven were burned alive. Ten of the victims were soldiers, and one was a woman.

In Charleston, the solicitor's office found "no clue as to who was responsible for the death of James Talbert, who died from gunshot wounds." Teddy, weary of this kind of news, signed his name to another editorial, this time one written by a committee of middle-class blacks. In the editorial, published in a local paper and headlined "Negroes Deplore the Mob Spirit," the writers pointed out how the violence had spilled over from the war.

> It has now invaded our own community, and is threatening the lives and property of Negroes, and thereby making this community unsafe for them as residents, whom the city, state, and federal government do not exercise their authority to protect. We know that in this instance of mob violence the Negroes were not the aggressors, and that they, innocent and unknowing, were the studied and intended victims of the mob, which took life and money, maimed bodies and destroyed property. We petition the mayor and the city officials to bring to justice all persons known to have been members of the mob, and to assure protection to all persons and property in the community.

In the fall of 1919, Elise and Teddy were finally able to put their plan in motion. The Emile Brunel School had an opening, and Teddy arranged to pay Elise's tuition. Elise boarded a train (they had stopped taking the Clyde Line) and left Charleston for New York.

"My dear Elise," Teddy wrote a few days later. "You looked so sweet and trim in your seat on the train I felt a thrill of sentimentality. You are perfectly at liberty to go where you will to stay—if to the Y.W.C.A. so much the better; and you shall have funds sufficient for your needs." Knowing how difficult it was to live in New York, Teddy tried a few jokes about city life. "Now I hope you will settle down to a genuine enjoyment of your stay in the metropolis, see the king and queen and the Prince of Wales and all the swells, have a real all 'round good time, behave yourself perfectly, keep in perfect health, make friends of worthwhile folks only."

Elise at first took a room at the YWCA in Harlem and then moved to Brooklyn, where she rented the upper floor of a house belonging to the family of Robert Elzy, a friend of Teddy's from Atlanta University who was director of a chapter of the Urban League.

The Emile Brunel School of Photography turned out to be a small operation run by its founder and namesake. Emile Brunel was a minor figure in the history of image making, and something of a dilettante. He not only made photographs but also produced abstract sculpture and dabbled in the movie business. Brunel is said to have developed a one-hour photography process, though he was unable to exploit it commercially. He is also said to have had a hand in early cinema, at one point making a silent film called *The Hand of God*. Later in life, Brunel would retreat to the town of Boiceville, his home in the Catskill Mountains about a hundred miles north of New York City, and surround his modest farmhouse with a collection of his own eccentric sculpture. Elise called this odd man "Mr. B."

In one of her earliest letters home from New York, Elise wrote:

> I went to the studio this morning where I was courteously received and invited to come in any day during the week and make myself at home. I told Mr. B. I'd be ready to start next Monday, so I am writing you as per instruction to let you know he said $100 down and a signed contract to pay $50 before the end of term. If it's all the same to you send money payable to E. Brunel.

In 1919, New York was a city inundated with immigrants. In the course of the previous twenty-five years, some ten million Europeans had passed through Ellis Island, and a good many had stayed, filling the tenements and rooming houses of the city's vast plain. Whereas in the South almost all of Elise's contacts had been with other black families, in New York she was suddenly surrounded by people of every background—and not merely on weekends in the protective company of Teddy, as she had experienced New York before. The experience had to have been a strange one. Most black Southerners, farmhands in remote villages and towns, had little knowledge of the world outside a radius of fifty miles from their cabin. Even the self-conscious members of the colored elite, with their appetite for education and cities, didn't travel a great deal out of the South until after World War I. For Elise to have been dropped into the turbulent stream of New York street life—where every day she rubbed elbows with Jews just arrived from Poland, Marxists fresh from the Bolshevik uprising in Russia, and Italians from the vineyards of Sicily—must have been startling to her senses.

Classes began, and Elise, at twenty-eight, again found herself a student.

> Dear Ted,
>
> This morning I went to school. I am the only woman. There is one other colored, a young man from Wilson, N.C., and a Jap. The others are Jews, Germans, and Irish. They are very polite, and today, every one wanted to show me something. The instructor is a very young man, German, I think. I can hardly understand him. He started me at retouching negatives to see if I have an artistic touch. The Jap is the model of the school, every one seems to try to retouch as well as he. Gee, how I do wish for you.

Elise had probably never met a person of Japanese descent, and on her lips, "Jap" was probably not the contemptuous term it would later become. Neither was "Jew." Teddy was enthusiastic about his girlfriend's new project.

"Find out for me, please," he wrote back, "every fine point about photographing a drawing and a painting for patent reasons. We may need it some day. When you have leisure afternoons and evenings stop up at the N.Y. Public Library and ask to see the collection of prints and photographs of paintings, and of course you will go to the Metropolitan Museum, Fifth Ave and 82nd Street, via the bus and see what artistic posing and lighting and arrangement mean. Is this asking too much of my Little One? When you go to the museum, buy a photo of 'Salomé' by Regnault (pronounced Rain-yo) for me, if you have the price. In other words study! Study! Study! That's the way to become a shark."

Elise inched into her new urban life, but she was less adventuresome than her boyfriend, and in every letter begged for news from home. By this time, three-year-old Gussie Harleston had moved in with Teddy, and some of her letters talk about the girl. Gussie had a nickname for Elise, who was her aunt: "Tantie," after the French for "aunt," *tante.* And so Elise wrote home, "Kiss Gus and tell her 'Tantie' longs for her, for the clasp of her dear little arms about my neck, and the warmth of her little soft body."

Despite her homesickness, in a few weeks, Elise managed to venture out a bit, and she began to use the spending money Teddy sent to explore the city. On one occasion, she went to a vaudeville show with friends who were visiting New York. The out-of-towners were light-skinned black women who were so pale that they were able to pass for white. Elise, who was slightly brown, felt at home with the three women until they got to the theater, when she began to feel that she stood out.

> I met the little Miss Grace Towns [a friend from Georgia] and her little friend at the Y the day I got over here. The other little girl had bought tickets for a vaudeville, and so they took me along. Of course when they saw the "brownie," we were told that there were not three seats together in the balcony, and I was sold a seat away up on the last row on the left. Nice Jim Crow, eh?

Teddy's occasional handouts weren't enough money for Elise to get by, so she found a part-time job at the YWCA in Harlem. Elise used her salary from the Y to go out and occasionally to buy photography supplies. She began to be lonely, and after a while, her letters filled up with complaints. Elise seemed to pity herself and didn't hesitate to tell her boyfriend she was unhappy. When Teddy wrote that he had been sick, Elise used the occasion to comment on their unmarried status.

> Dearest Heart,
>
> So sorry to learn of your illness. Please don't get ill enough to go to bed, as I'll surely come home. Then you'll have to marry me so I can take care of you, see?

Elise begged Teddy for more letters from him and scolded him when he fell behind: "It's been 18 days since I've heard from you. For the first time since Christmas I am unable to control my tears. As far as you are concerned, your promises are nil!"

Teddy answered: "Perk up and cheer up! What's the matter, cold feet? Your last letter was a sort of nightmare. It didn't have its silver lining, only the dark cloud. Cheer up for your own sake as well as mine, and don't court the blues."

Elise spent less and less time in the darkroom and more at her desk writing home. She asked Teddy to send her a ticket to Charleston for a holiday, and when he refused, she mentioned another man she had just met. He was a student at Union Theological Seminary by the name of Charles Winthrop.

> Teddy—Gee! Boy! But I missed you. The longing for you has become such that it is painful to have one of your sex get too near me. For instance, Miss Sims and I visited Charles Winthrop at his dormitory and on one occasion in his assisting me on a crowded car he had occasion to put his arm around me to protect me from the crowd; and so help me it was beyond me why I should feel as I did but it was merciful that it was only for a few seconds. Teddy boy—please hurry. Does it mean wait until Spring? Can I last that long and not blow up?

Teddy was not amused and told Elise to stay away from other men.

> You are such a poor little creature when you wish to be. I have very often had your image in my mind's eye. You are a sort of steady company with me. I judge folks by you, I compare you with them. If

you love me (and do you not?), you will make and keep yourself perfect for me, as I keep myself clean and wholly for you.

But Elise kept up the act.

I'm quite spoiled by the gentlemen. There are a half dozen awaiting a phone call, message or note to say when they may call, having been denied that privilege for over a week, they are quite anxious to see me.

After another letter from Elise, complaining that he wasn't writing enough, Teddy exploded.

We've been head over heels in the undertaking line. I know you must be glad to hear from home, and you ought to know that I appreciate the fact of your isolation. But I don't think you ought to be well nigh hysterical in your every other letter. I have a few nerves also, and an everyday schedule of from 9:00 a.m. to 1, 2, 3 a.m. is a little matter to put anyone on edge. Dry your tears and keep up the smile—and whistle—it helps. Don't reckon the extent of my love by the frequency or length of my letters. My heart is yours and all I have to give, but don't persecute me and stab me by continued peevishness.

Finally, Elise began to take her classes at the Emile Brunel School seriously. She wrote Teddy to say that she had gotten a few photography jobs, making portraits of people for barter and small fees. Perhaps the school had a bulletin board where Elise advertised her services, or maybe customers who were looking for a bargain used the school to hire inexpensive student photographers. In any case, Teddy liked this turn of events:

I am glad to learn that you are doing a little work sufficiently good to "charge" people for it. Keep it up, it is good practice. While you are at it, you will, of course, make every inquiry as to paraphernalia of every description, lighting appliances, furniture, screening, etc. I am

> hurrying this letter while a funeral is in the church. Let me know what you need. Maybe I can supply it. With love 'n everything, Teddy.

In early 1920, Teddy painted a picture of Elise, using his now-established practice of placing a photograph on the easel as a referent. He hung the painting on the wall, in a place where he could see it throughout the day. A short time later, he told Elise that he had begun to have "little chats" with her portrait. With the painting of Elise in front of his eyes, and with her apparent new seriousness about her work, Teddy began to think about a subject he had evaded for so long.

It had been seven years since he and Elise met, and during much of that time, Elise had tried to coax him into marrying her. Finally, in January, he relented. He wrote her in New York.

> Today I had occasion to look through a batch of my mother's old papers and letters and found some documents which I had read often. Some of them are of pleasant interest to me. Others are somewhat sad. But two in particular are of general interest. Of these one is the Bill of Sale delivered to my great-great-grandmother in 1804 when she bought herself and Flora her little daughter from slavery—brave woman. The other is the deed of emancipation and manumission, which she presented to her daughter Flora in 1820 that this daughter might marry then as a "free person of color" not being owned even by her mother. This was my great-grandmother. That was a hundred years ago. Nineteen hundred and twenty must be our year. It may seem like more than a hundred years to you since first we met, and it has been a fairly long time. But then you know we have not been in love that long—I didn't know you and you surely didn't know me, but love did come—it grew with me, which is safer than a flashing spurt. You must know that I regard it as a holy thing that has no relation to opportunity in the sense of furthering one's position either socially or materially, else I should have been married already and have had much of this world's goods.
>
> But you know I am not made that way and so I have waited and you came. Maybe the future holds something fine for us—maybe

not, but I will bet my chances that if we continue honest, we shall fare happily.

Of course I love you; of course I want you, of course, we will marry, and of course it will be this year and before Scorpio crawls into the sky, too. Meanwhile, keep busy and don't think of me too much.

Elise could not believe she had gotten him, and she answered:

There was a time when I could rapturously express my thoughts to you, but through your training I have learned to hold such thoughts in check, and simply say, "I thank you," while all the while my heart is bursting within my breast. I have always wanted to know myself in white satin and veil and orange blossoms. It has been one of my life's dreams.

By agreeing to marry Teddy, Elise knew that she was also agreeing to be the mother to her sister's child, Gussie Harleston. The engagement set, Teddy began to think about his apartment on Calhoun Street, where the three of them would live.

I salute you, the bride elect. I am exhilarated by the anticipation of our great day, and will advise you of what we may expect when I shall have gotten some of my plans under way. Meanwhile, will you not have some directions for me in the matter of furnishings? The trimming of the rooms? Etc.? Or will you let it all be a surprise box? If it can add any more to the happiness of my little brown lady, I send a kiss and reassure you of my perfect love.

And Elise replied:

I would most surely enjoy a "surprise box" for you are quite capable of pleasing me in every instance. So far as color scheme goes, who knows best the colors that harmonize with your "brown baby"? How I long for the feel of your arms about me. Fix your nest and then come get your bird. And "we'll live happy ever after" provided you promise not to fuss at me over the table during mealtimes. If you have made me a white kitchen, a dark, cool looking dining and living room, and a delightful bedroom why, I'm ready to come to, or rather with you, whenever you say.

Perhaps being busy with a funeral, Teddy didn't answer this letter as quickly as he had the others. In previous years, Elise might have sent another note scolding him for neglecting her. But this time, she remained uncharacteristically patient and wrote a second letter on top of the first:

> Having waited what seems months to hear from you, I feel it my Christian duty to remind you that I love you just the same.

Teddy enjoyed the duet and kept up his part.

> Now, Milady, tell me, as you are on the spot, where shall we face the preacher? What are your wishes and your ideas for the event—the place, the hour, the company, the aftermath?

Elise wrote back:

> The topic is open for discussion. As a matter of fact, I shall so cherish a decision from you that if it were that I should meet you in Chicago and that we go to the Klondike, I'd hail it as a mother would a long lost son whom she thought dead.

Teddy, on September 5, 1920:

> The refreshing of the flat is done and we are ready now (almost!) to fill it. It's a good thing I am not a millionaire, as you'd be utterly spoiled.

The lovers talked about their honeymoon. Because Teddy didn't have ready cash for a trip, he suggested a visit to Canton, Massachusetts, to the farmhouse of his friends, William and Ada Hinton.

> Mr. and Mrs. Wm. Augustus Hinton (Ada Hawes) are most likeable people, with two children, of Gussie's and Sylvia's ages, living

> in the country, on a little farm which she assists in working with pigs and all that. Plain, methodical living; no society connections. They invited me to return for my "honeymoon"—that meant *with you!*

Teddy and Elise decided to marry in New York and honeymoon at the Hintons' farm in Massachusetts. In the last letter he wrote before leaving Charleston to meet his bride, Teddy said:

> Think of it! Seeing you after a year. Of course I gaze at your portrait and have a little chat with her, but to see you in the flesh! My heart is no longer quite my own any more—I am giving it to you now. With all my love, Good night. Forever yours, Teddy.

The wedding took place on September 15, 1920, in the parsonage of Nazarene Congregational Church in Brooklyn. Only a handful attended. Rosalie Mickey, Teddy's cousin from Charleston, happened to be in town and came. Parson Jenkins's son Edmund (Jenks) was also in the city and showed up. Although Teddy was not keen on the relationship of his sister Ella to Jenks's father, the painter respected the young musician, whom he regarded as a fellow artist, and welcomed him at his wedding. After the service, Teddy and Elise boarded the steamship *Calvin Austin,* where they had a stateroom, and headed to Boston. It was an overnight trip, and darkness fell soon after they had come aboard.

At twenty-nine, Elise was still a virgin. That night, there was a conspicuous moon and clear sky. Later, Elise wrote in her diary:

> Our first night was spent on the Metropolitan S.S. *Calvin Austin,* Room 11. We stayed on deck until about 7 p.m. when we went in search of a bite. We had mackerel, eggs, potatoes, bread and tea. Then we went up to our room, where he left me to prepare for rest, for I was indeed worn out. About the time I had dozed, he came.

Teddy had a religious side, and so before getting in bed with her (according to Elise's diary), he paused to pray.

> Shall I ever forget how before he came to bed he knelt beside the bed, gathered me in his arms and prayed God's blessing on our union? Shall I forget the peace which stole over me when he laid down beside me, all the while speaking in terms of endearment to reassure me and put me at ease, to allay my fears?

A few days later, Elise added this note:

> I was not prepared for the abundance of joy. I was not prepared for the manifestations, the protestations, the adoration, for of one thing I am sure. My husband *adores* me. He loves me as I do him—for himself—*regardless*.
>
> Five days of married life—
> Five days of Bliss—
> Five days of Love—
> Five days of Extreme Pleasure
> Five days of living and loving
> Five days stolen from Eternity.

After the honeymoon, the couple settled into the apartment that Teddy had redecorated in Charleston. Moultrie Harleston had moved out, leaving Elise, Teddy, and Gussie the place for themselves. Gussie's parents, Bob and Marie Harleston, were still fighting tuberculosis and had evidently resigned themselves to having permanently surrendered their children. Bob fared better than his wife, however; Marie continued to decline rapidly.

Four months after the wedding, on January 14, 1921, Elise's sister Marie died of tuberculosis at age twenty-eight. Gussie Harleston was four and a half years old. She had been with Teddy for two years, and the little girl had lost her mother before the two could carry on a conversation. She says:

"I suppose I'm a child of sickness, death, and mourning—which about fits, because the Harleston family ran funerals for a living. My mother's wake was held in our apartment. They dressed her in white and put her in an expensive casket, which was open and in the middle of the room.

"I remember the wake was full of people, who were all sad. I couldn't feel anything. Uncle Teddy took me over to the casket. We stood there, and I could barely see over the edge to look at my mother's body. Uncle Teddy asked me if I wanted to be picked up so I could kiss my mother good-bye. I said, 'Yes,' and he picked me up. I leaned into the coffin and kissed her on the cheek. Her face was cold and rigid, and her eyes were a little sunken, because she was so thin.

"Across the room I heard somebody say, 'Oh, poor motherless child!' I don't know what got into me, but I looked over to them, and said, 'I'm no *poor child.* My grandfather is the richest man in Charleston!' I was kept home from her funeral, because it was thought the shock would be too much for me. Or maybe they were worried I would say something else inappropriate at the graveside."

Marie Harleston was buried in a simple ceremony at the cemetery of the Unity and Friendship Society, the funeral guild to which many of the Harlestons belonged.

Chapter Twelve

EFFETE, prickly Edmund Jenkins did not want to go back to America. Jenks wanted to stay in London and study music. The Reverend Mr. Jenkins had hoped his son would help manage the family's lucrative bands and choirs, but when the orphans sailed home from England at the start of World War I, Jenks stayed behind.

In September 1914, the twenty-year-old Charleston native applied for admission to London's Royal Academy of Music, where his father had reluctantly agreed to pay for tuition and rooms. The Royal Academy, founded in 1822, was the oldest professional music school in England. Housed in a great brick Victorian pile on Marylebone Road backing onto Regent's Park, it was England's leading school of the classical style and the antithesis of the loose orphan bands Jenks was used to. His credentials were the strangest of any applicant's, but he auditioned well and quickly registered as one of nearly eight hundred students.

Jenks studied piano, clarinet, and voice, and by December 1914, he had joined the student orchestra as a clarinetist and survived his first performances. The next year he filled out his repertoire of instruments with the oboe, the bassoon, and the organ, and added French lessons to polish his Continental airs.

The Parson's son was already a pretentious fop, but when he entered the Royal Academy, he doubled his outsider status: he was an American and a Negro in a school where neither attended. In the early 1900s, there were really only two models for blacks who wanted to write or play serious music: Samuel Coleridge-Taylor and Will Marion Cook. Coleridge-Taylor, a black Englishman born in 1875, was the composer, in 1898, of the immensely popular choral work *Hiawatha's Wedding Feast*. Based on *The Song of Hiawatha*, the poem by Henry Wadsworth Longfellow, *Hiawatha's Wedding Feast* drew a large following both in the United States and Britain, filling uncountable concert programs and inspiring Coleridge-Taylor clubs. The other black composer, Will Marion Cook, born in 1869, had studied violin at Oberlin Conservatory in Ohio and later at the University of Berlin. But Cook had given up classical training in 1898 to produce *Clorindy—the Origin of the Cake-Walk*, which he had written with Paul Laurence Dunbar, the African American dialect poet. The high-stepping show opened doors for hundreds of black performers who were willing to jive for white audiences on Broadway.

A white man—Czech composer Antonín Dvořák, strangely enough—was a third beacon for nonwhite musicians. Dvořák had come from Prague to New York in 1892 to lead the racially mixed National Conservatory. A Czech nationalist, he believed folk tunes—dances, ditties, and ballads—belonged in serious compositions. While in the United States, Dvořák is said to have acquired a liking for plantation spirituals and Native American music, which some thought led him to write his Ninth Symphony, *From the New World*. At one point during two stays in America, Dvořák famously declared (in *Harper's* magazine) that American composers would find their path only after they listened to their regional and folk music, including African American spirituals. Then in 1895, Dvořák went home to Europe, but he became a hero among colored musicians for his finger-wagging pronouncement.

The Royal Academy Orchestra, with Jenks on clarinet, put on a concert every two weeks. According to programs, in four years Jenks performed pieces by Ludwig van Beethoven, Georges Bizet, Johannes Brahms, Samuel Coleridge-Taylor, Claude Debussy,

César Franck, Edvard Grieg, Georg Friedrich Handel, Franz Liszt, Giacomo Puccini, Henry Purcell, Camille Saint-Saëns, Peter Tchaikovsky, Giuseppe Verdi, Richard Wagner, and Carl Maria von Weber. He won prizes in composition and received a scholarship that relieved his father from tuition bills. He also edited an issue of the school music journal, *The Academite,* and on May 18, 1916, Jenks made his debut as a conductor, leading the orchestra in a program with Franz Schubert's *Unfinished* Symphony (no. 8 in B Minor).

In addition to Schubert, that night there were two pieces on the program written by the conductor, "Rêverie" and "Esperanto Dance, No. 1." Intended for cello and piano, they were precious little works, striving for seriousness. In his second year as a student, Jenks wrote still more high-toned scores: three ballet movements, some music for wind instruments, and an overture for piano and strings to accompany Shakespeare's *Much Ado About Nothing*.

It was a strange time to be writing such music. Two hundred miles away in northeast France, the great mouth of World War I was devouring the youth of Britain, France, and Germany at a rate of five thousand and more every day. Because of the absence of men, women had begun to outnumber them in the school, and there was frequent news from the trenches about the death of former pupils in nerve gas attacks. As his fellow musicians were breathing their last in the mud at the Somme and Verdun, Jenks was polishing his melodies for *Much Ado*.

A year later at Queen's Hall, a concert stage in London, the son of the Orphanage Man premiered his piece for organ and orchestra, "Prélude religieux." Though it showed Jenks's mastery of academic taste, it was a timid invention, the kind of thing one would hear in a church before a Lenten service. Jenks had written a capable set of harmonies, but nothing at all dissonant; nothing, in other words, that would suggest he might have noticed the new subversive order that was then gathering at the edges of art and music.

Jenks was on the verge of becoming a conformist. But in the summer of 1917, his music began to grow. Without warning, he wrote a ten-minute piece called "Folk Rhapsody" that was unlike anything he had done before. At the center of that score, dated July 30, stands a little melody—just a ditty, really—that Jenks had pulled from his bank of childhood memories. The melody was that of a song once sung by black fishermen in Charleston, a tune with the very un–Royal Academy name "Hey, Brer Rabbit, what you do down dere?" Jenks made the song's eight or ten notes into the recurring theme of his "Folk Rhapsody." The piece had syncopation and an amusing lilt, flashy passages and stately ones, rhythm, and frequent shifts from major to minor key. It was a sophisticated work, anticipating in some ways George Gershwin's *Rhapsody in Blue*, which would premiere seven years later. It was as though Jenks had suddenly awakened to Dvořák's call for folk materials and gone home to his black roots.

There is no obvious reason why, in the middle of his classical immersion, Jenks would suddenly remember the black fishermen he was around as a child. Perhaps a maverick professor encouraged him. Or maybe he decided he wasn't really a British conductor in training, after all, but an American Negro.

Another explanation for the sudden appearance of "Brer Rabbit" is that Jenks was not getting along well with his father and was feeling defiant. In letters home, Jenks explains to Parson Jenkins that he is not about to give up his music and return to America (as the minister has asked), because he has too much learning to do. "It is very strange to me that you should continue to say to me come back home and all that sort of thing," he wrote. "Instead of encouraging me in my work, your letters full of lamentations tend to discourage me."

After "Folk Rhapsody," Jenks developed a split musical personality. On the one hand, he loved serious music, especially orchestral works and choral music. On the other, he felt the juices flowing in his own national music. He wanted not only the concert hall but also the sound of the street, which poured from the memory of his experience with the orphan bands.

The word "jazz" comes from the tough black streets of the American South, where it meant, in effect, "to fuck." In New Orleans brothels about 1910, money changed hands on the parlor level while sounds of "jassing" were heard up the staircase. (The spelling of the word shifted around before settling down.) To "jazz him" meant to screw him. It may be that "jism," slang for semen, comes from the same origin as "jazz," though this would be hard to prove. Something that was "jazzed" had sex in it, usually of the fast kind. And in the first years of the 1900s, in the alleys of New Orleans and Memphis, St. Louis and Charleston, when brass and drum bands began fooling around with their instruments, playing cakewalks and rags, the word "jass" was applied to their unnameable and deviant sounds. The bands "jazzed" with noises and beats other musicians lacked, and made a kind of music that was secret, taboo, and low.

In the same way that Louis Armstrong and Jelly Roll Morton were present at the birth of jazz in New Orleans, so too was the Jenkins Orphanage Band in South Carolina. They played fast and syncopated in two- or four-time, with the boys improvising in different directions. (As one band member put it, "When we played, it was every tub on its own bottom.") They used bent and tricked-up notes, hard riffs on the horn, and dancing snare drums. The orphans acted their part in some unlikely settings, from a world's fair in Buffalo, New York, in 1902, to the rich horse-racing country of Kentucky, and the Baptist churches of Richmond, Virginia. But undoubtedly the strangest venue where the Jenkins players tried their music was in the U.S. Army, on the blasted battlefields of northeast France.

In late 1917, near a French town called Givry-en-Argonne, a black American named James Reese Europe was training with a machine gun unit in the trenches. Lieutenant Europe was a thirty-six-year-old member of the Fifteenth New York Colored Infantry Regiment, but he was also one of the most successful black musicians of his day. Born in Alabama, he was raised in Washington, D.C., and had moved to New York, where he had founded a union for black musicians, the Clef Club. As a booking agency, the Clef Club helped raise the pay scale for working talent from vaudeville to saloons. It also fueled Europe's main career, which was as bandleader of the Clef Club Orchestra, an act with a strangely loose sound. In 1912, Europe's band, in a formation of more than a hundred musicians, had played at Carnegie Hall. Afterward, Europe had recorded several pieces, becoming one of the first black Americans in the recording industry. Then, with a pair of white partners, Vernon and Irene Castle, he had helped create a dance called the Castle walk, later known as the fox trot. By the time he enlisted in the army in mid-1916, Jim Europe, tall and commanding, had thoroughly mingled black and white music taste in New York.

Most white American soldiers refused to serve alongside blacks, so Europe's company had been dispatched to join a French division at Givry-en-Argonne. Here, a white commander, Col. William Hayward, asked Lieutenant Europe to form a band to help raise troop morale. Europe agreed, and in a few months had recruited forty-four musicians, all of them black soldiers. After the Fifteenth Regiment won a string of battles, the unit and its band were dubbed the "Hell Fighters." (Another colored unit, the 370th Infantry Regiment, got a juicier nickname from German soldiers for combat swagger: the "Black Bastards.")

Members of the orchestra who had come from the Jenkins Orphanage included the bandmaster, thirty-eight-year-old Sgt. F. Eugene Mikell (an orphan who had grown up to tutor Jenkins band #1), Amos Gilliard (a trombonist), and a pair of brothers, Stephen and Herbert Wright, who played drums. The two drummers had performed with the Jenkins band in London at the Anglo-American Exposition and had made their way to France after enlisting as privates in the U.S. Army.

The Hell Fighters band had a limber, loopy sound and played a repertoire of marching music, rags, and waltzes. In 1918, they set off on a tour through France into camps, hospitals, and war-torn towns. Their playfulness and rule-breaking kicks of melody and rhythm created a sensation. Soon the band was entertaining black soldiers, white officers, and French civilians by the thousands. They became especially popular with French audiences, because they sounded nothing like the square military bands of the European armies. Their music was so unhinged that one French conductor asked Jim Europe if he could borrow the sheet music to teach it to his own regimental band; when the music was loaned, but the same marches came out sounding flat, the conductor accused Europe of using altered instruments. Europe was successful enough that he could laugh off the charge and move on to the next town. Word of mouth around the Hell Fighters was unstoppable. In August 1918, their tour climaxed in a concert in Paris at the Théâtre des Champs-Élysées, with President Raymond Poincaré in attendance.

After the Armistice, the Hell Fighters returned to New York, where their fame had preceded them. On February 17, 1919, they marched in a victory parade through the city, leading their regiment beneath showers of cheers up Fifth Avenue, from Madison Square on 23rd Street to 135th Street in Harlem. Within a few weeks, Jim Europe had signed another recording contract, this time with the Pathé Talking Machine Company. (He was one of only a handful of black musicians to record more than once.) Pathé promoted the records as "Jim Europe's jazz" and "Eleven records of the world's greatest exponent of syncopation." The Hell Fighters then started on an American tour.

Stephen and Herbert Wright, the Jenkins drummers, must have been accustomed to cheering crowds: they had been hearing them for years, from Charleston to London to Paris. Still, the shock of emerging from an abandoned childhood into the light of international acclaim must have numbed their senses. After all, many of their old friends back home had come to the Jenkins Orphanage from jail, and their career ascents had stopped when they became newsboys.

Herbert Wright was apparently so rattled by the spotlight that he began to create problems. When Europe's band returned to the Pathé studio to record another dozen songs, Wright started to become disruptive. It's said that the former orphan so enjoyed the adulation that it became a prank for him to walk onstage when other performers were in the midst of their act. (The Hell Fighters appeared with warm-up bands.) This created a disruption and bad blood with rival musicians, who naturally wanted the audience's full attention. Several times, Jim Europe warned Wright to quit it, but the drummer remained intoxicated with the stunt and couldn't stop himself.

In May 1919, the band was out on tour again. They had recorded "On Patrol in No Man's Land," an eerie song written by Europe about his experience in a gas attack during the war, and were being welcomed by admiring newspaper reports in every town. On the night of May 9, the band was booked at Mechanics Hall in Worcester, Massachusetts. Herbert Wright again walked across the stage during someone else's act, and after the show, Europe sent word that he wanted to see Wright in his dressing room. Wright seemed to believe that Europe had been favoring his brother, Stephen, over him, and he was furious at this new reproach. Wright appeared in Europe's room, and the bandleader again dressed him down. When Europe told Wright to leave the room, the drummer pulled out a penknife and jumped at the bandleader, stabbing him in the neck. According to an eyewitness, Europe wrestled Wright to the ground. Bleeding steadily, Europe was taken to the hospital, where physicians couldn't close the wound. That night, Jim Europe slowly bled to death. A war hero and a jazz pioneer, he was buried at Arlington National Cemetery.

Herbert Wright was convicted of murder. Though there were calls to execute the man who had killed "the world's greatest exponent of syncopation," Massachusetts did not impose the death penalty. The drummer was sentenced to many years in jail, and the rough fame of the Jenkins Orphanage was lifted another notch.

When Herbert Wright was locked up for murder, a pall might have been expected to settle over the Jenkins bands, but instead the boys got more restless. The story of the Wright drummers who had taken off and found an audience was enticing; the killing made their escape mesmerizing.

Every couple of months, another musician would leave the orphanage to make his way to New York or Chicago, where he would try his hand in the black vaudeville or traveling band circuit. One orphan player remembered:

> When they put you in the orphanage, they signed you up until you were twenty-one, so running away was the only method to get out earlier. I started running away when I was fourteen, and must have run away six or seven times. We got in those little bands around town because somebody was always trying to steal you if

> they thought you were pretty good. And I'd go. But the orphanage would come and get me. The first time I really took off was in Jacksonville. My friend and me we lived on money we had knocked down. "Knocking down" was keeping a little of what we collected in the hat when we played in the street.

One of the first of the boys to get away and actually have a career was a songwriter named Tom Delaney. Thomas Henry Delaney was born in Charleston in 1889, and was delivered to Parson Jenkins a few years later. In the early 1900s, he studied piano and voice at the orphanage and was placed in one of the choirs. While a teenager, Delaney formed a singing group, the Springfield Minstrels, a troupe that evidently traveled with the regular bands, singing separately for tips.

Tom Delaney had an easy smile, sad eyes, and a large head. His voice was a big, sloppy baritone, and he wrote and played better than he sang. Escaping the orphanage, probably before 1910, Delaney went north. Merely by leaving the South, he was taking a risk. It was just the beginning of the mass migration of black people to the Northern cities, which meant there were few places he could play. The majority of working musicians were white, as were the parties and auditoriums where they got paid. Sources are thin, but Delaney seems to have been confident that as a blues singer he had something fresh he could sell. Arriving in New York, he got work playing the piano and singing in the gin joints and tin-roofed saloons of the black sections of town, which at that time meant the midtown blocks west of Broadway.

Jazz came from the brass bands in the Southern cities, but the blues came from spirituals and work songs in the rural sections. Black farm workers used songs of sorrow to breathe air into the suffocating routines of cotton planting. Blues was earlier than jazz, slower, and vocal. Jazz came later, and it was instrumental and fast. A regional source of the blues idea is said to have been the delta of the Mississippi River. But the entire South felt the same legacy of slavery and sang a similar blues reply: delta musicians just happened to be recorded first. By 1910, a recognizable blues sound was in place: the voice with one instrument (banjo or guitar), twelve bars per stanza, autobiographical lyrics, and three major chords with "bent" grace notes. Slow blues and fast jazz became the yin and yang of black music—performers oscillated between them, depending on their mood.

In 1912, a thirty-nine-year-old Alabama-born bandleader named William C. Handy began to pull blues out of the roadhouse saloons and clean it up for a wider audience. In those years, when "gramophone" disc players were still rare, the money in popular music came through music publishing. Success was measured not by the number of recordings sold but by stacks of sheet music. Moving to Memphis, W. C. Handy opened a tiny office, and when song publishers turned him away because he was black, he self-published the song "Memphis Blues." In 1914, he followed with "St. Louis Blues," which became a runaway seller. After Handy's success, other blues songwriters found it easier. James P. Johnson, the "father of stride piano," published his first song in 1917, "Mama's Blues," and twenty-five-year-old vaudeville pianist Perry Bradford followed with "Lonesome Blues" in 1918.

Songwriters with a good voice had another income stream in gramophone records. For fifteen years, gramophone discs had been raising salaries up and down the streets where musicians clustered. Some of the first recordings of music made for mass distribution were done in 1904. Opera singers, including the young Italian tenor Enrico Caruso, were persuaded to sing into trumpet microphones in return for royalties. The Victor Talking Machine Company earned its early profits on opera and became the industry bully, despite its cute logo of a fox terrier peering into a record player. Gramophone disc players came down in price and began to sell widely at the start of World War I, opening larger revenue streams from the records needed to feed them.

The popular music publishers had offices in New York City, on "Tin Pan Alley." At first just a nickname given to an actual New York City block (West Twenty-eighth Street between Broadway and Sixth Avenue) where start-up publishers had crowded together, Tin Pan Alley later became a term for the sheet music business. Tin Pan Alley employed composers and lyricists who earned salaries to create popular songs, or who sold them freelance. The business opened doors for black songwriters, because sheet music did not disclose the race of the composer.

The idea of making discs with black performers rarely occurred to white businessmen. Of the dozen or so colored performers who had made records, Jim Europe's band was the only act to have recorded more than once. Neither did it cross anyone's mind to sell records to black consumers. The explanation was seamless: there was no market for black blues and jazz, because recording companies turned it down; they turned it down because there was no market for it.

After years of singing to his own accompaniment on piano, Tom Delaney found his way into the subculture of Tin Pan Alley and became a songwriter. When he'd lived at the orphanage, Delaney's Springfield Minstrels had sung religious music. But now he wrote the blues, with lyrics about sex and struggle and drink.

Tom Delaney's breakthrough came when he was thirty-two, in 1921, with a song called "Jazz-Me Blues." The tune had a ragtime feel, which made it a little obsolete, but this was good for the mainstream. And it had cute lyrics. Though the word "jazz" was now respectable, Delaney's verses played with the sex still lurking behind it:

It sounds so peculiar 'cause it's really queer,
How its sweet vibrations seems to fill the air,
Then to you the whole world seems to be in rhyme;
You'll want nothin' else but jazzin', jazzin' all the time.

Every one that I ever came to spy, hear them loudly cry:
Oh, jazz me!
Come on, Professor, and jazz me!
Jazz me!
You know I like my dancing both day and night,
And if I don't get my jazzin', I don't feel right,
Now if it's ragtime, take a lick, play it in jazz time,
Jazz time!
Don't want it fast, don't want it slow;
Take your time, Professor, play it sweet and low!
I got those doggone, low-down jazz-me jazz-me blues!

Delaney was lucky, because the year before he wrote "Jazz-Me Blues," the lockout of blacks from the recording industry had ended. The door was first opened to a singer, Mamie Smith. In 1920, Perry Bradford, who had already published blues songs, partnered with vocalist Mamie Smith, who was then appearing in the musical revue *Maid of Harlem*. Bradford persuaded OKeh Records to let him cut several sides with Smith. Mamie Smith sang Bradford's songs "Crazy Blues" and "It's Right Here for You." In advertising the discs, OKeh promoted the newsworthy fact that this was a black singer and a black band. To African American listeners, the songs were like water after a drought. Newsboys sold the discs, as did door-to-door salesmen. They were peddled in railroad stations and barbershops. "Crazy Blues" became a hit, selling more than a million copies its first year of issue.

About this time, Tom Delaney met a young woman named Ethel Waters, a twenty-four-year-old blues singer on the rise. Born in Chester, Pennsylvania, in 1896, the child of a rape, Ethel Waters began performing as a shake dancer with the black vaudeville team Braxton & Nugent, using the stage name "Sweet Mama Stringbean." Waters preferred singing to shimmying, however, and she had a sultry voice and witty, intelligent phrasing. In 1920, Delaney and Waters were both angling for music jobs on Tin Pan Alley, with Delaney a veteran and Waters a latecomer.

Ethel Waters had her first recording sessions on Cardinal Records, but soon changed contracts to join the Pace & Handy Music Company, co-owned by W. C. Handy, the "Memphis Blues" songwriter who had since relocated to New York. Handy's company was trying to break into the new "colored" disc business and had started a label called Black Swan. Though there is no evidence about how Delaney and Waters met, it probably happened through a young employee of Pace & Handy named Fletcher Henderson. At the time, Fletcher Henderson was a twenty-three-year-old chemistry graduate from Atlanta University, the son of a mixed-race family, who had taken a job organizing recording sessions for Handy's acts. On March 21, 1921, Ethel Waters recorded two Tom Delaney songs, "Down Home Blues" and "At the New Jump Steady Ball." Fletcher Henderson played piano. The recording of "Down Home

Blues" took off, lifting Handy into profits and making Tom Delaney look like a good investment. A short time later, Delaney and Ethel Waters went on a tour together, from the eastern states to Chicago, and through the South, with Delaney playing the role of Waters's pianist and manager.

After Mamie Smith and Ethel Waters, the music business rushed to get Negroes into the recording studio. Every gramophone disc company in New York wanted black acts, and most created a separate product line called "race records" to sell jazz and blues. One advertisement published in a trade magazine by Ajax Record Company of Chicago ("the Quality Race Record") put things this way: "Wide-awake phonograph dealers all over the country are becoming more and more alive to the potential possibilities in the fast-growing Negro population and the musical demands of this special group. Wherever there are Negroes, phonograph dealers can tap an amazingly rich market with very little effort."

Two months after Ethel Waters made "Down Home Blues," the Emerson Phonograph Company approached Delaney about its new black blues act, Lillyn Brown. A veteran of vaudeville and musical theater, Lillyn Brown was part Iroquois, which led to her frequent billing onstage as the "Indian Princess." Brown's signature move was to disappear into the wings and return in men's clothes for a song-and-dance routine as a male impersonator. In May 1921, when an act that the recording company had billed "Lillyn Brown and Her Jazz-Bo Syncopaters" cut Delaney's "Jazz-Me Blues," the Charleston boy's success was assured.

In those years, it was common for a popular song to be recorded by several acts at the same time. Now everyone wanted a piece of Delaney, even the white companies. After Lillyn Brown, next in line was the Original Dixieland Jazz Band, a white five-piece act from New Orleans. Led by a cornet player named Nick La Rocca, the Original Dixieland musicians had the dubious credential of having recorded the "first" jazz record, "Dixie Jass Band One Step," four years previously, when black jazz bands were barred from recording anything. On May 3, 1921, the Original Dixieland Jazz Band did a version of Delaney's "Jazz-Me Blues" for the Victor Talking Machine Company. Six months later, Arto Records asked Delaney for his song. Arto had just signed a twenty-seven-year-old black vaudeville singer from Georgia named Lucille Hegamin. In November 1921, "Lucille Hegamin and Her Blue-Flame Syncopaters" was the third act to record "Jazz-Me Blues," giving the former orphan another boost.

With his music playing on gramophones around the country, Tom Delaney thrived. In these years before the jukebox, another source of income for songwriters was the piano roll. The Meloart Company, a piano-roll producer in Baltimore, released Delaney's "Graveyard Love—A Cemetery Moan." Delaney followed up with "Everybody's Blues" and "Nobody Knows the Way I Feel This Mornin'."

In 1925, a twenty-six-year-old bandleader named Edward Kennedy "Duke" Ellington asked Delaney if he could record "Jazz-Me Blues." Though Ellington's act, the Washingtonians, was a relatively obscure six-piece band, Delaney agreed. The result was one of the first recordings of Duke Ellington's career.

Delaney ultimately grew tired of placing his songs with others, and in 1925, he began making discs under his own name. With a solo piano and his own vocal, he recorded "Georgia Stockade Blues," which he had written in the voice of a prisoner.

Guards all around me with their guns,
Shoot me down like a rabbit if I start to run.
Five long years in a state stockade,
Working from sun to sun.
Evening goes, morning comes,
My silly task is never done.
Tipping boxes in a turpentine barn,
At night can't raise my arms.

Delaney's voice was large and off-key, and not easy to listen to. But his lyrics, sung against a strong backbeat, always told a story. Delaney had a writing range that appealed to all sorts of performers. In time, blues singers such as Bessie Smith and Alberta Hunter would record his songs. "Jazz-Me Blues" became a standard, recorded by Louis Armstrong, Earl "Fatha" Hines, Bix Beiderbecke, Count Basie, Jack Teagarden, Benny Goodman, and dozens of other musicians. "Down Home Blues" was done by Fats Waller, Eubie Blake, Charlie Parker, and, eventually, Etta James.

Tom Delaney's success traveled on the grapevine to the orphan boys, and he became a hero to the Jenkins kids. Everyone wanted to mimic his rise and make some real money. Delaney knew he had finally gotten away from Jenkins when, at age thirty-six, he made his first solo recordings. That year, he sat down and wrote a comedy tune that made him laugh about the old days back home. Accompanying himself on piano in 1925, he recorded a song called "Parson Jones (You Ain't Livin' Right)." If it reached the ears of Parson Jenkins, the Orphanage Man would not have been amused.

Parson Jones, you better watch your step
'Cause you sure ain't living right.
You better stop your slipping and dodging around
These sisters' house at night.
You ought to be ashamed to preach from that pulpit,
'Cause you ain't nothing but a hypocrite.
Parson Jones you better watch your step
'Cause you sure ain't living right.

In London after the war, the youthful classicist Edmund Jenkins increasingly thought about home. Jenks had actually begun to miss the orphanage band, with its chaotic rhythms and dancing kids. After five years in the Royal Academy, his taste for purely European harmonies had waned, and he had moved on. He was now supporting himself by playing in theater pit bands, in restaurants, and in cafés. He spent a summer in an orchestra at the Grand Theater in the resort town of Llandudno, Wales, playing for tourists on beach holiday. It was not the expected lifestyle of a clarinetist who had once carried a portfolio of Handel.

Despite his seeming distance from the war, Jenks had indirectly benefited from the fight. In addition to killing eight and a half million soldiers, the war had made Americans very popular in Western Europe, because the United States had helped bring about the Armistice. After the fighting stopped, a second invasion took place, that of American popular culture into Europe. The new black music began trickling into Britain on records made by white copyists of the black "jass" sound. It then streamed into the country as black musicians themselves started coming over to play their music. Will Marion Cook (the black composer who had broken the race barrier on Broadway some twenty years before) arrived in London with his band, the Southern Syncopated Orchestra, which included two former Jenkins band players: Jacob Patrick on trombone and Jacob's cousin Edward Patrick on trumpet. Jenks befriended Will Marion Cook, who was something of a celebrity, and appears to have played in some of his concerts. Cook played the big hotels, Philharmonic Hall, and even a garden party for workers at Buckingham Palace.

Like everyone else, Jenks was hit by the sounds coming from his homeland. But unlike his English friends in music hall bands, he knew what he was hearing. The music reminded him of the orphan boys, and it was as though the jazz tide was coming in for him alone.

Jenks devised a plan to rise on the wave of new black sounds. The Queen's Hall Roof Club was a barroom on an upper floor of the auditorium where Jenks had premiered several pieces. After the war, the Roof Club became a magnet for black music. Jenks sat in with the white bands and found that they liked playing with him, because as a Southern Negro his presence on the stage signaled the authenticity of their American-style music. Jenks became a regular and joined the barroom's house band, the all-white Queen's Roof Orchestra, which he was eventually asked to lead.

Jenks's plan worked, and jazz began to pay him handsomely. But the Queen's Roof Orchestra brought about a significant change in the Parson's son. He stopped paying attention to concertos and lost touch with his conservatory friends. The band kept him out late, and he had a large income for the first time. Jenks began to dress expensively and presented himself as a natural at the American sound. (With his background, it was undeniable.) And as much as a black foreigner was allowed, he became a debonair figure in London's burgeoning nightlife.

In the summer of 1920, for the first time in six years, Jenks decided to visit his home and family in Charleston. He wanted to show his family that he had made it—and furthermore, he seems to have been considering a move back to America, the source of the new music and Jenks's real home. Before he left England, he had written home to his stepmother, Ella Jenkins, for the first time in years, announcing he would be coming. ("My Dear Little Stepmother: What a disgraceful son you have who has utterly suspended correspondence!") On July 31, Jenks sailed from Southampton to New York. This time, instead of steerage, the way he'd come to Europe with his father, Jenks had paid for a first-class stateroom.

In New York, he visited Teddy Harleston's fiancée, Elise Forrest, who was attending the Emile Brunel photography school. He absorbed what he could about Tin Pan Alley and the songwriting business and listened to the changes in music appreciated by nonwhites. Jazz and blues were indisputably the hot popular sounds, spilling from the doorways of the colored barrooms of Manhattan. With the growing influx of black Southerners to the city, and the unlikely success of "Negro music," New York must have looked and sounded very different from the city Jenks had last seen as a twenty-year-old clarinet student six years earlier.

In contrast to the North, where a colored musician could find satisfactions, the Southern states were roiling when Jenks came home. The summer of 1919 had been dubbed "red summer" after the riots that killed colored people in dozens of cities, and the year 1920 was turning out to be a good one for the Ku Klux Klan. After staying dormant for forty years, the KKK had been revived in 1915, with new headquarters in Atlanta. The precipitating incidents had been the lynching of Leo Frank, a Jewish factory manager who was living in Georgia, and the runaway success of a Hollywood film that extolled the return of white supremacy to the South after the Civil War, D. W. Griffith's *The Birth of a Nation*. The leader of the new Klan, William J. Simmons, the forty-year-old son of an Alabama doctor, quickly attracted five thousand dues-paying members to his organization, the Invisible Empire, Knights of the Ku Klux Klan, Inc. In June 1920, Simmons hired an advertising agency to further recruitment. Within a year, membership rose to one hundred thousand, and the Klan was a mass phenomenon. The group's popularity meant that the beating and rape of blacks was once again considered an acceptable sport. The *New York World* kept count of the attacks: 152 of them in the year of Jenks's visit, including four murders, forty-one whippings, and twenty-seven victims who had been tarred and feathered.

It didn't take Jenks too long to become aware of all this, and he was not anxious to see the South, worried that home could offer him nothing. He wrote his father: "Though I was born in the South, you yourself have admitted it would be impossible for me to live in safety there now."

Nevertheless, after his immersion in New York, Jenks dutifully went to Charleston to spend time with his father, his stepmother, and the orphan boys. In Charleston, Jenks visited the old orphanage and looked in on the music classes for the new boys and girls. He looked up old friends and tried to patch things up with the Parson, who still wanted his son to move back to America. But looking around at the vast social straitjacket created by Jim Crow, Jenks was having doubts.

A friend of Parson Jenkins's, E. C. Lockhart of Arkansas, wrote the Orphanage Man about his son, after Lockhart had seen Jenks in South Carolina:

> During my visit to Charleston, it was my privilege to see Edmund in all the glory of young manhood. His face showed a peculiar radiance which reflected the artist. His affability was delightful. His English characteristics, which were no doubt unconsciously cultivated from contact, were superb. He was brilliant, scintillating—bubbling over with music, which seemed to hold his soul captive.

But Jenks couldn't return: he had tasted life in the metropolis and no longer wanted to live in the provinces, where the Klan was again beginning to throw its long white shadow, and there was no room for a cultivated person like himself to fit in.

After a few days in Charleston, Jenks took a train back to New York, on his way to London. As it happened, that week Teddy Harleston was being married in New York. (As the brother of his stepmother, Ella, Teddy was Jenks's stepuncle.) On September 15, Jenks attended the wedding of Teddy Harleston and Elise Forrest in a Brooklyn church. Then he sailed for England.

Back in London, Jenks marveled at how restricted his movements had been in America. He realized that he could no more play music with whites in Charleston—as he now did every night—than he could vote. In England, his lifestyle was far better than what he could expect to have at home.

Jenks resumed his gigs at the Queen's Hall Roof Club, where he earned more and more, and grew increasingly impressed with money. In late 1920, flaunting his cash, he bought a car. Although automobiles had been in mass production for some time, and he could have chosen an inexpensive sedan, he picked a pricey sports car instead, a Talbot-Darracq. The firm Talbot-Darracq was the product of a recent merger between two carmakers, the English company Clement Talbot and the French firm Darracq. Talbot was known for its performance cars, having made the first automobile to reach speeds of 100 miles an hour. Jenks's new purchase was an open-topped, two-seat sports model, which, at 695 British pounds, cost three times the price of a family car.

In 1921, with jazz selling so well, he received an offer from His Master's Voice, the British branch of the Victor Talking Machine Company, for the Queen's Roof Orchestra to make a recording. It was the breakthrough Jenks wanted, and the reed player agreed.

As it happened, however, the sound that His Master's Voice wanted was not exactly hot American jazz. The record company was looking for something tamer, something easy to sell to the English public. On May 28, 1921, eight musicians, seven of them white, went into a studio in Hayes, Middlesex. Jenks played clarinet and alto sax, while a former music hall entertainer named Jack Hylton played piano and celesta. The band recorded four songs, and each was a timid attempt to inflect a little colored rhythm into music hall ballads. The song titles give a sense of their tameness: "Idol of Mine," "Turque," "The Wind in the Trees," and "I'm Wondering If It's Love." The musicians were paid five pounds each and the recordings were issued on two double-sided, ten-inch discs under the name "Queen's Dance Orchestra."

Despite their blandness, the records must have sold a few hundred copies, because the band was brought back to the studio several more times. Their later recordings had more jazz, and they did dance numbers ("Oriental Fox Trot—Salomé"), novelty pieces ("Ilo—A Voice from Mummyland"), and Irving Berlin songs ("Beautiful Faces Need Beautiful Clothes"). They also occasionally cut loose a bit with blues songs ("Coal Black Mammy") and instrumental showpieces ("Trombone Cocktail"). In all of them, Jenks can be heard playing a wailing saxophone or clarinet, usually in the background, but sometimes stepping out for a brief solo. By the end of the year, Jenks had made some twenty records.

Though the Queen's Hall discs were "jazzed" by English standards, they were mild when compared with those of American bands. Every note is in place, every "spontaneous" rimshot carefully hit. But the reason the music was so tame had to do with who was playing it. Jenks's band mates came from the music hall: they had no feeling at all for the Jim Crow South and lacked any connection to the experiences of black Americans that were stored within the new style of music. Furthermore, the breakthrough of jazz into England had not been painless. In fact, many white musicians resented it. They disliked the good time that the black players had on the bandstand, and felt they were being made obsolete. The year Jenks made his first records, a London music magazine called *The Dancing Life* summed up the distaste and envy that greeted jazz this way: "The 'darkie' question, as it concerns dance orchestras, seems to be getting still darker. Everybody knows that Englishmen can play every bit as well as South Americans [meaning blacks from the American South], and sometimes better, and it is always annoying to see a coloured orchestra playing when there is so much English talent available."

Jenks probably knew his music was bland, and he probably wondered whether he could do it better elsewhere. At the time, Jenks had an English friend, Ted Toiros, who was spending some months in Paris. Toiros wrote Jenks letters about the simmering nightclub scene in France and the hot music being made there by Americans. Jim Europe had left an impression on French audiences, said Toiros, and a tiny but intense black jazz world was coming together in the cafés of Montmartre. There were just a handful of black

American musicians in Paris, maybe as few as fifty, but the French couldn't get enough of them. Parisians were paying good money for jazz and blues, which they didn't even quite understand, but which they wrote about in their newspapers as *le style primitif.*

Based very much on Toiros's impressions, Jenks decided to try playing in Paris in late 1921. He found an apartment in the Seventeenth Arrondissement, at 44 rue de Moscou, a one-block street off of boulevard Haussmann near the Parc Monceau. The apartment was a short walk to Montmartre, home of the black American expatriate colony. Jenks wanted to see just what these Negroes were doing in Paris. Many of them lived and worked in the streets just south of the place Pigalle. Soon he found himself walking into the night, listening for music coming from the doorways of the many cafés.

Chapter Thirteen

Gussie Harleston was a pretty child "with eyes like burning coals," according to one of the Harlestons. She remembers growing up on Calhoun Street in the 1920s:

"The Harlestons were light, and we didn't associate with people who were much darker than we were. Of course we didn't associate with white people either. We were a kind of in-between people. But we were Negroes all the same, and everyone in our circle was colored to one degree or another. In fact, I didn't know any white people, except for my friend Mildred Wieters.

"Her family lived next door to the funeral home. Charleston was like that, white and black families side by side, and no one thought about it—but no one talked to one another, too. Mildred Wieters's father was manager of the Coca-Cola Bottling Company, which stood on a corner of Calhoun Street two blocks from the funeral home. They made Coke right there, and you could stand on the sidewalk and look through the window and see the bottles coming around on the conveyor belt. Mildred and I used to do that, watch the Coke bottles turn the corner and swing by the window, and then we'd run along to play. A couple hours later we'd come back and watch the Coke some more. But something must have happened and one day Mildred snubbed me. I think her parents had probably had a word with her. I remember when I saw her downtown and she was with her classmates, and when I spoke to her she tossed her head and ignored me completely. We never talked after that. Mildred was growing up, and after a certain age, it was not acceptable for white children to have colored playmates, so she dropped me just like that.

"I remember Edmund Jenkins when he came home to visit from Paris. He had bought me a doll as a gift, but the doll was black and I thought it was ugly. Edmund told me it was going to last forever, because the store owners had made a point that the doll could not be destroyed. But I didn't like it, because it was black, and so I dug out its eyes, and then I beat on it until the stuffing came out. I proved that the doll wasn't indestructible, after all, and a little bit later I got a lighter one to replace it.

"My sister, Sylvia, had been sent to live with our aunt and uncle, Reverend Jenkins and Aunt Lou. I loved Sylvia, but I used to be jealous of her, because she had good hair and light skin. Her hair lay down flat when she combed it. Mine wasn't as straight and had to be tied up in bows. At that time, black people of some refinement didn't straighten their hair. Uncle Teddy wouldn't hear nothing about any of this from me. He used to say, 'Don't be telling me about anybody's color or anybody's hair. We're all God's children!' He drilled that into me, or anyway, he tried to."

W. E. B. Du Bois, Teddy Harleston's former teacher from Atlanta University, had left the South in 1908 and moved to New York. There, a year later, Du Bois had helped to found the National Association for the Advancement of Colored People (NAACP). The incident that launched the group was a lynching that had taken place a hundred years after the birth of Abraham Lincoln, in the former president's hometown of Springfield, Illinois. The banner on the NAACP's stationery announced that the association "proposes to make 10,000,000 Americans physically free from PEONAGE, mentally free from IGNORANCE, politically free from DISFRANCHISEMENT, and socially free from INSULT."

Teddy and the other Harlestons undoubtedly shook their heads about Jim Crow, which grew worse with each passing year, and wondered what, if anything, they could do about it. The Harlestons were among the privileged few, and if they stayed within black society, their lives remained fairly smooth. Teddy was one of the few non-whites who actually managed to defy the system and vote (as receipts for poll taxes that he paid demonstrate). But even so, he wanted more than the colored haven of his elite position and was sick of the locked doors of segregation. (Gussie remembers that for many years Teddy refused to allow her to go to the movies, because it meant the family had to sit in the "crow's nest," the colored seating section on the mezzanine.) Teddy Harleston had followed Du Bois's career for several years, until finally he decided to throw his lot in with his old mentor. In 1916, after a meeting with nine friends and relatives, Teddy contacted the NAACP staff at 70 Fifth Avenue in New York to let them know that he wanted to set up a South Carolina branch of the group.

The start-up committee consisted mainly of Harlestons, including Teddy's cousins Eddie and Dick Mickey, his brother Robert, and John Singleton, who was married to Teddy's aunt, Sister Sue. A letter came back from New York pointing out that at least twenty people were needed to start a new chapter. To drum up members, Teddy apparently asked Du Bois for help, because on March 2, 1917, the great propagandist and educator arrived in Charleston from New York to give a talk. Although no description of the event has survived, Du Bois probably stayed with his perennial subject, which was the heavy load laid on black people by enslavement, and the need to take human rights by demanding them rather than waiting for them to be offered. Afterward, twenty additional members were signed up, and a week later, Dick Mickey wrote New York to say the group had a starting membership of twenty-nine.

Though the NAACP was supposed to foster comradeship, it wasn't long before the old Harleston and Mickey rivalry divided the ranks. The two families disliked each other, and when the chapter took shape, Eddie Mickey, the gay undertaker, looked for ways to undercut his Harleston cousins. On the official roster of new members and their occupations, Eddie had his brother Dick type the word "Artist" next to Teddy's name, then typed over it "Art Student," although Teddy was nearing forty years old and had been out of art school for several years.

Teddy and his cousins began corresponding with James Weldon Johnson, who was the "field secretary" of the NAACP. Johnson was a black poet and critic who, like the Harlestons and Mickeys, had graduated from Atlanta University. When he pushed the new Charleston chapter to elect officers, the group met and elected Teddy Harleston as president and Dick Mickey as secretary.

The group's first campaign was an effort to get several hundred new jobs for black women. The Charleston Navy Yard, one of the largest employers in the city, had recently opened a clothing factory that would make uniforms for sailors. Six hundred jobs were available, but black women had been banned from the hiring pool, and in May 1917, Teddy circulated a petition to prominent whites protesting this discrimination.

"The available labor which we can supply is sufficient to meet the present demand," he wrote. "We can provide 500 or more women who can adapt themselves to the work."

The petition argued that if the jobs weren't forthcoming, there would be an exodus of black workers from the city. When local authorities ignored them, Teddy's group carried its appeal over their heads, to the United States Congress. Aware that politicians from the South would dismiss the complaint, the NAACP instead approached the (apparently liberal) delegation from Minnesota. The plan worked, and five representatives and both senators from that state sent letters to the Navy Department critical of the situation in Charleston. Suddenly besieged by inquiries from Minnesota about hiring practices in South Carolina, the secretary of the navy issued a statement implying that black seamstresses were incompetent, and in any case, it was a matter for local authorities in Charleston. However, when the Minnesota representatives refused to accept this evasive answer and continued to prod the military, the navy backed down. By the end of the year, the drive had paid off and the Navy Yard had hired 250 black women.

For his next campaign, Teddy turned to a pair of crimes that had recently occurred in Charleston. The first was a murder case in which the victim was colored and the accused was a white streetcar conductor; the other case was the attempted rape of a ten-year-old black girl by a white man. The Charleston solicitor's office had ignored both cases, but after gathering witnesses and lobbying the prosecutor with persistent calls for justice, Teddy managed to persuade the solicitor to investigate the crimes. Afterward, he wrote NAACP headquarters about the outcome.

"In neither case was the accused arrested until we had warrants issued," he told the New York office. "The state made no attempt to have either man brought to court, so we secured all available witnesses and had the great satisfaction of having the murderer brought to trial."

The murder case marked the first time in twenty years in Charleston that a white man had been tried for killing a person of color. By itself, this was a kind of victory, but Teddy knew he had no hope of a conviction. The composition of juries excluded both blacks and women, and the custom ran deep to acquit whites accused of crimes against nonwhites. In the end, as Teddy expected, justice remained symbolic: the white jury rendered a verdict of not guilty in the murder case, and in the attempted rape, a grand jury refused to indict.

As the months passed and Teddy showed increasing public defiance, he became a conspicuous figure in Charleston. Most blacks regarded his actions as brave, while some whites saw them as troublemaking. The fact that so few colored people had signed up with the NAACP (only about one in a thousand black residents of Charleston) was an indication of the fear that penetrated nonwhite society. Negroes, if they claimed any dissatisfaction with "their place," made themselves vulnerable to reprisals. The punishments didn't have to be violent to be effective. In fact, Charleston was not a center of Klan activity, and the tar-and-feather terror of counties in the upstate region rarely trickled down to the coast. But the majority of black wage earners worked for whites, and merely to join a subversive entity like the NAACP put one at risk of losing one's job. (In the case of skilled, self-employed workers—such as stonemasons, seamstresses, or carpenters, all of whom had white clients—one might be subjected to a mysterious boycott and be left without customers.) It was Teddy's good fortune, as a member of the colored elite, that the Harlestons were insulated from economic punishment. Teddy was able to take a leadership role in part because he, his family, and the Mickey family were all business-people, not employees. Their livelihood depended on a loyal black clientele who wouldn't abandon them, and who largely supported Teddy in what he was trying to do.

A third confrontation, involving the school system, absorbed more than a year of Teddy's time. When Elise had taught in the public schools, she had been forced to accept a job far from home, because according to state law, all instructors in "colored" schools had to be white. From New York, Du Bois was almost as irritated at this queer situation as Teddy and Elise. He denounced it with an editorial in *The Crisis,* the NAACP's magazine.

> Of all the cities in the South, Charleston is guilty of the meanest act toward colored folks. It keeps in the schools white teachers, teachers who do not want to be there; teachers who despise their work. These teachers are Southern whites and they are teaching little colored children with a cruelty of discipline that is shameful. Openly and persistently the white city gives two and only two reasons for keeping up this farce: first, that they want to teach black folk their place; and, secondly, that they want to supply certain white people with employment.

In early 1918, Teddy drafted a petition demanding an end to "the uncalled for, unnecessary, unusual and abnormal conditions in the public schools." For twelve months, Teddy and a friend, physician John McFall, collected signatures from adult heads of households, both men and women. By the end of the year, they had 4,551 names, each on an individual card. According to an affidavit, the petitioners represented approximately 22,750 blacks, or 70 percent of the black population of Charleston.

The petition went to the Charleston Board of School Commissioners, where Teddy expected a rejection. When it duly came, the stage was set for the main confrontation. On January 18, 1919, Teddy led a march of several hundred petitioners through the streets of Columbia, the state capital, carrying the appeal to the state legislature. For several days, Teddy waited in Columbia, and it looked as though politicians would dismiss the appeal. But finally, legislators passed a resolution favoring black teachers in "colored" schools. After September 1, 1920, jobs in the black public schools were no longer guaranteed to white teachers—only to African Americans.

Teddy was thrilled, and when news of the victory spread, membership in the "Harleston chapter" of the NAACP soared. By the time the appeal went through, the dues-paying membership had risen in two years from the original group of 29 to 646.

A few months after the march on Columbia, Teddy traveled up to Cleveland to attend the annual NAACP conference. Although at home he was one of just a handful of educated black leaders, at the meeting, hundreds of like-minded middle-class blacks surrounded him. Suddenly, he was a card-carrying activist in the thick of a national organization. A little bewildered, Teddy wrote Elise: "It has been very inspiring and reassuring to be here at this conference. Last night in half an hour $10,500 was subscribed to the Assn for the coming year. I have reason to be proud of the Charleston branch in comparison with other cities and the officers here appreciate what we have done."

Teddy's education as an activist peaked during this general meeting, when Du Bois asked him to stand up and tell the national membership the story of the fight over black teachers. Caught unprepared, Teddy jotted a few words on the back of an envelope, stepped to the podium in front of three hundred delegates, and summarized his two years of work from his notes:

> Welcome opportunity of reciting to this Conf. one of the campaigns carr. by Chas. Br[anch]. And to show that even in the far South it is possible to make a manly fight for our rights, without giving up our self respect.
>
> Chas. a city more than half col. [had] wh. Teachers in col. public schools, there being only 2 col. in system. 35 yrs ago attempt failed. Again 20 yrs. ago. Traitors. NAACP br[anch] determined to try again. Small memb. but yr. 16 public meetings, 1000 people. Victory.

When Teddy married Elise, he hoped for a creative partner who would help him with his art. But when his wife brought home her portfolio from the Emile Brunel School of Photography, he was less than impressed. No sooner had they finally moved in together than Teddy was making plans to send Elise away for more study, this time to Tuskegee Institute in Alabama.

Tuskegee was the best-known school in the South for African Americans. Its founder, Booker T. Washington, had died in 1915, but the school continued to grow. The head of the Tuskegee photography department was Cornelius M. Battey. At forty-seven, a veteran both of the studio and of location photography, Battey was hired at Tuskegee with money donated by George Eastman, founder of Kodak. After Teddy met with Battey, the photographer agreed to hold a place for Elise in his tiny studio class. Teddy wrote his wife back home in Charleston to give her the news: "Battey says he will be ready to receive you about the first of October, that I must just send you down and he will give you all he can. Tuskegee is a great place, but all else there is as dross compared to this artist and his equipment."

In the fall of 1921, to please her husband, Elise dutifully went to Alabama. "We are expecting Mr. Battey Saturday," Elise wrote to Teddy after arriving. "In the meanwhile every morning there is a class in chemistry, which I attend. Some reading, and then after dinner I visit the various buildings and classes. It is truly a wonderful place."

Teddy wanted his wife to help him build an enclave of art in his hometown, where art was largely lacking.

> Oh girl, if you will bring away some of the skill and surety that will give you the knack of turning out stuff much like that of Battey, we'll turn this town upside down. And so you are rusty, eh? Keep your eyes wide open. And don't forget to make those daily notes. And be on the job every minute of the time.

While Elise was away, in 1921, Captain Harleston decided to give his son and daughter-in-law a wedding gift: a new house. Across the street from the funeral home and next door to Emmanuel A.M.E. Church, the Captain made plans for a two-story building, with living quarters on the upper floor and storage at ground level for the funeral home. As the design went forward, Teddy lobbied for a large studio to be included in the upper rear of the building for him and his wife.

Hearing the house was going up, Elise had some ideas about decorating. "As to the studio," she wrote Teddy from Alabama:

> Get the floors painted, the windows cleaned, the shades up, and the rugs cut and down. You could even start moving before I come. Such as the machine and the living room things. See about the stove and the heaters for the various rooms. Then of course go right on with the electric fixtures. The shelves for dark room, the built-in closet under the sink, some shelves in the corner over the sink in the Studio. Put up the mirror in the reception room. I think beside the window is a good place. So there you are. Quite a full program.

The newlyweds' home was finished in January 1922, and the skylight took up the entire ceiling of the back room. The house was located at 118 Calhoun Street, directly across the street from the funeral home, which was at number 121. Gussie Harleston moved into the house with Teddy and committed the building to memory:

"There was a high ceiling on the first floor in the house, because that level was used for storage for the funeral home. Caskets were kept down there, and the carriages were parked in a livery stable in the back. They were landau carriages, with a high seat for the driver, and they were tall. After Captain Harleston got rid of the horses, the limousines went in there. Also on that level was a workshop where Teddy would frame and crate his paintings.

"We lived on the second floor. The way the house went, to get to our level, you entered the stair on the ground floor. Right there in the vestibule were some glass cabinets where some of Tantie's work was displayed. I called Elise 'Tantie,' which came from the French word that was easier to pronounce, because I couldn't pronounce 'Auntie.' Or maybe Uncle Teddy pressed it into me, I don't know. We had a lot of nicknames for each other. Uncle Teddy used to call Elise 'Sis,' and sometimes, 'my little brown thrush.' He called me 'my little heartstring.' Anyhow, it took three sets of stairs to reach our level. You'd go up six steps, then there was a platform, then some more steps, and another platform. At the second floor, there was a huge foyer, a kind of receiving room, with green wicker furniture. The chairs used to make a crunching sound when you sat in them. The foyer wasn't an enclosed room but an open area that gave into a long hallway. Off the hall on the right were four rooms. The first room we called the green room. It had a great big filigree mirror, full length, which had been my father's. The piano was there, and I also had my desk. Then came the living room, dining room, and kitchen, and off the kitchen was the bath. It was kind of bad having the bathroom next to the kitchen, but Captain Harleston was kind of a miser, and I guess he wanted to keep all the plumbing in the same area. Throughout the apartment, lining the hallway and in all the rooms, were Teddy's paintings. The house was a gallery of Harleston pictures—family members, people he painted in art school, Uncle Teddy's old girlfriends—and all of them with their eyes looking out at you. When Uncle Teddy finished a picture, he would put it at the end of that long hall, so it could dry. I remember when he finished one picture, a painting of Congressman Thomas Miller, and put it in the hallway to dry, I was afraid of the eyes. Congressman Miller was the last black congressman before the end of Reconstruction, and he worked with Uncle Teddy on the campaign for black teachers. Teddy's handling of eyes could be piercing, and Congressman Miller's followed me all around no matter where I walked, and gave me the shivers.

"When you got to the end of the hallway, you were in Elise's and Teddy's domain, the studio. Teddy had his easels and paints, and he used to put on an artist's smock when he worked, which he kept there. Tantie had a huge camera that she used to stand behind with a black cloak over her head. There was an elevated platform with a black backdrop, and she also had a darkroom closed off by a door in the corner. Over your head, the skylights had shades to change the light.

"Of course, Uncle Teddy was still in the undertaking line, and he was back and forth to the funeral home. Sometimes he brought business back to the house with him, and once I had a run-in with a little thing he took home from the office. In Uncle Teddy and Tantie's bedroom there was a dresser, and on top of it I found some little bottles. I thought they were perfume, because they smelled nice. I was a little girl and was going to go to a party that day, so I sprinkled some of the perfume on my dress and put it behind my ear. When I got to the party the people kind of screwed up their noses and moved away from me, and I didn't know why. Later on I asked Uncle Teddy about that perfume he had in the bedroom. He told me he had been wondering why the bottle was half empty, because it had been a sample of some new embalming fluid."

Elise came home from Tuskegee in the early part of 1922. About the same time, Gussie's father, Robert Harleston, got married for the second time. Since the death of his first wife, Robert's health had stabilized and he had resumed a normal routine. His new wife was a twenty-eight-year-old woman named Louise E. Ransier, the granddaughter of a Reconstruction-era politician. But neither Gussie's father nor Louise Ransier wanted to take Gussie into their house, so the five-year-old girl stayed with her aunt and uncle. After that, Gussie saw her father only infrequently, and he seems to have accepted the arrangement.

Now, with Elise's new training in hand, Teddy and his wife decided to launch their business. A couple of years earlier they might have bought some champagne, but the Volstead Act, which outlawed all alcoholic beverages and enforced Prohibition, had gone into effect in 1920. No matter: Teddy had a private cache of homemade wine that he kept in five-gallon kegs in the downstairs storage. Perhaps Teddy and Elise raised a toast with their home ferment before they started work.

Teddy had new visiting cards printed. In an era of scarce telephones, the old Victorian practice of leaving a business card after an unannounced visit was still practiced by the Harlestons and their ilk. Although he had previously carried a card with his name and the phrase "Licensed Embalmer," he now had it redesigned and the lettering changed to read "Edwin A. Harleston—Portrait Painter." He also hired a print shop to make some brochures:

The Harleston Studio—Portraits—From life or from photographs, Oil Painting, Pastel, Charcoal, Crayon—Free Hand Machine Enlargements—Highest Artistic Standards—118 Calhoun Street.

When the Harleston Studio opened in the spring of 1922, it was probably the only workshop of its kind in the country—certainly the only one in the South—a studio run by a pair of black artists. Elise's work had improved at Tuskegee, and she started shooting again. Her pictures were formal and posed, but they had a serene quality. One of her early photographs was a portrait of Teddy's cousin Rosalie Mickey. The sister of Eddie Mickey, Rosalie posed in a three-quarter profile, wearing a wool coat, with her hand resting on the lapel. The picture feels somewhat remote, but it was nicely finished in a way that Elise's earlier photographs were not.

Elise's prospects were likely to come from her own tiny mixed-race middle class, but in Columbia, a hundred miles to the west, a skilled black photographer named Richard Roberts had just opened a competing studio and begun to attract clients. Roberts's pictures were beautiful but stiff: they flattered sitters in their fine clothes. Elise was less skilled as a technician than Roberts, but she was better able to compete with him artistically. And she was at her best when she photographed strangers.

"Tantie made portraits of people she didn't really know," Gussie remembers. "She would get to talking with folks in the street, invite them up to the studio, and take their picture. Uncle Teddy was kind of standoffish. He would fight for the masses in the NAACP, but he wasn't intimate with common people. Tantie was much more outgoing and warm, and she found deep relationships with a lot of different people. She loved people in the particular and in the general sense. Some of her subjects were poor, and she probably paid them a little something if they looked like they needed it, which was pretty bad for business."

One of the people Elise persuaded to sit for her was an elderly black man named Smart Chisholm. Chisholm was both a former slave and a veteran of the Civil War. He had enlisted in the Union cause after being freed toward the end of the fighting, and had become a second sergeant in the Thirty-third U.S. Infantry, Company C. In Elise's photograph, made more than fifty years after the war, Chisholm appears in uniform, a wiry old man with a deeply lined but contented face. On his hat is the insignia "G.A.R.," for Grand Army of the Republic, the official name of the Union forces.

Though Teddy was the more public figure in the couple, it was harder for him to get subjects to paint, and this was not merely a matter of personality. Portrait painting was the signature art of white Southern society, and every respectable family was expected to have one or two (preferably dead) relatives hanging above their fireplace. But white people would never hire a nonwhite artist to paint them or their children. Teddy knew that the phrase "colored artist" was enough to lift a sneer on the lips of white patrons, and that nine-tenths of his natural market was closed to him.

He faced equally thin prospects in black society. Teddy's obvious patrons should have come from the colored middle class, but in Charleston, with some thirty-two thousand blacks in 1920, perhaps as few as fifty nonwhite families lived as comfortably as the Harlestons. Even this handful had tiny resources when compared to those of whites. The black bourgeoisie held none of the type of clients—railroad barons, social climbers, dissipated English dukes—that allowed a portrait artist like John Singer Sargent to wear silk shirts. Moreover, the average black South Carolinian had never been to a museum, much less haunted the galleries of great collections, as Teddy had done. Most black businesspeople, though sympathetic to Teddy's situation, didn't understand him. He was a completely new creature: a Negro Southerner with a classical artistic background.

Not everything was lost, however, because Teddy started to get commissions from people he knew. He painted the young daughter of a friend, Robert Wilkinson, president of South Carolina State College. (The official name of the school filled Wilkinson's stationery: the South Carolina Colored Normal, Industrial, Agricultural, and Mechanical School.) He did a full-length portrait of businessman Alonzo Herndon, founder of Atlanta Mutual Insurance Company. He got a commission in St. Augustine, Florida, through his sister Kitty Fleming. But when that client moved away without paying his fee, Teddy got a taste of what might await him if he ever became a busy freelance artist.

On top of commissions, there was the additional problem of exhibition. Although most portrait work was for private use and not for galleries, portrait painters needed occasional exhibits to help drum up work. Even if Teddy got the job, black artists could not get shows. Art galleries barely existed in the South, and museums refused to admit a black person through the front door. (Exhibitions by nonwhite artists were out of the question.) In the North, art dealers did not represent black painters, which meant the collectors' market was closed, and neither did museums buy them. As a result, Teddy and other black artists were compelled to show their work like amateurs: at churches, club meetings, schools, and libraries.

With so many obstacles, Teddy was forced to become not only a painter but a traveling salesman hawking his own art. In August 1921, he rented a booth at a convention of the National Negro Business League, which was being held in Atlanta. His white peers from art school would have been humiliated by such a move. At the trade show, Teddy's booth of paintings was wedged between the peanut display of George Washington Carver, from Tuskegee Institute, and a case of hair products from a black-owned cosmetics company. For a full week, Teddy stood in the aisles, mingling with insurance adjusters, barbers, shopkeepers, and dressmakers, trying to interest someone, anyone, in a portrait.

"Black Swan Records has a booth," Teddy wrote Elise, naming the music company that had recorded the Jenkins orphan Tom Delaney. "Also Tri State Casket Co., several insurance companies, and toilet article mfgrs." The show ended, and still Teddy had no new commissions. Dejected, he packed his paintings and returned to Charleston.

Teddy decided to concentrate on people he knew, and let the undertaking business pay his bills. He approached his friend Rev. Caesar Ledbetter, pastor of Plymouth Congregational Church, and asked if he could paint him.

"Reverend Ledbetter was a kind of philosopher," Gussie Harleston remembers, "and a brilliant preacher. He had a deep baritone voice, and his sermons were full of allusions to literature as well as to the Bible. He and Teddy were friendly. Teddy and Tantie sang in the choir at the church, and the portrait Teddy did of Ledbetter did capture the man."

In *Portrait of the Reverend Caesar S. Ledbetter,* a commanding, seemingly egotistical man looks down at the viewer, with one hand on his hip and the other holding a book. Like most of Teddy's paintings, this one cropped the body at the waist. One art historian has pointed out that Ledbetter's pose resembles that of the young nobleman in Il Bronzino's *Portrait of a Young Man.* In the Bronzino painting, made in Florence between 1535 and 1540, a princely youth wearing a smug expression holds his hand on his hip and props his other on a book, in precisely the same way as the Reverend Mr. Ledbetter. As it happened, Il Bronzino's painting had been shown at the M. Knoedler and Company gallery in New York during April 1915, and it was later hung at the Metropolitan Museum of Art, one of Teddy's frequent stops when he was in New York.

One of Teddy's ideas for the Harleston Studio was that Elise would photograph people and he would paint them. After the Ledbetter portrait, the couple did the first of their collaborations. Among the acquaintances Elise had gotten to pose for her was another elderly black man, a Mr. Wigfall. His first name has not survived, but Wigfall was about the same age as Smart Chisholm, though with a less wizened face. In the pose Elise arranged, this white-haired man with a gentle expression sits with his head bowed, reading a large book. For his painting, based on Elise's photograph, Teddy left the pose intact but added a shawl to the figure's shoulders. The result was *The Bible Student,* one of Teddy's most emotionally direct portraits.

Teddy and Elise struggled along, picking up small jobs, earning practically nothing from their work. "I have an order from Mr. James Frasier for two of the wedding group. Mr. Mickey for three of his," Elise told Teddy. "Had two jobs last week but only $5. Have a sitting tomorrow." The couple kept a sense of humor in their frustration. "Trust by this time you have your commission to do the next portrait because I will have to help you do it," Elise wrote Teddy, half seriously. "You know you cannot attempt so important a job without my supervision, advice, and instruction, plus constant criticism."

In July 1923, a letter arrived that promised to bring an end to their troubles. It came from the typewriter of a woman named Ernestine Rose, an administrator in the 135th Street branch of the New York Public Library in Harlem. "Your name has been given me by Mr. Russell of the *New York Age* as a painter who might be willing to let us have some of your work to exhibit," the librarian wrote. "Our plan is to have several exhibits by individual artists."

Despite her obscure job, Ernestine Rose held a position of considerable power in the nonwhite world. The Harlem branch housed the so-called Negro Division of the New York Public Library. A focus of black New York society, the building on 135th Street had been growing in importance for years, ever since the trickle of black Southerners moving to New York had begun to turn into a flood. After World War I, an economic boom stirred industries in the North, and by 1920 hundreds of thousands had left home for jobs up the East Coast and in the upper Midwest. The black population of New York City tripled, and upper Manhattan was fast becoming a colored enclave—larger, more progressive, and richer than any of the black cities of the South. Black New Yorkers interested in "high culture" eventually found their way to the 135th Street library. Teddy knew that showing his paintings there would mean tremendous exposure, and that this letter was like a little saving grace.

He wrote Ernestine Rose to accept, and in October Teddy packed *Portrait of the Reverend Caesar S. Ledbetter, The Bible Student,* and several other paintings, as well as photographs of his paintings' models by Elise, and took a train north.

Stopping in Washington, D.C., Teddy went for a quick look around the art museums. (Although in the South this errand was almost impossible, Teddy, as a nonwhite patron visiting the capital, could actually look at art.) "Spent several hours in the National

Gallery today," he wrote Elise, "and think still better of my things!!" He was high on the prospect of New York. He also seemed to miss sleeping with his wife, because he told Elise he felt aroused in front of one painting. "Just saw 'Portrait of a young woman' and her shoulder and breasts were so much like you I was almost for kissing her on 'them' as I love to do." Expecting he might meet a commission on the road, he asked Elise to send him some brushes and gear. "You might get my box ready for me right away—all the good brushes, 2 sq. tin bottles, 1 oil cup, little hammer, pastel box, a few charcoal sticks. Have box wrapped only in heavy paper. Love—lots of it for you and Gus."

Teddy reached New York and hung his paintings at the Harlem library. Looking around at the competition, he allowed himself to gloat. "There is work by Laura Wheeler, Al Smith Scott, and others, but for solid stuff our dear little attempts *have them*. I am making some headway toward publicity and my pictures are something of a 'hit' to all who have seen them."

A few days later, two magazines asked him for interviews. One was *The Crisis,* run by his old friend W. E. B. Du Bois, and the other was *Opportunity* magazine, a new monthly published by the National Urban League. In a letter from the editor of *Opportunity,* Charles S. Johnson, Teddy read, "One of the most delightful surprises of my visit to the art exhibit at the 135th Street Library was the personal discovery of your work. I was immediately struck with the fact that I was looking at a development of artistic expression which quite easily transcended racial lines."

The magazine wanted to do a feature about him. He wrote to Elise:

> It almost seems that I am dreaming. I have seen Mr. Johnson of "Opportunity" and his photographer is to come tomorrow to make some copies for a later number of the magazine as well as an "interviewer," if you please, to make a little write-up of yours truly. A formal "opening" is being arranged for the first week of December, when the critics of the daily papers and magazines will see all of the pieces shown. If I just had some capital to wait around and work it up for a few weeks here I am confident of turning out some hits.

Teddy languished in New York as long as he could, visiting friends, looking at art, hoping someone would see his work and hire him. He lounged in the museums and tasted the kind of nightlife he couldn't get back home. "Saw 'Runnin' Wild' last night," he wrote Elise, "and was thrilled."

Runnin' Wild was a black musical then playing on Broadway. It was a big success, not least because of its inclusion of a new hit tune, "Charleston." The show's producers had hired several of the older Jenkins orphans to play in the jazz orchestra, and evidently the orphans had taught the band some of their riffs, as well as their street dances. Immediately noticing the song and the jerky dance that went with it, Teddy laughed that a New York choreographer had pilfered the comical steps he sometimes saw at Harleston Hall, the party room above the funeral home.

> They have some beautiful folks in the show [*Runnin' Wild*], and more lovely music. They have taken a step out of the last figure of the quadrille as it is done at the famous Harleston Hall and called it "the Charleston." It makes a grand hit, closing the first act in a whirlwind of action amid great applause.

As the weeks passed, Elise, back in Charleston, pined for Teddy. "What am I to do to pass the time until you come?" Elise wrote. "My love, I am so nervous you'd think I am waiting my groom instead of a three year old husband. With happy thoughts of coming bliss, Your own."

The longer they were separated, the more seductive their letters became, especially if they were written on a Saturday. Teddy and Elise had a ritual of stripping down and bathing each other on weekends, what Teddy called "our little Saturday evening institution of the Bath." In his weekend correspondence, he teased Elise about what it would be like to sleep with her.

> This is my Sat. night you know, and no one to dry me. Wish I could just kiss you good night, right on your neck and your chest, and then put my face against my warm little ricebird. So sweet!

> Within a few days I shall be back to pester and worry and harass and annoy and exasperate and fret and tease you, but also to love and cherish and adore, fondle and caress you, my dear.

Elise also craved her husband's touch. "The longing to hold something you lately touched was so strong I could not resist it. The shirts that you had worn I took to bed with me. Gus says 'Send that kiss to Uncle Teddy and tell him hurry up and come on home.'"

But Teddy stayed away. He told his wife he was denying himself the pleasure of her bed because he had his eye on something else.

> I am continually thinking of you, saying, "hold on, hold out, girl." It could be and it can't have been much worse, but better things are just beyond my fingertips. Set no date for my coming home—I want money! Don't you?

Being in New York seemed to make Teddy materialistic. He saw the urban Negroes of Manhattan, who were twice as rich as Southern blacks, or at least twice as conspicuous about spending. Teddy was tired of being the overgrown son who depended on handouts from his father. But New York also made him ambitious. Though he missed Elise, Teddy stayed on in the metropolis as long as he could, looking for just the right commission, the one big job that seemed always to dangle just out of reach.

Chapter Fourteen

OF all the Harlestons, Ella seemed to have the most money. In the spring of 1918, Ella and the Reverend Mr. Jenkins moved into a mammoth old house. The brick manse stood at 34 Magazine Street, a half-block from the orphanage and across the road from the Charleston city jail. It was one of the largest houses in the neighborhood, a four-story Federal period mansion with high ceilings, formal rooms, a great central hall, two tiers of porches, and some five thousand square feet of living space. The house had been built a century earlier, in 1815, and Parson Jenkins had bought it from an estate.

Gussie Harleston spent much time in her aunt's home.

"You entered it from the porch, through a screen door into the foyer. To the right was a huge living room, with a bust of Beethoven on the mantel. Aunt Lou loved that sculpture and used to make the servants dust inside Beethoven's ears. Also in that room was a record player that you had to wind up with a crank, and across the hall was a huge dining room. On the second floor, one of the parlors had been converted into the Jenkins Orphanage office, and across the hall from that was Aunt Lou and Uncle Joseph's bedroom. I called Reverend Jenkins 'Uncle Joseph.' They had their own sink in the room to wash before bed, which I thought was very unusual.

"On the third floor were the servants' quarters. The head housekeeper was a stocky woman who wore aprons with a big bib, named Jane Dickerson. Everyone called her 'Aunt Jane.' She was actually a distant relative of Reverend Jenkins, and she was very dark, like he was. Aunt Jane guarded Reverend Jenkins's belongings as though they were the king's jewels. We children couldn't touch anything of his if she was in the room. There were a couple of younger women under her, girls from the orphanage, one called Lottie Belle and the other, Ruth. They never seemed to graduate from the orphanage school, because they worked for Aunt Lou for years.

"The fourth floor was the attic, which had been made over into a kind of playroom. My sister, Sylvia, and I had a clubhouse up there where we would disappear for hours. The Jenkins house stood across the street from the city jail, and from the attic window we could look down into the jail yard. We'd watch the men in their striped uniforms, moving around, and the paddy wagon would come and go. The prison was one of our entertainments."

In a photograph from the period, Ella Jenkins stands in the parlor that had been given over to the orphanage. The usual filing cabinets, office furniture, and typewriters surround her. But on the wall, Ella has hung a charcoal drawing of her father, Captain Harleston, that was done by her brother Teddy. Although the Captain had vilified her for having a child outside marriage, and even Teddy had shunned her, Ella kept the picture where she could look at it every day.

In 1920, after eight years of marriage, Ella Jenkins was thirty-six and the Reverend Mr. Jenkins, fifty-eight. Family memory is divided about Ella. Some say she was serious and gave her all to the orphanage, but she also had a wry sense of humor, and she could be uninhibited. One time, Ella and her husband were eating in the formal dining room when the Reverend Mr. Jenkins said something Ella didn't like. Ella had an ear of corn in her hand, and without a pause, she threw the half-chewed cob across the table, just missing the Parson's head. Jenkins reared back, smiled, and said, "Bless Jesus," because he liked his wife's fiery side. Other family members say Ella was a cold woman who held herself above other people and starved her family of love. Where Jenkins was outgoing and personable, Ella was often silent and judgmental.

Though opinions about Ella conflict, everyone agreed about their marriage. All who knew them say that Jenkins adored Ella and held her up on a pedestal. The Parson's endearment for his wife was "Madam." The Reverend Mr. Jenkins knew what he had done by impregnating Ella when she was little more than a girl and wrecking the Harlestons' love for her. He felt remorse, and with equal parts guilt and love, Ella's seducer supported her in high style.

Ella Jenkins had a taste for fine clothes and accumulated racks of them. Even though the temperature in semitropical South Carolina rarely dipped below freezing, she also owned several furs. In the Jim Crow South, it was no simple matter for a black woman to buy expensive clothes. The few fine shopkeepers in Charleston didn't carry the New York styles Ella liked, and even when they did, Negroes were not allowed to try on clothing in stores. If a colored customer put on a hat or stepped into a dress, it was as if she had bought it, because no white woman would ever wear clothing touched by black flesh. Ella disliked being treated like a pariah, and so with her husband's money, she developed her own method of shopping. White clothiers knew of Mrs. Jenkins's generous spending, and whenever a new shipment came in, they would have some of the best things delivered to her house. Ella would try the clothes on, keep the pieces she liked, and send back the rest. This was how Mrs. Jenkins acquired the wardrobe of a proper rich lady, while also avoiding being humiliated.

In addition to their mansion, the Jenkins family also owned real estate in New York. The Parson bought a town house in Harlem at 147 West 132nd Street so that the bands and their managers would have somewhere to stay for extended trips. He also bought for himself and Ella a summer home in Asbury Park, New Jersey, a beach town with a breeze and a busy amusement pier. The couple had developed an annual ritual. In June, at the start of the summer band tours, Ella and her husband would move to New Jersey, taking housekeepers and a cook with them. Using Asbury Park as a springboard, Jenkins ventured out by chauffeured car from Virginia to Massachusetts, following the travels of the various bands. Ella either stayed at the beach or visited friends in New York. When the temperature cooled down in the fall, the Jenkins household returned to Charleston.

Every year, the scale of the Jenkins operations grew and the luxuries enjoyed by Ella and her husband became more visible. In addition to houses and servants, the family owned several pieces of rental property in Charleston and kept various bank accounts. Respectable black citizens struggling to make ends meet gossiped at church about the Jenkins family's high living. No other women could afford the clothes Ella wore, least of all the impoverished mothers who sent their children to Jenkins's care. The Parson traveled with a car and driver at a time when few black citizens could afford even a bicycle, and the diet at the orphanage consisted of corn bread and molasses while Ella supervised her cook's preparations of fish and game. To some, the orphanage looked less like a charity and more like a little empire. Ella and Parson Jenkins dismissed the complaints as jealousy and inadequate appreciation for their difficult work. But the criticisms never rose above the level of disgruntled remarks or became organized enough to wound their target. In theory, an annual audit by an accountant provided oversight, but Ella was the bookkeeper for the orphan house, and she made sure the ledgers were as correct as they had to be.

Edmund Jenkins, the Parson's son, was settling into Paris. Jenks had moved to France in 1921 and was trying to find a life in the tiny black subculture of the capital.

To many Americans, Paris in the 1920s is regarded as the city of the Lost Generation. Dozens of American writers made bereft by the war were drawn to France, where they lived precariously and spent years honing their art. The unknown beginners from those years included Gertrude Stein, F. Scott Fitzgerald, Henry Miller, and Ernest Hemingway. But the French cared nothing for scribbling foreigners, having ample writers of their own, and Parisians did not think of their city as an American literary salon. In the 1920s, Parisians were far more likely to pay attention to American music than American letters, and for Jenks, this difference would be a boon.

The first radio broadcasts in France were made in 1921, from the Eiffel Tower, causing thousands to go out and buy expensive radio sets. Crowded around tinny, unreliable receivers, Parisians tuned in to all sorts of music, but the searing notes of "le jazz hot" were easily the most modern (like the radio itself). The city's nightlife, dormant for years, bristled again, this time with jazz and cabaret. Memories of Jim Europe's Hell Fighters were still fresh, and young people who had survived the war were hungry for more bands "like them." The dingy cafés in the cramped streets of Montmartre, beneath the white dome of Sacré-Coeur Cathedral, provided the setting, and a small group of black jazz players provided the sounds. American musicians working in Paris included Louis Mitchell, who fronted an act called the Seven Spades; Palmer Jones, leader of the International Five; Elliot Carpenter, pianist and head of the Red Devils; and Will Marion Cook, with whose Southern Syncopated Orchestra Edmund Jenkins had played in London.

Arriving in Montmartre in his Talbot-Darracq roadster, Jenks cut a swank figure. He was black and could play the new music as well as anybody. Jenks soon found work as a clarinetist in a white six-piece band at the Club Daunou. The cramped barroom stood at 7 rue Daunou, next door to the more famous Harry's Bar (at number five).

Although he was personally debonair, Jenks had stepped into a raunchy scene. By the 1920s, Montmartre was no longer the titillating bohemian quarter that Impressionist painters had idealized in the late nineteenth century. It was a working-class slum and the center of the city's vice industry. Submerging himself in a decadent world, Jenks swam through a subculture of drunken Americans, cocaine and opium, and ubiquitous prostitutes. A well-heeled clique of Russian expatriates—thrown out of their country by the Bolshevik Revolution and pining for St. Petersburg over pitchers of Pernod—provided the odd grace note. After their gigs, the black Americans often retreated to their own hangout, the Flea Pit, a pool room, bar, and cigar stand on the corner of rue Pigalle and rue

Bergère. Only rarely was Jenks home before 3:00 or 4:00 in the morning.

He had come a long way from the sleepy American South. Commanding high fees for his clarinet and sax work, Jenks became even more comfortable with money and luxury, and he had put his "serious" composing back on the shelf. After a few months of this, Jenks's old friends at the Royal Academy in London began to wonder if he had lost his way. In July 1922, an English friend named Randall Lockhart wrote him a scolding letter.

> It is with great concern that I noticed the thinning of the chances of you beginning the work, which was your first ambition. I have always felt, Jenks, that you never had that happiness and the feeling of purposefulness as in the days when you had little money and many ideals, few acquisitions and little worldly experience, and much religious faith and moral earnestness.

Lockhart was a white English classicist who could little understand what it was like to be a black American—for whom any professional life, no matter how loose, must have seemed a small miracle. In Paris, Jenks saw the chance to exploit a burgeoning French fascination with black people. Surrealist poet Guillaume Apollinaire called the fad "melanomania." The newspapers termed it *le tumulte noir*, "the black chaos."

When work at the Club Daunou dwindled, Jenks found himself playing in the Art Hickman Orchestra at L'Ermitage, a chic nightclub in the Bois de Boulogne. Art Hickman was a white American bandleader with a following back home who, like his black fellow countrymen, had come to Paris to cash in. Something happened, however, and Hickman was called back to America. When the bandleader chose Jenks to take over his job, the twenty-eight-year-old clarinetist became the black conductor of a twelve-piece orchestra of white jazz players.

Now he was flying. He rented a more luxurious apartment, at 27 rue de Lecluse, and played his music into the night. To add to his income, Jenks began making trips to Belgium, a five-hour train ride to the northeast, where he played jazz in Ostende, a resort town on the coast, at the Kursaal, a theater for wealthy vacationers. Club owners in Italy invited him on more concert dates, and soon he was also hopping trains to Milan. The European appetite for Jenks's American style seemed insatiable.

Ella Jenkins had only one natural child, Olive, the girl she had abandoned who was growing up in England, three thousand miles from her mother. Not long before he and Ella abandoned Olive, the Reverend Mr. Jenkins had had a daughter with his first wife, Lena. The girl's name was Mildred, and when Ella finally married Jenkins, she took Mildred in as her own child. The arrangement worked both ways, and Mildred called Ella "Mama." Then, in 1919, Ella became a stepmother two times over when Gussie Harleston's infant sister, Sylvia, went to live at the Jenkins mansion.

With her giant house and two adopted children, Ella seemed to have found symbolic compensation for having given Olive away. Tradition says that the idea of sending for Olive in England originated with the Reverend Mr. Jenkins, and that Ella didn't think it would be wise. Although she knew she wasn't going to have more children, she wasn't eager to have her real daughter back, because the danger of being discovered as the child's true mother would be too great. But when her husband pressed her, Ella relented, and she made plans to make room for Olive in her life.

In 1920, Olive turned fourteen, and that year, the Reverend Mr. Jenkins sent word to the family of Alice Layland, the midwife in England who had raised her since birth, that he and his wife were ready to bring Olive "home." According to a childhood friend of Olive's, the Layland family was stunned and at first resisted, but eventually accepted that the girl would be "better off" in America with her prosperous, and black, family. Olive herself must have been frightened by the news that she now had to go. It would be heartbreaking for her to leave the only family she had ever loved for an unknown destiny overseas. Whatever her childhood had been like as the only Negro for miles around, the town of Wigan was still the place she called home.

Although she had been known as Olive Layland in England, when she took a steamship to America (alone), the teenager reverted to the name on her birth certificate, Olive Harleston. The Reverend Mr. Jenkins went to New York to meet his daughter, while Ella stayed in Charleston. The girl's British citizenship meant that officially Olive was an immigrant and had to be processed at Ellis Island. When she arrived, thousands of Italians, Poles, Greeks, and Russian immigrants were also clamoring at America's gates. Eventually, Olive got through the tests and questions and took the ferry to a dock in Manhattan, where her father was waiting.

According to family memory, Olive had a pronounced English accent and carried herself like a British lady. Until coming to America, the only black people she had ever seen were her parents, and they had visited Olive just once. Jenkins had come to bring his daughter to South Carolina, a state where the majority of the population was black, and he wanted her to mingle with black Americans before heading down South. So when the teenage girl reached the dock, Jenkins greeted her and then walked her to the subway, which they rode straight to Harlem. As Olive came above ground into the teeming masses of black America, she was speechless. It was a frightening shock and a flash of self-recognition that forever marked her.

Olive and her father then took a train to Charleston. After so many years, Olive would have had reason to expect some kind of celebration to welcome her into the family. As a mother bringing a lost child back into the fold, Ella might have wished to embrace Olive with the warmth of her long-denied love. But the stain of her daughter's birth had never washed out of Ella's mind, and the warm homecoming soon turned into an emotional freeze. Although Olive's half sister, Mildred, called Ella "Mama," Ella told Olive to call her something different. Ella's full name was Eloise Chapman Harleston, and many of the orphanage staff called her "Miss E. C." Ella let Olive know that "Miss E. C." would be the right name, both in public and in private.

After that, things deteriorated into an even stranger charade. To control the gossip and keep up appearances, Ella decided to pretend that Olive was a foundling. She insisted the English girl was an orphan, like the Parson's other orphans, and vowed that she and her husband had no prior relationship to her. To prove this, Ella went to family court and formally adopted her own daughter in a show of "charity." Olive was astonished that she, too, was required to observe this pretense. But Ella could never admit to anyone, not even to Olive herself, that she had given birth to a love child.

Gussie Harleston says: "Olive changed her name from Harleston to Jenkins. Reverend Jenkins always acknowledged her, but Aunt Lou couldn't do it. If you asked her, she would always call Olive 'my adopted daughter.' That was true, as far as it went. She did adopt Olive, but she also birthed Olive."

Olive, torn from a loving home in England, was now heartbroken a second time as she was thrust into the cool care of her real mother. Although eventually she came to idolize her father, the Parson, and saw him as free from blame, Olive developed a sense that her mother had misled and betrayed her. In school, and then later in music, Olive spent years trying to rack up impressive achievements, hoping that these might lure her mother closer to her, but Ella could never be moved.

There is no obvious explanation for Ella's behavior. Perhaps at some level she wanted a reunion with her daughter, but when it came to the event, the fear of being publicly judged overwhelmed her. Ella was an esteemed member of the community: she enjoyed being recognized for her work as a pious soldier in a great charitable enterprise. It turned out this image wasn't compatible with acknowledging an out-of-wedlock child, not to mention adultery, in the distant past. In the end, Ella withheld her love from Olive, even as she demanded her daughter's loyalty and love in return.

Fortunately, when she came to Charleston, Olive discovered that she had a half sister, Mildred Jenkins. Though they had never met, Mildred and Olive were nearly the same age, and were now united under the same roof. The sisters attended school together and grew fond of each other; when it came time to go to college, they both left Charleston for Hartshorn Memorial College, a small school in Richmond, Virginia. Olive's friendship with Mildred became a kind of reward, a small consolation for her mother's chilly indifference.

By 1923, Edmund Jenkins had lived abroad for nine years. The Parson wrote him a stream of letters asking when he planned to move back to America—letters that Jenks answered or ignored, depending on his mood. In Paris, Jenks began seeing more of Will Marion Cook, leader of the Southern Syncopated Orchestra, and evidently he asked Cook to help him negotiate with his meddling family back in Charleston. Obligingly, on March 7, 1923, Will Cook wrote Parson Jenkins a letter intended to bolster Jenks's status in the eyes of the family.

> Dear Rev. Jenkins,
>
> Want to congratulate you on your son Edmund T. Jenkins with whom I had a most wonderful association while in Paris. He is possibly the best Musician in the colored race; the very best instrumentalist in any race, and one of the most perfect Gentlemen I have ever had the pleasure of meeting.
>
> —Will Marion Cook

Will Marion Cook and Edmund Jenkins were very much alike. The more time they spent together, the more they egged each other on with big plans for black music and their careers. But after several months in Paris, Cook said good-bye to Jenks and went home to New York, where he started work on a new musical. Between 1900 and 1910, Will Cook had practically created a franchise for "colored musicals" on Broadway. His shows that decade included *Jes Lak White Folks, In Dahomey, In Abyssinia,* and *In Bandanna Land.* But by 1910, producers had stopped backing black acts, and after ten dry years of whites-only Broadway musicals, black performers had finally returned to the boards with a show called *Shuffle Along,* in 1921. The reason Cook left Paris for America was to raise interest in a new musical of his own.

In July 1923, four months after Cook had written the Parson, Jenks received a letter from C. A. Parker, of G. Lewis Theatrical Enterprises, a talent agent in New York. Parker said that Will Cook was planning a new "colored show" to open on Broadway, and that he wanted Edmund to come to New York to play in it. Having just seen Cook in France, Jenks was probably expecting the offer. Business was good in Paris, but Jenks could only go so far in nightclubs. The centers of black American music were New York and Chicago.

Cook was offering $125 a week, a fraction of what Jenks could make in France; but Edmund had been hearing about the burgeoning jazz scene and wanted to angle for bigger fish back home. He had talked with Cook about forming a black orchestra, maybe even a black school of music. Will Cook's Broadway show could be a means to an end—to get him there and get him started—so he accepted the offer.

Jenks wrote his stepmother, Ella, to tell her he was coming to Charleston. In the letter, he boasted about various jobs he had had in Europe, and then said he would be home for Christmas. In December, he bought first-class passage on the *Maryland,* bound for New York. Determined to make a show of himself in America, he had paid an exorbitant premium so he could bring his car. With Edmund Jenkins in a stateroom above, and the Talbot-Darracq in the cargo hold, the *Maryland* left Le Havre for the States.

Will Marion Cook's new musical was scheduled to open on January 27, 1924, at the Times Square Theater. Jenks first spent two weeks in Charleston, bringing gifts to his long-remote family, including expensive French pastels for his cousin Teddy. ("The pastels!" wrote the artist, who was out of town. "Oh, fine. Thank him!") Then, in mid-January, the musician headed for New York. The white producers of the Will Cook show had given it the name *Negro Nuances*. Billing announced that singers Paul Robeson and Alberta Hunter would front an "orchestra of twenty-five, under the direction of Edmund T. Jenkins and Will Marion Cook."

When Jenks arrived in New York for rehearsals, Teddy Harleston was in town. "I am going to look into Edmund this aft—through Will Marion Cook," Teddy wrote Elise. But Jenks proved hard to locate. Teddy looked all over town, but his nephew had seemingly vanished. Several days later, Teddy wrote home, discouraged. "I have just about run my legs off hunting Edmund—without success."

Jenks's disappearance turned out to have been staged. The day before the opening of *Negro Nuances,* the *New York Age* ran an item announcing that the show had been moved from the Times Square Theater to the smaller Forty-fourth Street Theater, around the corner. On February 1, Edmund wrote his father to explain that the musical had poor advance ticket sales, and to escape embarrassment, he had backed out. "The show was not at all a success, and I was lucky in being able to get out of having anything to do with the program just at the last moment," Jenks said.

His Broadway debut having failed, Jenks pushed other plans. His first idea was to launch a music publishing company, to be run by and for black musicians. Jenks met with another black bandleader, Will Vodery, to talk about the launch. He wrote his father to ask whether he would be willing to invest in the project, but Parson Jenkins demurred. The Orphanage Man felt from experience that black-owned companies had little chance of success in the white business world. By April, the music publishing scheme had come to nothing.

Jenks's next idea was for a black music school and symphony. He left New York for Chicago, Baltimore, and Washington, D.C., searching for partners. But he had lived abroad for many years, and his contacts were thin. Jenks tracked down a former teacher from college, Benjamin Brawley, and laid out his plans. Brawley had no suggestions. He talked to white music producers and was reminded that dance music, not overtures, was the main thing white people wanted from black musicians.

After nine months, Jenks despaired of realizing his plans. He wrote his family that everywhere he went he had encountered "dubious smiles." He said, "I found that in America the dollar talks," Jenks said, "and no one is found wanting in rallying to its stentorian tones whatever may be his or her professed ideals."

In late 1924, Jenks left Charleston for New York, where he planned to take a ship back to France. Along the way, he stopped in Virginia to give a recital at Hartshorn Memorial College, the school in Richmond where Jenks's sister Mildred and his half sister Olive were both students. The Richmond concert was probably his first and only meeting with Olive, after which, saying good-bye to his sisters, Jenks headed for New York.

Jenks was running low on money. Accustomed to stylish travel, he was chagrined that he had to book second-class passage. Worse, Jenks realized he couldn't afford to bring back his car. He left the roadster with a distant cousin in New York and asked the man to do what he could to sell it. Then, in a cramped, windowless compartment on a liner called the *Paris,* Jenks shipped out for "home."

In the year Jenks had been gone, the French jazz scene had heated up to a fever. In 1924, a thirty-year-old black vaudeville singer from West Virginia, Ada Louise Smith, began singing at Le Grand Duc, a nightclub at the meeting of rue Pigalle and rue Fontaine. Smith had a freckled face and red hair, evidently the harvest of interracial sex somewhere in her family's past. Nicknamed "Bricktop" for her coloring, the American caused a minor sensation. (Langston Hughes, a twenty-two-year-old poet from the Midwest, was working at the time as a dishwasher and busboy at Le Grand Duc and observed the singer's electrifying effect on audiences, an experience he would later recall in his memoir, *The Big Sea.*) Ada Smith soon opened her own club at 26 rue Pigalle, called Bricktop's. Boosted by regulars like songwriter Cole Porter and American movie actress Fannie Ward, Bricktop's became a way station for the fashionable that helped fuel the spread of jazz. Meanwhile, a jumping dance called the Charleston swept through the nightclubs. Black Americans began opening restaurants and clothing boutiques, and the dribble of new acts from the States turned into a stream.

In 1925, a nineteen-year-old chorus girl named Josephine Baker, a veteran of Broadway and New York nightclubs, arrived in Paris with *La Revue Nègre,* a twenty-five-member troupe of American dancers. Under the direction of a French choreographer Jacques Charles, Baker created the "Danse Sauvage," a nude act that took the sex content of hot jazz and moved it to center stage. On opening night for *La Revue Nègre,* October 2, Baker made her entrance entirely nude, except for a pink flamingo feather between her legs, held aloft and upside down on the shoulder of a male dancer, while doing a split with her legs. When her partner turned her right side up and placed her on the boards, she danced and swung her hips in a way that unleashed a cavalcade of newspaper coverage—and boosted the appeal of black Americans another notch.

Returning despondent from New York, Jenks was probably not surprised at these developments, but not happy with them, either.

Something in him had changed, and now he was less interested in the jazz scene and its money. The trip to America seems to have chastened him and reminded him of his early ambitions. After his failed tour of the States, Jenks did only enough nightclub dates to pay the rent, and again he began to write music. He started an operetta, which he called *Afram,* and in February 1925, Jenks wrote his father with a description of the new piece.

Afram was to be set in the American South, Jenks told the Parson, amid cotton fields and palmetto trees. Simple black people would be the mainspring of its plot. The libretto was to have colorful cameo characters, such as gin-runners and revenue officers, but always it would come back to its central theme, the longing and survival of black people. The piece was serious and not necessarily commercial, but he thought it was his best composition so far.

Jenks also turned to instrumental work. From his files he pulled out an experimental piece, "Folk Rhapsody." This was the orchestral work he had composed in London in 1917, the one based on the folk song from Charleston "Brer Rabbit, what you do down dere?" Jenks expanded and orchestrated the piece, and renamed it "Charlestonia." (Because the Charleston was an internationally famous dance, it's possible he thought the name would click with audiences.)

In July 1925, at the Kursaal, in Ostende, Belgium, Jenks conducted "Charlestonia" with a sizable orchestra. The performance magazine *Comedia* called the concert "an unprecedented success" and "a faultless performance" by "composer Mr. Edmund T. Jenkins, a Negro." It was a strong review of a serious concert far from the dives of rue Pigalle and truly signaled that Jenks had turned a corner. Encouraged, he revived his idea for the music publishing company, which had failed to materialize in America. But Jenks had stayed away too long from the bandstands of Montmartre, and his income was dropping. He wrote his father and asked for a loan of three thousand dollars to help him get the business on its feet:

> That is all I need to start with, just a real decent office in the proper district, and I would have the field practically to myself in the way of arranging French tunes in the popular American fashion.

> If you could possibly help, I assure you, you will have given me the lift I need just now to get in the way of making some really big money.

Parson Jenkins seems to have said yes, because Jenks was soon renting an office at 23 rue Pasquier and printing up stationery for a new venture, the Anglo-Continental-American Music Press. Moving quickly, Jenks contacted the French copyright society, SACEM (Société des Auteurs, Compositeurs et Editeurs de Musique), and registered some fifty works, his own compositions and some belonging to others. The pieces included jazz, serious music, and the operetta he had been working on, *Afram*. Now he was not only composing but also publishing music, and he wrote his father excitedly, "Europe is the only possible place for me to live!"

Though Jenks was writing orchestral music, he kept an ear open to commerce and wrote some blues tunes just in case. Among the songs he wrote and copyrighted was "Levee Lounge Lizard," blues in the dirty American style. Like the songs Jenks had lately heard in New York, this one was four minutes long, built on bent seventh chords, and dressed up with roadhouse lyrics.

> *I ain't no movie star,*
> *I ain't no desert shark.*
> *But when I get behind the women folks*
> *I shows them where they are.*
> *I don't need no Rolls Royce*
> *I can't use no Ford*
> *Cause when I get down to talking*
> *My words am the best-est road.*
>
> *I am the Levee Lounge Lizard*
> *Sho Sho*
> *And I ain't yet let a woman make a fool out of me.*

On February 12, 1926, Jenks copyrighted *Afram,* the operetta more to his own taste, with a libretto in French. The piece was a narrative in three acts about the love and losses of an African prince and princess. It opened with a scene in Dahomey during a war between two tribes. A prince and princess are in love, but when the war intervenes, they are captured into slavery. Separated by slave traders, the two are delivered to different plantations in the United States. The story picks up in the American South, with slaves at work, singing their laments. From their separate farm prisons, the prince and princess pine for and sing about each other. After much plot digression comes the happy ending, and the couple is reunited in America.

Jenks was thrilled with *Afram,* which he regarded as his breakthrough composition. He decided to make France his permanent home and began shopping for a house. In late April 1926, he wrote his father once again for a loan: "If it were not for the fact that you have already very severely scored me in one of your earlier letters about asking you for money, I would suggest that you let me have a thousand dollars so that I could get the house right away and take advantage of the extremely low exchange of the franc (the lowest in its history)." Jenks told the Parson that he could realize his ambitions if he could live in the style to which the Jenkins family had been accustomed, "with a respectable housekeeper or valet who can cook." Back in Charleston, Parson Jenkins wondered whether he should help his son with money this one final time.

On July 15, 1926, three months after this last communication, Jenks was admitted to the Hôpital Tenon in eastern Paris. The diagnosis was appendicitis. The musician underwent surgery and was returned to his hospital bed. According to family tradition, Jenks was recovering in his room when he fell out of the bed and lay on the floor unnoticed for some hours. After that, he is thought to have contracted pneumonia. Jenks's condition worsened, but for some reason, his physicians sent him home to be cared for by friends. Several weeks later, the coroner of Paris issued a report:

> The twelfth of September nineteen hundred and twenty-six at 21:15, there died at no. 4 rue de la Chine, Edmond Thornton Jenkins, living in Paris at 27 rue de Lecluse, born at Charleston (North America) aged 32 years, composer of music, son of Daniel Jenkins

and of Lena James, couple without any profession living at the said Charleston, bachelor. Reported 13th September 1926 at 13:10 on the declaration of Pierre Rivassoux, age 40, employee at 4 rue de la Chine, Paris.

The coroner's office had not been required to list a cause of death since 1912 (when the law had been changed), so it is impossible to say what truly happened. The American consul in Paris cabled Parson Jenkins to inform him of his son's death. Ella and Jenkins were stunned and unable to function for several days. When a second cable followed with the news that it would cost the Jenkins family six hundred dollars to have the body embalmed and shipped to America, the Reverend Mr. Jenkins sent the money to Edmund's bank in France.

A funeral service was held at the American Church in Paris, and Jenks's body was then loaded onto the steamship *Leviathan*. The remains arrived in Charleston and a second funeral was held on Thursday, September 30, 1926.

The Jenkins Orphanage Band marched solemnly through the cemetery and at the gravesite struck up a dirge. Some of the musicians had known Jenks from the time he played with the band in London, a dozen years earlier. Parson Jenkins stood and wept over the casket. He had buried most of his own children and had had highest hopes for this one. The gravediggers started to lower the body of Edmund Jenkins into the ground, next to the grave of his mother, Lena James. But as the casket descended, Parson Jenkins asked the hired men to bring it back up. He could not accept that this child, his last surviving son, had been taken away on the verge of a great musical career. While Ella looked on, the Parson asked the gravediggers to bore a hole in the casket large enough to see inside. An opening was drilled, and the Parson used his fingers to pull out the padded lining on the inside of the lid. He bent down and pressed an eye against the hole, so he could look at the face of his boy one more time.

Part V

~

The Orphan Dancers

Chapter Fifteen

TEDDY Harleston was in New York, basking in the publicity around his show in Harlem, when he began to have abdominal pains. He wrote Elise nervously, "Have had to pass up some very tempting things to eat so far on account of the pet stomach." The pain got worse and finally he realized it might be serious. Teddy couldn't trust medicine in the South for a severe condition: hospitals in Charleston were segregated, and the "colored" wards less reliable than the white. Rather than go home for negligent care, Teddy decided to head to New England, where one of his friends, William Hinton, was a physician at Harvard Medical School.

William Hinton worked as a bacteriologist for the state of Massachusetts and taught public health at Harvard, where he specialized in research on syphilis. Teddy and Elise had spent their honeymoon at the Hinton family farm in Canton, Massachusetts. In early November 1923, Teddy arrived in Canton in sharp pain. Hinton had already arranged a meeting with a colleague, an internist and surgeon named Hilbert Day. Afterward, Teddy told his wife, "Dr. Day was very positive in advising an operation with a stay of 10 days to 2 weeks in a hospital. It will cost money for a bed, in addition to the fee for the operation, which he has not yet disclosed."

Teddy was forty-one at the time, and although his letters don't name the illness and family tradition has forgotten it, the diagnosis seems to have been appendicitis. In the 1920s, even simple surgeries like an appendectomy could be life-threatening. Teddy knew he might not survive, and so the day before he was admitted to the hospital, he sat in the guest house at his friend's farm and wrote Elise a farewell letter.

> My love to all. My kisses to you and my little heartstring Gus. By the time this letter reaches you I shall have entered the Roxbury Hospital, Vernon St., Boston, and had my little operation. If I do not come out, take my insurance, doll up and catch yourself a real, sure-enough regular guy for an A-1 Husband. Up to the present I have not filled the bill, but forgive me, please, I have loved you so. I'm positively not much afraid of anything, but you will appreciate the expected anxiety of the situation. If you could just hold my hand. Good night, with all my love.
>
> —Your "first" husband (and of course your last).

At home in Charleston, Elise waited nervously for word about the outcome. Two days passed, and the telegram Teddy had promised to send as soon as he was alert did not come.

"When Uncle Teddy was away from home," Gussie remembers, "Tantie minded the Firm. Tantie disliked the funeral office as much as Teddy, but for different reasons. Captain Harleston had all his children under his thumb. He didn't pay Uncle Teddy a salary, but just doled out money when it suited him. If Teddy was out of town, Tantie had to beg the Captain for money, even to buy me shoes, and she found it humiliating.

"The other thing that bothered Tantie probably was the Mickey family, who weren't easy to get along with. One of the Mickeys, Ellen, made kind of a nuisance of herself. Ellen was the daughter of Teddy's aunt Hannah, and the sister of Eddie Mickey, who was in the NAACP with Teddy. Like Uncle Eddie, Ellen Mickey slept on the far side of the bed—I mean she was homosexual. She was sort of manly, and she wore her hair pretty short, but I didn't know she was a lesbian until much later. Ellen was kind of the black sheep of the family. Heaven knows what Aunt Ellen did to find girlfriends—or what she did with her girlfriends once she found them. But she had a temper and was always having confrontations with someone.

"Anyhow, Ellen drove the hearses for the Mickey Funeral Home. Our house stood next door to one church, Emmanuel A.M.E., and across the street from another, Zion Presbyterian. Both the Mickeys and the Harlestons used to run funerals out of those churches, so the hearses outside the door were constant. Whenever the Mickeys had a service, Ellen would park the hearse while the funeral was going on and come into our house just to sit around. I think she got in Tantie's hair. Ellen used to take the milk bottle out of the icebox and drink straight from it, without using a glass. She didn't do it in front of Tantie, but I spied on her.

"My grandfather Captain Harleston had some wineglasses that had come from England, which he had inherited from his father, William Harleston. They were old crystal and shaped like an inverted cone, with a stem. Tantie and Uncle Teddy had a lot of them in the house, and I used to take a spoon and chime them, because they made a beautiful sound. Well, the glasses started to disappear. I remember seeing Ellen in the kitchen one day after she had come in from a funeral, and she glanced around and reached for one of those stemmed glasses, and put it inside her jacket. This must have happened several times, because the glasses became fewer and fewer, until they were all gone."

Another family member who might have bothered Elise was Captain Harleston's mistress, Mamie Randall. At forty-three, Mamie Randall was the mother of the Captain's second family and got much of his money and attention. About the same time he built a house for Teddy and Elise, the Captain had also put up a new cottage for Mamie. The four-room house went in the backyard of the funeral home, and Mamie lived there with the two daughters she'd had with the Captain, fourteen-year-old Alethia (known as "Leetie") and Catherine, twelve.

Gussie Harleston, who was seven years old at the time, remembers Mamie's presence.

"We used to call her 'Mama Sister,' which was what the Captain wanted us to call her, I guess. She wasn't my mother, nor was she my sister, but Mama Sister was a personable lady, and always very good to me. She looked good, too, all the time."

Mamie, or Mama Sister, was around the house quite a lot. The Captain doted on her, which was a difficult adjustment for Elise, who felt her father-in-law owed more to Teddy. "Mamie Randall has a new stove, come in from Detroit, so we are invited to dinner tomorrow, Gussie and I," Elise wrote Teddy. But Mama Sister was not entirely a kept woman, because she held down various jobs. Having given up her first line of work, dressmaking, she had worked for two or three years as a maid at a local theater. But in the early 1920s, she changed tack again, prompting city directories to list her as a "nurse." Mama Sister ran a unique "nursing" business from her cottage in the backyard. According to family tradition, the Captain's lover earned her living as an abortionist.

In the early 1900s, even though abortion was a felony, with jail time guaranteed, it was nevertheless widespread. Many women could not face the prospect of having a child without a husband—or, even with a mate, an endless chain of babies. Mamie Randall's clients are said to have consisted chiefly of white women; they could afford her, while most black women couldn't. The women came at all hours—some of them "in trouble," others just tired of children—walking back behind the funeral home to the freshly painted house of "Nurse Randall," the amateur surgeon. Mamie's method of inducing miscarriage is no longer remembered, but it's not likely to have been scrupulous. (Did she scrape out the embryo with a knife? Did she borrow the Captain's embalming pump and suction out the fetus?) But whatever the details, thinned by secrecy even during that time, it's curious to imagine Captain Harleston and his girlfriend working in their separate careers—the Captain embalming a body in the rear of the funeral home, while just a whisper away, his girlfriend performed an abortion in her kitchen.

After two days spent worrying about Teddy, Elise received a letter from the patient in Boston. The handwriting was shaky.

> Well, I am here by a large majority. Thank God. Doing fine in every way, except devilish bowels. Look fine, talk all right, sleep

> pretty well now, but can't turn over alone without great effort. Very little pain or soreness below, but I am disjointed from the world.

Teddy's stitches came out at the end of November. He complained that the remaining bandage was like a corset, and he couldn't bend over. The pain made him weak, so he spent most of his time around the Hintons' farm. Soon he found he had plenty of time to write home from the guest house where he and Elise had spent their first week of marriage. "I'm back at *our* old home," he wrote, "the place so hallowed by our golden moon—that ne'er forgotten moon—spent beneath its roof three short years ago."

To pass the time, Teddy began sketching William and Ada's children, and after regaining some strength, he went into Boston, where he reminisced about his life there as a penniless art student. In depressed moments, he worried about the funeral home and wrote dreary instructions about the inventory to Elise:

> Tell Moultrie to write H.S. Eckels, Arch St., Philadelphia, for 3 cases of Reconcentrated Dioxin and 25 lbs of Hardening Compound. And please remember to see that if a Birmingham salesman comes in, to have an assortment of silk robes and dresses ordered costing up to six or eight dollars each, and to replace any plush caskets sold with similar ones from Florida or Charlotte only.

Elise played along, responding with business reports: "There were three bodies in state in the chapel and three children buried in the past week," she wrote.

By Christmas, Teddy was strong enough to go home. He was relieved that he had cheated death and been given a second chance. On New Year's Day, 1924, he wrote Elise with a list of things for which he was grateful:

> Thank God that we chose each other to join our lives, our flesh, our destinies together, and that for three years now—three beautiful years, we have so enjoyed our union. Thank God that our happiness has been so almost completely unalloyed—no great domestic jars—no clouds but those sent by Nature—one great grief, your sister

> Sweet Marie, but no demons like Jealousy! Thank God for that, no Jealousy. Thank God for Gussie and her sunshine, her cheer, her comfort. The little cares that make us happy, her health, her fine mind, her wonderful disposition, her love for us and the good reasons for our great pride in her.

Gussie was a source of joy, but there was one sadness Teddy could not overlook. After more than three years of marriage, he and Elise had had no children of their own. The Harlestons lived in an age before fertility tests and treatments, when women without children were described as "barren" and wives took much of the blame for a couple's childlessness. In the biases of the time, for Teddy to mention the lack of children might be taken as an accusation against Elise. And so throughout his letters, he avoids the subject, except this single time. In the letter home to Elise, Teddy mentions having children, framing the idea in the most positive way he can: "One of these moons there'll be no visitors and our great enjoyment of our love will be changed only in kind," Teddy wrote, "but I can't see that it can be in extent or in purity, for I love you for yourself and nothing shall come between us, not even childlessness." And that was it: no mention of the subject ever again.

Elise also worried that she and Teddy weren't going to have any children. She wanted more in the house than Gussie and had set her sights on a new adopted daughter. Elise's brother, Tom Forrest, had a young daughter he couldn't take care of named Doris. Doris's mother seems to have had a breakdown, and Tom Forrest was willing to let Elise take the three-year-old girl. "Teddy, will you let me have Doris?" Elise wrote. "I mean are you willing that we take her just as we have Gus? You'll love her sweet heart equally as well as you love Gus." After Teddy came home, and he talked it through with his wife, Doris moved into the house on Calhoun.

Starting with the exhibition at the New York Public Library, Teddy's art career took off. Stories about his work appeared in both the *Crisis* and *Opportunity* magazines, and he began to get commissions from perfect strangers who had read about him. "Please give price on the grades of oil painting, portrait 16 × 20 inches," wrote Solomon Johnson, editor and publisher of *The Savannah Tribune*. *Opportunity* had put a reproduction of Teddy's painting *The Bible Student* on its cover, and an offer came from Chicago from a man who wanted to buy it. His name was Jesse Binga, president of a bank for black depositors on the city's South Side. Teddy sold him the painting and shipped it. From Atlanta University, a request came for a portrait of the retired president of the school. "I wish you would let me know if it is possible for you to paint a picture of President Twitchell Ware in Montclair, N.J.," wrote a professor in charge of the project. "I know the graduates would be willing to pay a decent price for it."

Teddy wasn't prepared for his new good fortune. For the first time, he was able to accept some jobs and turn away others. Then, in July 1924, there came a letter that floored him. It seems some people in Delaware had approached the artist's old friend W. E. B. Du Bois, asking for the name of a black portrait painter. The group of black teachers wanted to honor the industrialist Pierre S. Du Pont, who had taken an interest in the public schools and who had personally financed the building of eighty-three "colored" schools in Delaware. Du Bois forwarded the inquiry to Teddy and enclosed the reply he had sent to the group. "Answering your letter of July 14," Du Bois had written the teachers, "I beg to say that there is but one colored man resident in America, whom I would unhesitatingly recommend to paint Mr. Du Pont's picture and that is E. A. Harleston, 118 Calhoun Street, Charleston, S.C. If you will write him and mention my name he will write you and tell you of his training and let you see specimens of his work."

Pierre S. Du Pont was a household name. At fifty-four, he was chairman of E. I. du Pont de Nemours & Company, an explosives manufacturer with a history reaching back almost 125 years. The company had begun selling stock only in the early 1900s, after which it mushroomed into a monopoly, making record-setting profits by selling deadly munitions during World War I. Pierre Du Pont also owned a one-third stake in General Motors, of which he had been president for several years. His fortune stood in the tens of millions of dollars. If Teddy landed a portrait of the robber baron Du Pont, he might find himself with a nice batch of commissions from the white ruling class.

The artist followed up and asked a high fee. "It was thrilling to have the Delaware letter and I have answered it," Teddy wrote Elise, "from $1000 to $2000 according to size and what they want in it. Hope they come across." The reversal of fortune seemed complete. From the brink of death only a few months earlier, Teddy was now able to contemplate giving up the funeral business for his art career.

As an artist of color in the South, Teddy was an anomaly. Art made by African Americans could be found easily enough in music, dance, and material culture: from jazz to ironwork, spirituals, masonry, blues, tap dancing, woodwork, "ring shouts," basketry, and a dance called the Charleston. But as for fine artists—printmakers, sculptors, oil painters—Teddy had no peers in Southern black society. His equals (the ones of color) lived in the North, and he seldom visited or corresponded with them. Meta Warrick Fuller, five years older than Teddy, was a sculptor who had graduated from the Pennsylvania Academy of Fine Arts and had studied at the Académie Colarossi in Paris with Auguste Rodin. However, this first African American woman to become a professional artist lived in Framingham, Massachusetts, and Teddy saw her only when he went to Boston. Laura Wheeler Waring, who was five years younger than Teddy and lived in Pennsylvania, had also studied in Paris. Her classically styled portraits shared the same reserved intensity as Teddy's. William Edouard Scott, after studying at the Art Institute of Chicago and in Paris until 1911, became a muralist and painter of small canvases in an Impressionist idiom during a long career in Illinois and Indiana. Though they had had little to do with each other until the 1920s, Scott was soon to become Teddy's rival. Archibald Motley was a painter from New Orleans (very much of the same light-skinned Creole class as Teddy) who lived in Chicago, while Augusta Savage was a sculptor and art teacher in New York City. And that was about it.

Teddy's natural colleagues should have been among white artists in the South, because they were more likely to practice the European tradition of portraiture, and almost as important, they lived nearby. But in South Carolina's climate of racial loathing, white artists would have nothing to do with him.

For many years, the fine arts were moribund in the city of Charleston. Modernism did not penetrate the South, and even its milder forms, such as Impressionism, were considered recherché. Painting had not changed much since its documentary phase before the Civil War, when a coterie of portrait artists created flattering images of the slave-owning class. In the early 1900s, portrait painters in Charleston who were still working in the old traditions included two women, Leila Waring and Jean Robinson Flemming.

But the 1920s were a turnaround time for the arts in South Carolina. In 1905, a museum had opened in Charleston, the Gibbes Art Gallery, and beginning in 1915, its drawing class, known as the Charleston Sketch Club, breathed air into the stale scene. A few years later, in 1921, a thirty-eight-year-old printmaker named Elizabeth O'Neill Verner helped found the Southern States Art League, intended to promote art and artists in the region. Shortly after that, the Charleston Museum propelled things further with another art salon, the Charleston Etchers' Club, founded in May 1923. The Etchers' Club confirmed a growing local interest in printmaking. This group consisted of seven women and two men, all white, the most accomplished being the native-born artist Alice Ravenel Huger Smith. At forty-seven, Alice Smith had been working in wood-block prints and watercolors for several years and was in a position to share her skills.

The Etchers' Club bought a printing press and set it up at the Charleston Museum. Some members came in from out of town for meetings, including forty-six-year-old Alfred Hutty, sometime of Woodstock, New York, and Anna Heyward Taylor of Columbia. The Etchers' Club became an instrument for the launch of several regional careers that soon escalated out of the state. The prints of Alice Smith, Alfred Hutty, and Elizabeth Verner, in particular, began to sell in Charleston to tourists from the North, and then to collectors. The focus of the nascent movement remained local, which made it popular with Northern art buyers. Subjects were usually Charleston scenes: classic architecture moldering from neglect, natural vistas of the bygone plantations, and pictures of dilapidated buildings with black people sitting on the steps.

If it were not for Jim Crow, Teddy would reasonably have taken part in the salon life of these printmakers, who, after all, lived a short walk from his house. Whereas the white artists were often self-taught, Teddy was the best-trained painter in the city. (The Gibbes Art Gallery showed the work of his former teachers Frank Benson and Edmund Tarbell.) Members of the Etchers' Club knew of the Harleston Studio, but they did not or could not allow Teddy (or Elise, a trained photographer) into their circle. Segregation and condescension kept the door locked.

In the 1920s, Leila Waring, Alice Smith, Elizabeth Verner, and Anna Taylor all opened "studios" on Atlantic Street, a two-block street off the harbor. The Harleston Studio, already established, may well have been a catalyst. These were closer to drawing room stage sets than they were to working artists' lofts. The women mixed hospitality and art, opening their doors on Sundays to offer tea in silver service to visitors in hopes of encouraging sales.

By 1925, these developments were recognized to be part of a local progression in the arts that later became known as the "Charleston Renaissance." But Teddy had to watch the action from afar. Other artists began to arrive in town, bringing their talents and preoccupations with them. Lilla Cabot Perry, a seventy-seven-year-old Impressionist painter from Boston who had lived for ten years as a neighbor of Claude Monet's at Giverny, France, came to Charleston to paint landscapes. Another American Impressionist, Childe Hassam, appeared in the city in 1928, and the following year, Edward Hopper, in midcareer at age forty-seven, visited South Carolina, where he made drawings and watercolors, choosing as a frequent subject the battered cabins of black sharecroppers. In the work of most of these artists, black people appeared as anonymous decor, not as individual subjects but as ornaments in the landscape. (Two notable exceptions were a pair of etchings by Elizabeth

O'Neill Verner, one of Rev. Daniel Jenkins and the other of Jenkins's daughter Mildred.) Ignored by the white "Renaissance," Teddy, self-reliant, soldiered on.

In 1924, Teddy and Elise decided to use Teddy's new commercial renown in the North to try to leave Charleston for good. But while Teddy had a batch of new commissions, the years of struggle had made him realistic. He knew that even with steady painting work, he and his family could not live on portrait fees. Few artists could make their clients pay on time, and (Du Pont aside) his work started at about one hundred dollars per painting, not enough to support a middle-class lifestyle. Teddy wanted to leave the funeral business and get to a city where he would be noticed, but he and Elise knew they needed a job with a reliable income. A new occupation would support the family—which included eight-year-old Gussie and now three-year-old Doris Forrest—while the couple developed their art careers out of reach of Charleston.

"You have worked for ten years to help your father keep his business alive," Elise wrote her husband. "Today it is worth something to him—so much so he will not turn it over to you nor take you into it as a partner, but it does seem he should be willing to give you sufficient means to care for your family. The Captain gives me $5 every week."

Teddy sometimes despaired at his situation with his father. Captain Harleston made promises to change, but to share the wealth would mean a loss of control, and so the Captain remained neglectful of Teddy's needs.

In Washington, D.C., Teddy had apparently talked to a man named William Nixon, an arts administrator in the public school system, about a job opening up for an African American artist. "The job in Wash? Can't tell you—yet," Teddy wrote Elise. "I saw the folks, but the politics, oh the politics. Pays nicely, though." The exact nature of the position has been forgotten, but it seems to have been that of an art teacher at Dunbar High School, a new "colored" school attended by the city's mixed-race elite. Teddy told Elise, "At the present the superintendent at Washington insists upon teaching experience plus training, with preference to the former over the latter."

According to family tradition, Teddy and Elise wanted to move to Washington, but there was an unexpected problem. In the previous decade, Impressionist landscapes had come into vogue, especially the plein air genre, or pictures painted on-site in the open air. Teddy made portraits, but the job in Washington required that he teach landscape painting. In order to qualify, he had to show he knew plein air technique. He could give up the undertaking business if he could paint water lilies like Monet.

As it happened, a strong program in outdoor painting had developed at the School of the Art Institute of Chicago, in a city that was home to several of Teddy's old college friends. (T. K. Gibson, or "Gib," was an insurance executive in Chicago, and Inez Canty, an Atlanta University alumna who had been in love with Teddy for years, also lived there.) Inez Canty and Teddy exchanged letters in which Canty suggested he come to Illinois to study. The courses at the Art Institute lasted just seven weeks, and he could spend time with old friends while he learned about outdoor painting.

In June 1924, Teddy took a train to Chicago and enrolled at the Art Institute. "Your gray and bald-headed husband is a boy again—a schoolboy—a student," he wrote home. "I registered yesterday. I shall need my little sketch box, and please send my rubbers."

Like New York, Chicago had become a boomtown for black Americans. It was the beginning of the movement known as the Great Migration. Between 1916 and 1920, some 110,000 black Southerners had moved to the city, nearly tripling the number of nonwhite Chicagoans. At forty-two, Teddy had never been to the city. "This is a spacious, breezy, drab town, an overgrown frontier city with sooty buildings—not dingy, inky black buildings, and wonderful avenues full of flying motors going 30 to 40 miles per hour," he told his wife. Though he was a frequent visitor to Harlem, all the colored people in Chicago took Teddy aback. "There are Negroes by the tens of thousands—intelligent, refined, thrifty, coarse, vulgar, filthy, crooked—a motley group such as I've seen nowhere else."

According to the records of the art school, Teddy found an apartment at 4439 Calumet Avenue, where his phone number was Oakland 2164. He paid tuition for the summer, forty-five dollars, and then enrolled in a course in "Outdoor Sketching." The class met five days a week and was taught by a local landscape artist named Carl Krafft. Teddy was one of twenty-seven students—five men and twenty-two women—and the only black person.

The outdoor class met at 9:00 A.M. in Humboldt Park. Although plein air painting worked well enough in Provence or Giverny, it was another matter in gritty, industrial Chicago. After the first week, Teddy wrote Elise about the dangers of taking his easel outdoors. He reported that one morning he had been painting near the Illinois Central Railroad, with its coal-powered locomotives, when the wind "brought a drizzle of coal dust, most of which seemed to land either in my eyes or on the canvas. A cinder lodged under my upper eyelid."

Despite the smokestacks, Teddy managed to produce some credible work. His *Landscape with Painters,* Impressionist in style, shows two artists wearing smocks and working at their easels, with a woman convincingly placed in the foreground. Light filters through the trees, giving the painting the evanescent quality that art collectors in the 1920s seemed to want.

Though Teddy applied himself, as the weeks passed, news trickled in from Washington that the job might be less of a sure thing. In fact, the grapevine suggested that it might have already disappeared. "Perhaps you may be interested to know," he wrote Elise, "that I was repeatedly reminded of the politics at work in the system, and that someone already in the system would most likely be promoted for it." But still Teddy stayed on in Chicago.

Teddy's friend Inez Canty was an unmarried bank secretary working on Chicago's South Side. She had sent Teddy love letters for years before he married Elise and continued to pine for him. Canty worked for Jesse Binga, one of Chicago's richest black citizens and the art collector who had bought Teddy's painting *The Bible Student*. Teddy did not want to sleep with her, but he didn't mind that Inez Canty was in a position to help him get a commission from her boss.

Jesse Binga, a former Realtor, had opened a private bank for black clients in 1908. In 1921, he obtained a charter for Chicago's first black-owned deposit institution, the Binga State Bank. The bank stood at 3633 South State Street, on "the Stroll," an eight-block stretch of road that was the city's black main drag. When Binga and his wife moved into a house in a white area, at 5922 South Park, their home was bombed. The banker hired guards and started to carry a gun, announcing to the press that he would not be forced out—and as a consequence becoming a hero to black Chicagoans. When Teddy arrived in town, Binga was sixty years old and at the peak of his power. From time to time in Binga's office, Inez Canty would drop Teddy's name in hopes of convincing the banker to commission a portrait. The Binga State Bank was building a new headquarters, and finally, in July, Binga hired Teddy to paint his picture to decorate the new office.

"I am just back from the Bingas', and I have just been whisked home in their real Lincoln limousine," Teddy told Elise. "Their house is very nicely furnished, showing nice taste with a number of paintings, mostly landscapes, but the one 'Harleston' [*The Bible Student*] holds its own. He wants a half-length portrait much like Rev. Ledbetter's only not so 'posey.' I think he will be fussy and a little irksome, but one-hour sittings won't kill him. In all the conversation there was no word of price. What shall it be?"

Because the banker had a busy calendar, Binga asked Teddy to work on him early in the morning, before office hours. For a month, Teddy had to get up and out by 5:30 A.M. in order to reach Binga's house via public transportation by 7:00. After the hour-long sitting, Teddy took a bus to the art school, where classes met throughout the day.

"I have settled down to work and it is a sort of strenuous business," Teddy said, "getting there after finding a breakfast and carrying on from 8:30 to 4:00." Elise wrote back to encourage him. "Hope you are getting on with the 'Great Mogul's' portrait. You have worked so very hard it is time you had some measure of success." Teddy was relieved that Binga, despite expectations, was easy to get along with. "He talks business all the time while sitting but is not a bad sitter and our meetings are very pleasant."

The painting, thirty-two inches by thirty-six, was finished on August 9, and the next night Teddy went out to celebrate. It was evening along the South Side's showiest blocks, and Teddy was a little wowed by the money into which he had stumbled. "I played cards at one of the 'palaces' in which these folks live. For miles the houses look like something in *Vanity Fair* or *Vogue*. You know I told you I'd begin to make some money this year. When we travel again, who knows? We may go in state class."

In the previous eight months, Teddy had hardly been in Charleston, and Elise had begun to resent her husband's absence.

"Did you think I would miss you most in the studio?" she wrote to Chicago. "I do. It makes my heart hurt to come in here and am always ready to cry when I look around on the many reminders of your dear presence. I wanted you so this morning at 7 A.M."

Elise told Teddy she had been dreaming about making love to him.

> I dreamed of you last night. Such a vivid dream. You remember how we used to carry the papers to bed and how somebody would get sleepy and leave t'other one reading? Well just such, only I remember distinctly imploring you to "love" me so I could go to sleep. Had a terrible "hankering" for you today. I am quite positive you're not to leave me this long again.

Teddy tried to make amends. "I will bear in mind your statement that this is my last long trip without you. I'm looking forward to our meeting when I can tell you how after seeing all these beautiful, dashing, charming, well dressed, painted flappers hereabouts, it will be a welcome relief to embrace a real true blue little woman, beautiful, affectionate, dependable—my wife!!"

Elise was not as diligent as Teddy, but she tried to keep up her studio work when her husband was away. Often this meant taking pictures of family members. One of the most interesting of these shows Parson Jenkins's daughter Mildred. (Having also sat for printmaker Elizabeth Verner, Mildred seems to have been a busy model.) In her photograph, Mildred wears a cotton shift and a scowl, and her hair is an unkempt, ratty nest. Though in reality Mildred was a soprano vocalist with a cultivated manner, Elise had her pose as an uneducated waif indifferent about her appearance.

Elise took herself seriously only half the time. Many of her photographs are casual, little more than family snapshots, and sometimes she liked visual jokes. When she and Teddy opened their art business, Elise took a photograph of Gussie as a toddler standing behind the old Brownie camera and wrote on it in pen, "Greetings From the Harleston Studio." Because of her casual approach, Elise's photography career, compared with Teddy's painting commissions, was weak. Photography was something she did to please her husband, and she was disorganized about keeping her negatives and prints. Though Teddy would never say so, Elise was turning out to be a dilettante.

In August 1924, Teddy finished the course at the Art Institute of Chicago and headed home. When he arrived, he found a letter even more surprising than the one about Pierre Du Pont.

The fall of 1924 was a presidential campaign season, and the Republican incumbent, Calvin Coolidge, was running in the election. Calvin Coolidge was the former vice president who had succeeded to the presidency in 1923, when Warren Harding died in office, but he had not yet won the top job on his own account. In the candidate's home state of Massachusetts, a black political guild called the Coolidge Republican Club had hatched an idea to help Coolidge along. Black people (and for that matter, most whites) thought Calvin Coolidge, "Silent Cal," was close-lipped and uncharismatic. In an attempt to get more Northern blacks to vote for Coolidge, the chairman of the Coolidge Republican Club, Benjamin Robinson, suggested to the Republican National Committee (RNC) that Coolidge hire Teddy Harleston to paint his portrait. (The suggestion may have come from the NAACP, or perhaps Teddy's notoriety was beginning to spread.) Robinson wrote to William Butler, head of the RNC: "We are in a position to have for the first time in our history the services of an artist worthy and capable of doing justice to so distinguished a sitter as our President: Mr. E. A. Harleston, trained in Boston, and widely known as the foremost portrait painter his race has yet produced in America." On September 24, the Republican chairman sent his reply: "I am advised that the President is favorable to this suggestion."

Teddy got wind of this in the mail, and he was stunned. The commission to paint Pierre Du Pont was set, and now he was on the short list to paint the president, the most powerful white man in the land.

While he waited for the next word from the Coolidge campaign, Teddy decided to do a little advance research. He had not painted white people since his days in art school, fifteen years earlier. For a quick primer, he took a train to Washington, D.C., to look at the paintings of his former teacher Edmund Tarbell. If anyone was an expert at depicting white skin, it was Tarbell, one of the most successful society painters of his day. Arriving in Washington after an overnight ride, Teddy went to the National Gallery, which had two Tarbell paintings he wanted to study. One was his portrait of President Woodrow Wilson and the other Tarbell's painting of Herbert Hoover, secretary of commerce under President Coolidge. A little delirious from all the travel, Teddy wrote Elise to explain:

> I am sending this greeting from the National Gallery where I am now sitting. I'm sleepy till I'm drunk but if I fall asleep here I know I'll be arrested or shot or something like that for not attending to duty, my duty being to see how Mr. Tarbell manages little things in the rendering of white folk that I might refresh my memory in the same thing when I begin the President's portrait.

Still awaiting word on Coolidge, Teddy turned to his next job, a painting of the former president of Atlanta University, named Edmund T. Ware. Though he lived among African Americans and had run a school for black students, Edmund Ware was white as a Southern politician. In November, Teddy began painting the elderly man at his home in Montclair, New Jersey. During the work, the two talked about their common thread, Atlanta University, and after a couple of weeks, Teddy finished the painting and made his way home to Charleston, leaving the portrait behind with its subject. The retired college president apparently had time to study the painting, because a month later he wrote Teddy to complain that there was something wrong with the face.

"I miss your daily visits, and our discussion of the old days in Atlanta," said Edmund Ware in a letter. "And by the way, we and the friends who have seen the portrait feel that the light and shadows on the nose make it appear as though there was a bump at the bridge which as a matter of fact is not there. The line from the brow to the tip of the nose is nearly, if not quite, straight in life. The portrait makes it appear more like a Roman nose."

Teddy must have smiled at the appeals of his subject to improve on his face, which he ignored, and the painting was unveiled in Atlanta.

Fresh from this commission, Teddy went to Pennsylvania to start the picture of Pierre Du Pont. Sittings were to take place in the Philadelphia office of E. I. du Pont de Nemours & Company. Pierre Du Pont, a much-painted man, was accustomed to artists. He already owned a painting of himself by John Singer Sargent, the most sought-after portrait artist in the land. But Du Pont did not present an impressive subject. Bald and of small stature, he had a pinched expression permanently etched on his face. Teddy decided to paint him in as direct and traditional a manner as he could. He put a book in the magnate's hand, sat him in his desk chair, and improved a bit on his countenance.

The painting was done in October. Pierre Du Pont eyed the results, and then told Teddy he wanted to have his wife take a look. Mrs. Du Pont took a train in from New York to give her opinion. After studying the canvas, she told Teddy that she felt the picture made her husband look too grim. What's more, she said John Singer Sargent had made the same mistake. Teddy wrote home:

> Well, the portrait is done. It is now drying out waiting to be varnished. Everybody seemed to develop considerable interest toward the end. Mr. DuPont said that he would like to have Mrs. DuPont see it. Of course I was glad and told him so. She came on Wed. and liked it then, except she said that I had made his mouth just as "Mr. Sargent" had done in the one he made of Mr. DuPont—giving him pronounced shadows under the lower lip, and making him appear too severe. I told her that was the way I saw it but that if in softening it I did not weaken it, I would try it; otherwise I'd put it back.

The painting stayed as Teddy made it, and it was unveiled on December 5, 1924, at one of the high schools the Du Pont money had built. "Went down to Delaware," Teddy said, "to DuPont day celebration and wished so for you and others who think well of me."

Pierre Du Pont apparently thought better of the result than his wife; he wrote Teddy a thank-you note for his work:

> My dear Mr. Harleston:
>
> When you were first announced, I had thought you intended using the photograph for copying and that one or two sittings on my part would be sufficient, but I became so interested in your work that it was a pleasure to give the sittings, and the result has made it well worthwhile.
>
> Sincerely yours,
> Pierre S. Du Pont

Elise wrote back to her husband with loving compliments. But despite the acclaim from both family and client, Teddy felt ashamed. He thought he still wasn't doing well enough and told Elise to bear with him.

> It made me feel very fine indeed to have you reiterate in your letter what I have known all along, that you are with me heart and soul in my little ventures in my work—in my life. I must admit it is a poor and barren sort of affair, having no big or great things accomplished to my credit, but at least it is all that I ever promised you and something worthwhile yet may come out of it.

At the end of 1924, the *Chicago Defender,* one of the largest-circulation black-owned newspapers in the country, ran a piece about Teddy and his accomplishments. The *Defender* had subscribers as far south as Alabama and Florida and east to New York. From St. Augustine, Florida, Teddy's sister Kitty Fleming wrote to congratulate him on a year of victories.

> Do you remember how you felt and how dejectedly you wrote me when you decided to take up embalming? You figured this was the end of your cherished hopes! Now, you're surely in the limelight as you longed to be. 'Twas with a deal of pride I read of you in *The*

> *Defender*, and I'm writing to congratulate you and wish you success, Success, *Success*.

Teddy's moment seemed finally to have arrived, but he felt pulled between two worlds. When he showed up for appointments in Northern cities, he would modestly describe his background, only to find people had already read about him and could name his paintings. How strange it must have been to then return to Charleston, where he was still a cultural nonentity. Among blacks at home, Teddy was a businessman and the esteemed founder of the NAACP branch. Among whites, he was merely one of the Harlestons, just another "colored embalmer" from the local undertaking family.

Chapter Sixteen

ELLA Jenkins had conflicting emotions about the children at the Jenkins Orphanage. Although she felt empathy for the wastrel boys and girls, she saw her own status as several levels above the dark, loud, and rough Negroes in her husband's institution. It was true that Ella had married a dark minister who was born a slave, and spent most of her life surrounded by tough urchins, but she had come from a sheltered family. Ella could not have grown up the child of Captain Harleston without absorbing some of his sense of entitlement and command, and at least some of her family believed they were better than other nonwhites (they were certainly better off). The Harleston family legacy gave Ella superior airs and allowed her to condescend to the abandoned children even as she helped care for them.

She enforced a strict household rule at the Jenkins mansion: family members did not fraternize with the waifs. Sylvia, the five-year-old sister of Gussie Harleston, sometimes had to play alone because games with the girls from Jenkins were forbidden. As for her daughter, Olive, Ella kept the orphan boys as far as possible from the eighteen-year-old girl, who was now curious about men. Ella remembered what she had done in her own youth, and believed the adolescent boys to be a sexual menace.

Even the Jenkins band couldn't escape Ella's judgmental gaze. The music program might be an important fund-raising tool, but as for the music itself (especially jazz), she wanted nothing to do with it, her bust of Beethoven being a visible sign of her tastes. The musicians might be cute, and they might keep the Jenkins family comfortable, but Ella knew that grown-up jazz players were debauched. They were often drunkards (despite Prohibition) and drug users (marijuana was the stimulant of choice) with loose morals. (Many in Ella's day thought the saxophone aroused the libido, especially among women.)

Occasionally, the Parson and Ella would hear that a former orphan had gotten work with a jazz band in the North, but strangely, this was not welcome news. On hearing of a boy's "success," Ella would nod knowingly, wondering to herself how long it would take before the young man fell afoul of the law or was consumed by his own corrupt appetites.

The black exodus from the South peaked after World War I. By the early 1920s, between five hundred thousand and one million black Southerners had left the cotton fields of the former slaveholding states and moved to Chicago, Washington, D.C., Philadelphia, and especially New York's Harlem. The neighborhood of Harlem occupied about two square miles of the upper third of Manhattan Island, north of 116th Street on the west side, north of 96th Street on the east. It was a district of town houses and apartment buildings formerly occupied by a white middle class that scurried off as the Negroes arrived. In 1925, Harlem was predominantly black and dominated by newcomers, who were sometimes called "Russians" because they had "rushed up from the Carolinas."

Around this time, the Reverend Mr. Jenkins and Ella were having trouble keeping band members at the orphanage, because whenever a teenage musician escaped and made his way north, stories trickled back about the jazz clubs where he was pulling down a big salary and the speakeasies where he had picked up flagrant women. One of the Jenkins's toughest cases was a trumpet prodigy named Cladys Smith. Ella and the Parson wanted to help Smith into a reliable and subservient life as a shoemaker, but hobnail boots were the last thing on the child's mind.

Cladys Smith was born in Pembroke, Georgia, forty miles west of Savannah, on December 24, 1908. About the same time, the boy's aunt had a daughter named Gladys, giving his mother inspiration for his unusual name. Cladys never saw much of his father, a barber, and his mother cleaned Pullman railroad cars for a living. When he was four years old, his family moved from Pembroke to Savannah. Cladys proved hard to control, and in May 1915, when he was six, his mother put him in the Jenkins Orphanage.

In an interview decades later, Smith recalled that at Jenkins, he became one of the "yard boys." A yard boy was an orphan without a skill. "They try to give you some sort of trade like carpenter or shoe maker or baker," Smith told a jazz magazine. "A yard boy is what you are when you're not doing anything—you're just out there picking up sticks in the yard."

When Cladys came to the Jenkins compound, the Parson welcomed him with the usual intimidations.

"Jenkins was a stately man with a beard, and he looked seventy or eighty to me," Smith told Whitney Balliett, a writer for *The New Yorker*. "There were four hundred kids at Jenkins, and . . . Jenkins was strict. At six o'clock in the morning, there would be a prayer meeting, and after that the roll would be read, and if you did anything bad the day before you'd be called up and tied to a post and whipped with a rope or a piece of rein."

One day, when Cladys was eight years old, an orphanage staffer named Alonzo Mills, a cornet player and former Jenkins inmate, picked Cladys out of the yard and put him in a music class. On his first day out with the band, Cladys had the job of holding up one end of the marching banner. His first instrument was the trombone, but he didn't take to it, nor it to him. A few weeks later, in the practice room, another boy played something on the trumpet that grabbed Cladys's attention. In the next few weeks, he picked up a trumpet the moment any player put one down. When the older kids didn't whack him on the head, Cladys showed he could play better than boys who were band veterans.

> They'd start you off playing, after you learned your fingering, hymns, like "Nearer My God to Thee." Then you'd graduate to marches, and overtures, you name it. On the street corners, it was every kid doing his own thing. All you needed to know was the melody, and then you'd take off from there. There was always a little drum major in the middle doing his antics. We'd play and then pass the hat.

The boys could tolerate money passing under their noses only so long. Eventually, the better musicians looked for ways to abscond so they could play for tips of their own. When he reached his teens, Cladys began running away to earn a few coins from adult bands around Charleston, but the orphanage staff would always track him down and bring him home.

A turning point came when a trumpet player he admired, Gus Aitken, ran away to New York and got music work. The long tentacles of the Reverend Mr. Jenkins, who had connections in black churches straight up the East Coast, located Aitken in Harlem. When he was brought back to South Carolina, Aitken played up the role of captured hero, and Cladys was in awe.

The first time Cladys escaped town was in Jacksonville, Florida, with his friend James Reddick, a trombonist. On the march back to the boat that was to carry the band home, James and Cladys, both fourteen, ducked down a street and disappeared; Jenkins had to leave without them. The boys lived on money they had scooped out of the collection hat and then got work with Eagle Eye Shields, a Jacksonville bandleader. In Charleston, the Parson reported them missing, but three months passed before Florida truant officers caught up with them. Cladys spent three nights in jail. When he was brought to the orphanage, he was stripped, tied to a pole, and whipped with a horse rein.

Cladys Smith hated his own name and wondered how he could lose it; finally, in Philadelphia, he found a way. He was in his teens and playing with the orphanage band in Pennsylvania when he got away for a session with a local act. The bandleader asked if there were any more down in Charleston as good as he was. Cladys said he had a friend back at the orphanage, James "Jabbo" Reddick, the trombone player. When the bandleader laughed at the nickname, Cladys explained that his friend had been tagged Jabbo by the other boys after an Indian character in a western movie. Within minutes, Cladys found himself called Jabbo, which he didn't mind, as long as it meant the end of Cladys.

Jabbo Smith grew up into a thin, medium-size man with an oval face and deep black complexion. He had a quick laugh and nervous demeanor and was always ready for the next prank. Jabbo ran away five or six more times, and on each occasion was dragged back. The Jenkins staff realized that sixteen-year-old Jabbo, while difficult, had become a rare horn player. The idea occurred to Ella and the Reverend Mr. Jenkins to send him to college for deeper study. After Ella made inquiries, Jabbo was dispatched to the music program at South Carolina State College, a historically black school in Orangeburg, ninety miles from Charleston. But soon after he arrived, Jabbo had another pratfall. He told an interviewer:

> When I was sixteen, I got in real trouble. I accidentally shot myself in the leg with a pistol I'd picked up—just fooling around, and not knowing there was still a bullet in the chamber. They fixed me up and sent me back to the orphanage, and the Reverend Jenkins took me up on the veranda and said he'd done all he could for me—that they couldn't keep me any longer, because I was too wild. He gave me nine dollars in an envelope to go to Savannah, where my mother still lived. But I went to Philadelphia. I played with Harry Marsh's band at the Waltz Dream Ballroom for three months. Then I went to Atlantic City and ran into [trumpet player] Gus Aitken, who was with the Drake and Walker Show, and I joined them for a month. I went back to Atlantic City and met Charlie Johnson, and he said he wanted me. It was 1925, and Charlie Johnson had the best band in New York.

Jabbo Smith was seventeen. His signature style concentrated on the high register, with runs and furious trills. He was not a subtle player, but he electrified a bandstand. His first big break was performing with Charlie Johnson, a thirty-four-year-old pianist from Philadelphia, one of the most popular bandleaders in a booming

Harlem music scene. When he auditioned with Johnson, the kid from the orphanage floored the bandleader with his feverish runs. Jabbo knew Johnson's group was handsomely paid, so he told the bandleader that he had to have the exorbitant salary of a hundred dollars a week. The novelty of an adolescent who could front a barrelhouse jazz band got him the money.

In October 1925, Charlie Johnson began headlining at Smalls' Paradise, a new Harlem nightclub on Seventh Avenue at 135th Street, with Jabbo at center stage front. It didn't hurt Jabbo's billing that Ed Smalls, owner of the nightclub, came from South Carolina and liked the young prodigy. The crowd was mixed, as hundreds of Harlem blacks mingled with a few dozen whites who had come uptown on the subway to drink bootleg liquor and watch the Negroes dance. In addition to Johnson's band, Smalls' Paradise featured an all-gay male chorus line that performed under the name "the Internationalists." But the Jabbo Smith act, now dubbed the "Paradise Ten," stole away all the glory.

When Jabbo Smith arrived in Harlem, American culture seemed to be entering a fresh moment, one in which black voices suddenly filled the air. In 1925, a twenty-two-year-old preacher's son named Countee Cullen published *Color,* his first book of poems, stirring the Harlem strivers and piquing the interest of whites. A year later, Langston Hughes, twenty-four, published his own first collection, *The Weary Blues,* with its simple ode to black memory, the poem "The Negro Speaks of Rivers." Previously, black Americans had spread thinly across the rural South; but with the huge new concentration of black families in tiny Harlem, a chain reaction could be set in motion by a couple of books.

In 1925, a sociologist and critic named Alain Locke published a book of essays, *The New Negro,* about the increasing urbanism and sophistication of black America. The "new Negro," according to Alain Locke, seemed to be not unlike Locke himself, a Philadelphian with degrees from Harvard, the University of Berlin, and Oxford, and the first black Rhodes Scholar. In reality, there were few blacks with degrees of any kind, but in New York there was certainly a critical mass of nonwhite creativity, both "high" (poetry) and "low" (jazz).

Like Montmartre in Paris, Harlem was a working-class neighborhood that was becoming a playground for the voyeuristic rich. Whites came to Harlem to "slum" at nightspots like the Cotton Club, where, after 1923, black musicians wailed for an all-white clientele at the corner of 142nd Street and Lenox Avenue. (Sightseers who didn't want to venture uptown could visit the Plantation Club, on Broadway and Fiftieth Street, where singers like Ethel Waters performed amid log cabin decor, and waitresses dressed as black mammies served pancakes.) Harlem was full of "ofays," pig latin code for whites, distilled from the word "foe."

Jabbo Smith was living a fantasy life. He was a poor orphan from the South who had landed on a cosmopolitan stage, surrounded by fans, with a wad of money in his pocket. Musicians of the era recall that the lavish attention gave Jabbo delusions of grandeur. He was handsome and tireless with women, drank a good bit, and had trouble getting to the show on time.

In 1926, Duke Ellington was playing with his six-piece band, the Washingtonians, at the Kentucky Club, a speakeasy on Broadway and Forty-ninth Street. Ellington's regular trumpet player, Bubba Miley, had proven himself to be unreliable, failing to show up for work several nights in a row. Ellington was frustrated and wondering where to turn when he saw Jabbo Smith play in Harlem. Ellington was aware of the Jenkins musicians, because the Washingtonians had recorded a song by orphan Tom Delaney. By this time, Jabbo's reputation as a teenage phenomenon covered black New York. Ellington fired Bubba Miley and asked Jabbo Smith to sit in.

Jabbo topped himself at the Kentucky Club, running up trills into the high register, then playing slow, muting the trumpet to make it fit over Ellington's calm, sultry arrangements. Ellington knew he had a good soloist, and a few weeks later took his band (along with Jabbo) to the studio of OKeh Records. In two or three days, some of the classics of early jazz were captured in recordings. Jabbo Smith and the band made a version of "Black and Tan Fantasy," a signature Ellington song, and several other numbers. On the fast pieces, Jabbo does a couple of wild trumpet solos that are effortless in their paces. But on "Black and Tan Fantasy," with its slow beat and spare arrangement, Jabbo is a different musician. He plays a muted trumpet that wanders and howls over the band, creeping to a finish in a heavy dirge. To jazz aficionados, Jabbo's version of "Black and Tan Fantasy" is one of the purest pieces of early Ellington that has survived.

"The night before I recorded with Duke, somebody stole my horn," Jabbo said later in an interview. "I had to go to a music store and get a replacement, and the mouthpiece was way too big. I had a hell of a time hitting that opening high C in my solo."

After the recordings, Ellington offered Jabbo a permanent place in his band, but the trumpet player turned him down. "He wasn't one of the big boys at the time," Jabbo said. "He wasn't paying enough money, and only offered me $90 a week. I'd look like a fool to jump from a hundred and fifty a week."

The same season Jabbo and Ellington were cutting records, another trumpet player with a name, Louis Armstrong, was getting around Harlem. Louis Armstrong, born in 1901 in New Orleans, had a background remarkably similar to that of the Jenkins orphans. Raised by his single mother in the prostitution district of New Orleans, Armstrong was arrested for theft at age ten, when a judge placed him in the so-called Colored Waifs Home, a reformatory and orphanage for hard-luck kids. The home had a band consisting of black boys dressed in military-style uniforms, and it was with that ensemble that Armstrong began playing cornet. Released from the reformatory at fifteen, Armstrong spent several years playing at parties and saloons before linking up with Joe "King" Oliver's Creole Jazz Band. In 1922, King Oliver invited Armstrong to Chicago to play with his act at Lincoln Gardens, a nightclub on East Thirty-first Street. When the Oliver band made a breakthrough record called "Chimes Blues," Armstrong's improvisatory genius glowed through the static. In September 1924, after two years with Oliver, Armstrong moved from Chicago to New York to join Fletcher Henderson's band, then playing at the Roseland Ballroom in Times

Square. From time to time Armstrong did solo gigs, and it was on one of these that he first met Jabbo Smith.

It was jazz practice (and a way to raise a crowd) for star musicians to duel each other onstage, each bringing his hottest material to the contest, and to let the audience choose the winner with applause. Once or twice in New York, Jabbo and Armstrong had a showdown, or "cutting match." The first took place at an Easter Monday dance at the Rockland Palace. A ballroom at 155th Street and Eighth Avenue with a blue ceiling and crystal chandeliers, the Rockland Palace competed with the gay chorus line at Smalls' Paradise by staging a periodic drag party known as the "Faggots Ball." During the Jabbo-Armstrong showdown, however, the atmosphere was entirely macho. Jabbo appeared with the Paradise Ten; Armstrong fronted a band from another dance hall, Connie's Inn. At twenty-five, Louis Armstrong was the top draw at the event, leaving Jabbo, still a teenager, to play the princely pretender. A couple of thousand fans packed the house, many of them "Russians" from South Carolina, rooting for their favorite son.

The two trumpet players showed off in sharply differing styles. Where Armstrong's solos were lyrical and searching, mournful and expressive, Jabbo's performance was about technique. The Jenkins orphan played crackling bursts of melody and arpeggios, attacking the trumpet. Armstrong floated melodic and beautiful lines; Jabbo countered with his rapid, machine-gun passages. When it was over, opinions differed, but musicians who had attended seemed to think that Louis Armstrong had gotten the better of the kid from Charleston.

Ella Jenkins would have heard about the success of Jabbo Smith, and she possibly shook her head with regret. "The only thing that will save us as colored people," she once wrote in a letter, "is the church." But there was another orphan musician whose personality seemed the antithesis of that of the reckless, fast Jabbo—a drummer named Tommy Benford. Benford worked more than he relaxed, took his music seriously, and had no time for what Ella thought of as the jazz low life.

Tommy and Bill Benford, two brothers who played the drums and tuba, respectively, entered the Jenkins Orphanage about the same time as Jabbo Smith. Thomas P. "Tommy" Benford was born on April 19, 1905, in West Virginia; brother Bill was three years older. The boys never knew their mother, Ann, and their father, a carpenter and musician, died when Tommy was five. Afterward, Tommy and Bill were sent to an aunt in Charleston, who handed them over to Parson Jenkins.

The younger boy, Tommy, started on baritone horn and trombone but soon switched to drums. (For a time he studied under drummer Herbert Wright, the orphan who would murder bandleader Jim Europe.) Bill Benford started, and stayed, on tuba. Tommy Benford was short, with a square face framed by eyeglasses, and a gravelly voice. Like his drumming, his manners were precise and direct. He sat straight as a post behind the drum kit and outshone his brother Bill on the grunting tuba. Tommy Benford later told Whitney Balliett of *The New Yorker:* "When I was sixteen, I ran away from the orphanage. We knew about the money that musicians were making on the outside, and we wanted some. Bill and I ran away, and we got ourselves to Virginia, where we joined a minstrel show called Green River."

The traveling variety act "Green River and His Transcontinental Tributaries" was owned by a promoter named Charles Collier and toured the South with a roster of comedians and musicians.

> Then we went with a doctor show, where the man sold patent medicines. We'd play two or three tunes, the man would talk about his medicines, and we'd play another selection while the people paid up. He had his own bus, and we traveled in that. The orphanage caught up with us in Georgia. We stayed two or three months, and ran away again. I got a job in a hotel in Charleston as a bell-hop, and we joined a circus. We played in the band and stayed six months, until it reached Cincinnati.

Tommy Benford arrived in New York in 1920, when he was fifteen. His older brother had preceded him and was working in bands around town, but it was Tommy who would have the memorable career. Tommy's first job was at the Garden of Joy, an outdoor cabaret situated on top of a rock shelf that occupied a block at Seventh Avenue and 139th Street. At night, Japanese lanterns lighted the dance floor, which was enclosed by canvas sheeting. Tommy Benford played in a band that followed a more famous act, Mamie Smith and her Jazz Hounds. Mamie Smith, a thirty-seven-year-old blues singer from Cincinnati, was the breakthrough vocalist who had recorded "Crazy Blues," which made her a small fortune and first threw open the door for black musicians in the record industry. With Mamie Smith as the headline, the Garden of Joy stayed full.

Soon Benford moved to a taxi-dance club, the Rose Danceland, on Seventh Avenue at 125th Street. Men bought dance tickets from the cashier for a dime and used the chits to pay women employed by the club to fox trot with them. If the song ran too long, the drummer (Benford in this case) hit a wood block on his kit, at which point the men had to stop in midstep and give their hired partners another ticket.

> I shifted over to Marie Lucas at Goldgraben's, at 133rd and Lenox Avenue. She was Will Marion Cook's niece, and she played piano and trombone, and arranged and conducted. We jumped out of New York with her and into Washington, D.C. After a couple of months, we went to the Smile-a-While Café in Asbury Park and then to the Tent in Atlantic City. In the fall we were hired into the Everglades Club, at 48th and Broadway. Ethel Waters was there, and Adelaide Hall was her understudy. Red Fletcher took over the band, and we went on the road to places like Binghamton and ended at the Metropolitan Burlesque, on 14th Street.

Tommy Benford was a workaholic jazzman, for whom no job was too short or pay too little, as long as he could keep playing. In 1925 or 1926, he started at the Rhythm Club, a basement room under the Lafayette Theatre, on Seventh Avenue at 132nd Street. Then he moved to another club, which led to his making the first of uncountable records.

> I jumped to my brother Bill [Benford]'s band at a place on 125th. Jelly Roll Morton started coming in, and he'd ask my brother if he

> could sit in and sing. He took a liking to us and he'd bring his music and we'd run it down right there. Jelly Roll still had that diamond in his front tooth. He bragged a lot, but I liked him. We had heard of him, but we didn't know how good he was until he asked us to record with him for RCA Victor. I guess we made about a dozen records between 1928 and 1930.

Jelly Roll Morton, thirty-eight at the time, was a New Orleans piano legend who alienated nearly everyone he met with his egotism, which included the claim that he had personally invented jazz. But his small Chicago band, Jelly Roll Morton and the Red Hot Peppers, had undeniably recorded some of the finest jazz songs. In the late 1920s, Morton moved from Chicago to New York and put together a second version of the Red Hot Peppers, which included the two Jenkins orphans, Tommy and Bill Benford.

On June 11, 1928, the Benford brothers, twenty-three and twenty-six, went with Jelly Roll Morton to the offices of RCA Victor in New York, where they made six recordings: "Boogaboo," "Georgia Swing," "Kansas City Stomp," "Mournful Serenade," "Shoe Shiner's Drag," and "Shreveport Stomp." At the start of "Georgia Swing," Tommy Benford leans heavily on the high-hat cymbals, and then retreats to cool timekeeping, using the snare on the upbeat. In "Mournful Serenade," a blues at slow-walk tempo, Benford is nearly the only accompaniment to each instrument in turn, starting with Morton's piano, moving to the clarinet, and finally the trombone. Circling between drumrolls, rim shots, and cymbals, Benford's playing is hypnotizing. About a year later, the Benford brothers came back with Morton and made another half-dozen records. After working with Jelly Roll, Tommy Benford resumed his peripatetic career. He moved to Europe for ten years and then came home to the United States, where he played for several more decades in dozens of acts. Benford made hundreds of recordings, and along the way, he became the opposite image of the dissolute jazz musician.

The "Harlem Renaissance" is the term applied retroactively to the jazz movement, literature, and poetry of 1920s New York. In 1926, the Charleston Renaissance met the Harlem Renaissance at the Guild Theatre, just off Broadway at 245 West Fifty-second Street. The occasion was the rehearsal of a new play called *Porgy*. The script for *Porgy* adapted the 1925 novel of the same name by DuBose Heyward, a white, forty-year-old Charleston poet. Several dozen black actors and actresses had been hired to fill roles, and the director of the play insisted that they share the stage with the fourteen-piece Jenkins Orphanage Band #1.

Porgy, the play, differed from *Porgy and Bess,* the George Gershwin opera, which Gershwin would not compose until eight years later. After the commercial success of his book, author DuBose Heyward and his wife, Dorothy Kuhns Heyward, collaborated to write a stage version. They hoped for a New York venue, and sent their work to several companies. The seven-year-old Theatre Guild had gained a reputation with the plays of Eugene O'Neill, but its managers wanted fresh material, and so they bought the Heywards' pitch. The Charleston couple insisted on a black cast, and *Porgy* was to be the Guild's first production in which whites were confined to cameo parts.

Set in Charleston in the early 1900s, the story of *Porgy* follows a disabled black beggar, Porgy, and the woman he loves and loses, Bess. (A porgy is a mullet that runs thick in the summer; Charleston blacks often ate it for supper.) The action unfolds at a black tenement on the city's waterfront, Catfish Row. Porgy loves Bess, the cast-off lover of a violent dockworker named Crown. One night after a craps game, Crown kills another gambler and escapes. In his absence, Porgy woos Bess with his kindness, and the two are happy together for several days until Crown reappears and tries to take Bess back. But Porgy kills Crown with a knife to the back. When the police call Porgy in for questioning, his friends back home at Catfish Row protect him with their silence. Meanwhile, Bess is lured away by another man, a flashy gambler named Sportin' Life, who entices her to New York using cocaine ("happy dust") as a lure. In the end, Porgy is released, sees he has lost Bess, and sets off for New York to find her.

The author of *Porgy* came from an old Charleston family and was well aware of the Jenkins bands. In chapter two of his novel,

Heyward inserts a scene in which an "orphan band" appears on a corner while Porgy looks on:

> Bare, splay feet padded upon the cobbles; heads were thrown back, with lips to instruments that glittered in the sunshine, launching daring and independent excursions into the realm of sound. Yet these improvisations returned to the eternal boom, boom, boom of an underlying rhythm, and met with others in the sudden weaving and ravelling of amazing chords. An ecstasy of wild young bodies beat living into the blasts that shook the windows of the solemn houses. Broad, dusty, blue-black feet shuffled and danced on the many cobbles and the grass between them.

In August 1927, the director of the stage version, Rouben Mamoulian, took a train from New York to Charleston to soak up some of the city's atmosphere in hopes of infusing his show with authenticity. Mamoulian was a twenty-nine-year-old immigrant of Armenian descent, born in Tiflis (now Tbilisi), Georgia—and if anyone needed to get in touch with the American South, Mamoulian did. DuBose Heyward, then in London, asked a friend and fellow writer in Charleston, John Bennett, to show Mamoulian around town and introduce him to sights that had inspired his novel. Sixty-two-year-old Bennett, originally from Ohio, had a fascination for black Southern life. He was a white folklorist steeped in local tradition. In a letter reporting on his doings, Bennett wrote that he took Mamoulian

> [t]o find the odor and grip of our colored democracy in its lairiest lairs. We navigated the most picturesque and disreputable sections of the city, where we found much richness, scenery and personalia, boys gambling under the fig-trees, Negro men loafing in indescribable purlieus, and women in costume. We ended at Jenkins's Orphanage, where we arranged for a private hearing of one of the famous Orphanage Brass Bands. Mamoulian was so charmed with the infinitesimally small darkey boy who "led" the band, and by the melodious discordance of the band itself, that he was inspired by a desire to hire one of the Orphanage's traveling bands for use during the production of PORGY and perhaps for actual introduction upon the stage.

Mamoulian loved the band's act. Three weeks later, Heyward, back in New York from London, told Bennett that he and Mamoulian wanted the orphans on Broadway:

> Mamoulian and I are pulling like the devil to get the Jenkins Orphanage band here in the flesh for the show, and have got the Theatre Guild to authorize an offer for them for the lot a sum of $250 per week, and traveling expenses. The Rev. would make a good thing in cash, and get an enormous publicity among these rich "Yankees" for his old orphanage.

By chance, Jenkins Band #1 had just finished its summer tour and was already in New York, staying in one of the Jenkins town houses at 67 West 131st Street. A few days later, Rouben Mamoulian was directing the band in rehearsals.

Mamoulian was said to have been amazed that the "high yellow" members of his cast were reluctant to fraternize, even onstage, with darker Negroes, but he pushed ahead over this unexpected obstacle. The Jenkins boys' star moment was supposed to come when the band led the cast in a song, "Ain't It Hard to Be a Nigger," which appears in the novel and which the script called for in the play. But another song had to be substituted in rehearsals, because despite Mamoulian's pleading, the black cast in New York flatly refused to sing the lyrics:

> *Ain't it hahd to be a nigger!*
> *Ain't it hahd to be a nigger!*
> *Ain't it hahd to be a nigger!*
> *'Cause yo' can't git yo' rights when yo' do.*
> *I was sleepin' on a pile ob lumbah*
> *Jus' as happy as a man could be*
> *When a w'ite man woke me from my slumbah*
> *An' he say, "Yo' gots fo' work now cause y' free."*

On opening night, October 10, the performance of *Porgy* was segregated. Black New Yorkers sat in the balcony (as a joke, Jim Crow seating was called the "crow's nest") while white people took the orchestra level. In act 1, the Jenkins band did nothing. But midway through act 2, the fourteen boys appeared onstage, leading a procession of Catfish Row tenants to a picnic. (This was the moment where the play had called for the controversial song, and it seems Rouben Mamoulian replaced it with a hymn.) According to John Dowling, a manager of the band in those years, after the picnic scene the boys had nothing to do until a burial scene in the fourth and last act; they then reappeared at the end of the show for the curtain call.

Initial reviews of *Porgy* gave it faint praise, but within a week, strong word of mouth had made it standing room only. The play ran for 217 performances, went on tour, and then returned to Broadway the following year for 137 more performances. In the spring of 1929, *Porgy* was revived and went to London, where it played at the Pavilion Theatre. The Jenkins band went along. The group had played in London in 1895, and then again in 1914, so the *Porgy* trip in 1929 was their third overseas tour. Jazzing the audiences of the West End, the band spread its fame a little more. But to everyone's surprise, the London show closed after only six weeks. It failed to catch on, people said, because English audiences could not follow the Gullah dialect from black Carolina, and the Jenkins band was simply having too much fun.

In 1928, Jabbo Smith, never an easygoing star, had a falling-out with bandleader Charlie Johnson and left the Paradise Ten. He joined the orchestra pit band for *Keep Shufflin'*, a black musical then playing down the street from *Porgy*. For his new band mates, Jabbo had Fats Waller on the organ and James P. Johnson on stride piano. The *Keep Shufflin'* band recorded several songs taken from the show, including "'Sippi" (as in the state and river). On that track, Fats Waller plays a wavering organ in the background, while Jabbo's horn, muted with a metal cup, weaves in and out with a beautiful bell-toned melody. *Keep Shufflin'* began touring, but when the musical reached Chicago the show fell apart. It seems the show's financier, a gangster named Arnold Rothstein, who smuggled liquor from Canada, had run afoul of his enemies and been shot dead in New York. With Rothstein's demise, *Keep Shufflin'* summarily closed in Chicago.

Stranded in Illinois, Jabbo picked up a job as the house trumpeter at the Sunset Café. The Sunset was to Chicago what Smalls' Paradise was to New York. Located in a former automobile garage—315 East Thirty-fifth Street at South Calumet Avenue—the Sunset Café had been remodeled in 1921 to resemble a modernist building by the likes of Le Corbusier, with a completely clean white facade. Under the hand of Al Capone's gangster syndicate, the speakeasy became one of Chicago's early jazz venues.

As it happened, Louis Armstrong was a frequent headliner at the Sunset, and after Jabbo took the house job, he and Armstrong once again faced off. More showdowns were arranged. Sentiment among musicians was that this time, Jabbo got the better of Armstrong. In 1929, a music producer at the Brunswick Company, one of the large recording firms of the day, noticed the rivalry between Jabbo and Armstrong. Brunswick wanted someone to challenge Armstrong, who was then recording for Brunswick's competitor OKeh Records. The Brunswick producer asked Jabbo to headline, and in January 1929, Jabbo put together his band for the session, the Rhythm Aces.

Jabbo Smith's Rhythm Aces recorded nineteen songs. Their range is virtuosic, and Jabbo, at twenty, shows himself to be a swaggering prodigy. On "Let's Get Together," Jabbo sings taunts to a woman he's trying to seduce, then steps away from the microphone to show off his horn tricks. On "Rhythm in Spain," a fast instrumental with Latin coloring, the full orchestration makes the Rhythm Aces sound more like a big band from ten years later than a hot jazz ensemble. "Take Me to the River" is a showcase piece with scat trumpet and Jabbo's own scat singing. And "Rub Me Some More" is suggestive and funny, with a trumpet solo tossed off like a joke, and Jabbo singing the refrain: "Baby, rub me some more / Rub me till I'm on the floor."

The songs were released two at a time, on flip-sided seventy-eight-speed discs. But the records didn't sell, and Brunswick dropped Jabbo Smith after six months. The boy from Pembroke, Georgia, was just twenty years old, and he had reached the peak of his career.

Chapter Seventeen

THE little girl named Doris Forrest moved in with Teddy, Elise, and Gussie Harleston in 1924. She had come from Elise's brother Tom, who could no longer take care of her. Not surprisingly, eight-year-old Gussie was unhappy that she would have to share the attention with her three-year-old cousin and began a campaign of intimidation.

"Doris had fair skin, and her hair was real fine, and when she came to live with us, I was jealous of her. I used to grab Doris and straddle her with my legs on the floor. I held her down on the rug and pinned her arms, then I would let a string of drool out of my mouth and drip it down to her face before slurping it back up on my lips. I was good and mean. Uncle Teddy made a pencil drawing of me from that time that shows a distinct bald spot on my head. I had given myself a haircut with a pair of scissors, because I wanted hair like Doris's, and I sort of messed up and cut away too much hair. Uncle Teddy thought that was funny, so he showed the bald spot in his portrait.

"Doris could have passed for white. There was a church for the real light people, St. Mark's Episcopal, not far from the funeral home. It was for the old free families of color, many of whom were just about white. They had this great big building with columns, and beautiful stained glass windows, and it looked kind of like a temple. The legend was that they had a vestryman standing by the entrance, and they had one of the pillars painted a certain color yellow. If you were too dark, darker than that yellow, they wouldn't let you in. I guess that's what they meant by high yellow. If you made it through that, you passed the color thing, then they had a comb that was fixated on a string nailed to the wall. That was the last barrier. If you tried to comb your hair and the comb wouldn't go all the way through your hair, then you were told to think about going to another Episcopal church, the one for the darker people.

"A lot of families who could do it started passing. People would disappear overnight from Charleston, just vanish from the city. Their names would drop from the phone book. When they took the census in 1910, and then again ten years later, just gobs of people were unaccounted for. They were tired of being Negroes and all it meant, and they didn't want anybody to trace them, so a lot of them went to Brooklyn and became white people. They wouldn't tell their friends they were going to New York to start a new life, and the population of Charleston slowly shrank.

"Some people called it 'dyeing up.' I had a friend who lived on the same block as we did, John MacBeth. He was real smart and real light-skinned. John was my classmate, and we were kind of competitors in school, so much so that we used to do our homework together and compete for grades. There were two branches of the MacBeth family. They used to call them the 'white MacBeths' and the 'black MacBeths,' although both branches were Negroes. John was from what we called the 'white MacBeths.' John's father owned a dry cleaning place on King Street, and one year John's family just vanished. They had packed up overnight and gone to New York, and were never heard from after that. All of them went but one, John's younger brother Bob, who was light enough to pass but didn't, and who stayed in Charleston.

"The rules of passing were that if you knew somebody was colored, you didn't give them away. When you saw them in white company, they would signal to you with their eyes and you would know and respect their charade. I knew other families who did it, like a family called the Dawsons, and the family of Dr. Harry Boston, who left, too. There was a real conspicuous case, a man named Thomas

Grant, who owned several groceries and butcher shops. He was a member of the Republican Party and even attended one of the Republican conventions as a delegate. Then, poof, he crossed over to white, and must have just locked up and left his stores."

While tens of thousands of African Americans began to live as whites, even more colored people began openly celebrating their black identity. In the mid-1920s in New York, the "new Negro" movement was in full flower, as artists and writers of the Harlem Renaissance explored what it meant to be a nonwhite American. While Teddy wanted his canvases to be indistinguishable from "white" art, the new tendency in Harlem was to stuff the artwork—from music, to painting, to literature—up to the brim with black filling.

Most of the unusual women and men making "black art" in the 1920s came from a new generation. The signature painter of the Harlem Renaissance, a Midwesterner named Aaron Douglas, was seventeen years younger than Teddy. Born in Topeka, Kansas, in 1899, Douglas studied at the University of Nebraska and at Columbia University Teachers College before turning to painting. After a stint at the Académie Scandinave in Paris, he returned to the Midwest to teach in Kansas City. Douglas moved to New York in 1924 at the urging of Charles S. Johnson, director of the National Urban League and editor of *Opportunity* magazine. Although he had trained in painterly realism, Douglas now studied with Fritz Winold Reiss, a German folk painter whose influence helped to bend the twenty-five-year-old's style back to its African American roots. Douglas illustrated the cover of *Opportunity* in 1925 and supplied drawings for Alain Locke's essay collection, *The New Negro*. When he illustrated James Weldon Johnson's book of poetry, *God's Trombones: Seven Negro Sermons in Verse*, in 1927, his reputation as a "black modernist" was made. Commissions for murals and portraits flowed to Aaron Douglas and away from the likes of Teddy.

Harlem in the 1920s was the birthplace of hip. Liquor became illegal at the start of the decade, which meant unregulated cocktails of all strengths took their place alongside marijuana and sundry stimulants. (A "speakeasy" was a saloon where patrons kept their voices low so as not to draw attention, but with a band blaring, conversation hardly mattered.) Between the jazz clubs and speakeasies—with names like Hayne's Oriental, Lybia, the Catagonia Club, Happy Rhone's, Basement Brownie's, and the Bucket of Blood—the older Negro sophisticates began to look wooden.

Teddy paid attention to the new mood but couldn't quite get it. When he was in New York, which seemed to be every few months, Teddy stayed with his friend Jim Cartier, whose apartment stood in the thick of Harlem at 159 West 145th Street. Cartier was a radio enthusiast who tried to keep Teddy in sync with the times. "We are searching the heavens with the radio," Teddy wrote home. "Just heard Chicago. Lots of N.Y. jazz on now." But Teddy showed how much he was out of it when, on a trip to the North, he found he couldn't do the country's most popular dance. He wrote Elise after a party, "It was queer that I, of all the people, couldn't do the 'Charleston' when the orchestra stepped on the gas!"

Signs of black consciousness were everywhere. There were a dozen national newspapers and magazines for black readers, including the *Amsterdam News*, the *New York Age*, *The Crisis*, *Negro World*, and *Opportunity*. Blackness could be embarrassing to a light-skinned undertaker from South Carolina. In 1916, a former civil service worker from Jamaica named Marcus Moziah Garvey arrived in New York, where, from a street-corner lectern, he single-handedly launched a back-to-Africa movement. Marcus Garvey's organization, the Universal Negro Improvement Association (UNIA), put out a weekly, *Negro World*, to help agitate its doctrine of an African renaissance and the spread of a new kind of black power. Attracting tens of thousands of followers, many of them working-class migrants from the South, the UNIA peaked in the early 1920s with a series of mass rallies that were the first loud public declarations of black pride. Teddy's friend W. E. B. Du Bois looked down his nose at Garvey, and Teddy probably did the same.

While hip nightlife helped to loosen the masses, a new verbal culture simultaneously spread through the ranks of the high yellow class. In 1922, Jean Toomer, a handsome and effete young man from an old Creole family, published *Cane*, a strange and beautiful prose poem that paired images of the new Negro of the North with the "old Negro" of the Deep South. Meanwhile, the boyish Langston Hughes appeared on the brand-new literary scene and assumed a position of leadership. Hughes moved to New York from Paris in November 1924 and published *The Weary Blues* two years later. His poems replaced the stilted phrasing of the black bourgeoisie with the plain speech of ordinary Negroes. Zora Neale Hurston, a thirty-something folklorist originally from northern Florida, did something of the same by gathering and publishing the peasant lore of Southern blacks. Hurston wrote a series of beguiling books, including *Mules and Men* (1935) and *Their Eyes Were Watching God* (1937), which applauded the reality that black Americans were one generation removed from slavery and still sounded like it. Showing a talent for neologism, Hurston coined a word for the boring, respectable people who worked for black uplift: "Negrotarians." She claimed to come from new and superior literati, which she called the "Niggerati." Such a group even had a kind of clubhouse. When Aaron Douglas moved into a rooming house at 267 West 136th Street, where Langston Hughes and Hurston herself already lived, Hurston dubbed the building "Niggerati Manor."

There were other Harleston family members in New York, but they seemed to be having as much trouble as Teddy fitting in. "I saw Alethia yesterday on 7th Avenue," Teddy said, speaking of one of Captain Harleston's daughters by his mistress, Mamie Randall. "Tells me her mother has not been well since being here, that they don't like N.Y. and expect to be home the last of this month."

The New York new Negroes took the old sore points of black life and made jokes from them. Teddy, as an undertaker, was spoofed as an "underground specialist." The texture of hair acquired all sorts of prank names. Slang for kinky hair included "naps," "tight head," "mailman hair" (every knot was said to have its own "route"). A "jar-head" was a person with straightened hair (because chemical relaxers came in a jar). A "kitchen head" was a person who had straightened his front locks but had left the back of his head nappy.

Sexual life was getting looser, and it became easier for women of all classes to have sex. New kinds of brothels opened, and "buffet flats" were the most libertine of them. These apartments had different sex acts (hetero- and homo-) on offer in each room: fellatio here, anal sex there, and straight coitus down the hall. The best-known buffet flat in Harlem was the Daisy Chain, run by a bon vivant named Hazel Valentine.

At least some of all this was documented in the extraordinary pictures of James Van Der Zee, a photographer with a studio at 109 West 135th Street. Van Der Zee captured the highs and lows of black life after the exodus from the Deep South, including the garish wealth of a tiny few. In 1928, A'Lelia Walker, daughter of the hair-straightening magnate Madam C. J. Walker, opened a salon called the Dark Tower in her twin town houses at 108–110 West 136th Street. Walker stenciled Langston Hughes's *The Weary Blues* on one wall and proceeded to play hostess to a stream of flush entertainers, well-dressed strivers, and their hangers-on. Teddy was probably never seen there, but then again, neither were most black New Yorkers. Despite a sprinkle of new money and much posturing, the majority of people remained destitute. Harlem became famous for end-of-the-month "rent parties," where hosts charged admission as a way of making the rent. Food was served—okra gumbo, red rice, sweet potato pone—so that people felt they were getting something for their money, and the dancing and seductions went on until sunrise. One year, a woman named Cora Jones, of 187 West 148th Street, distributed a printed invitation that must have brought at least a few strangers to her rent party:

> Let your papa drink the whiskey
> Let your mama drink the wine
> But you come to Cora's and do the Georgia grind.

With its swagger and cool, the Harlem Renaissance was passing Teddy by. He was aging, and his work also seemed stylistically out of step. Teddy had matured before modernism had penetrated the art schools; after it did, he stayed loyal to academic painting and found it hard to change. The avant-garde was a powerful new force in the circles of "white" art, with apostles like Alfred Stieglitz, the photographer and art dealer who imported abstract experiments to the American scene. White painters such as Marsden Hartley, Georgia O'Keeffe, and Arthur Dove, and photographers like Man

Ray, adapted European formal innovation to their work and acquired followers of their own.

Similar energies filtered into the colored world. To some black leaders, modern art was doubly suspicious, first as a "white" intrusion into black life and second as wasteful leisure. Art made little difference when race towered over everything. Despite his friendship with Teddy, W. E. B. Du Bois actually had small patience for art, which he thought distracted from the great struggle. Speaking to an NAACP conference in late 1926, Du Bois said, "How is it that an organization like this, a group of radicals trying to bring new things into the world, a fighting organization which has come out of the blood and dust of battle, struggling for the right of black men to be ordinary human beings—how is it that an organization of this kind can turn aside to talk about art? After all, what have we who are slaves and black to do with art?"

Teddy made a weak effort to stay current. In the fall of 1925, he came to New York to try to drum up some portrait work and looked around at the new art. "I am doing the exhibitions this week to get some ideas of what the painters from all parts are doing," he told Elise. But mainly he stayed with the old beauties: the application of paint, a source of light, and resemblance of the image to the sitter.

Back at home in Charleston, Teddy demonstrated his old skill in a painting of his half sister Alethia, or "Leetie." The Captain's mistress had sent Leetie to Teddy's studio when the girl was still in her teens. In the painting, Leetie is fancifully dressed, looking more like a character in an Edith Wharton novel than a bastard child, and her eyes reflect Teddy's special talent. Leetie's face is bright and ethereal, but she looks at the viewer with a distant gaze, her eyes bereft, as though she had been thrown into the world.

Once when Teddy was away from home, eight-year-old Gussie Harleston received a visit from a prince of the Harlem Niggerati.

"Our world was so small, the colored world, that artists of any accomplishment knew about each other even at great distances. I remember the visit of Langston Hughes. Teddy was out of town, Tantie was out of the house, and I was the only one there. This was after Hughes's poetry had first been published, but before his books

Union Army Veteran, *Elise F. Harleston, photograph, ca. 1921. The plan for the Harleston Studio was that Elise would photograph portrait subjects and Teddy would paint them. Managing the funeral business, Teddy didn't have weeks of free time for sittings. So a client would sit for initial drawings, and then Elise would make a photograph of the pose, from which Teddy could finish a portrait. One early collaboration involved an elderly black man named Smart Chisholm, a Union veteran of the Civil War. In Elise's photograph (the only surviving print is torn), Chisholm wears a hat with the insignia "G.A.R.," for Grand Army of the Republic, the name of the former Union forces.*

Union Army Veteran, *Edwin A. Harleston, French crayon, ca. 1921.*

Tanti, *Edwin A. Harleston, charcoal, 1922. A portrait of Teddy's wife, Elise, so tactile that one can almost smell their bedroom. Elise was nicknamed "Tantie" by Gussie, who couldn't pronounce "auntie" and opted instead for a diminutive of the French for aunt,* tante. *Teddy scrawled the name on the picture and dropped the "e" from the spelling.*

Sue Bailey, *Elise F. Harleston, photograph, 1930. When a young New Yorker named Sue Bailey, then an administrator for the YWCA, came to Charleston on business, Teddy asked if he could paint her portrait. She first sat for Elise, who draped her with a shawl.*

Portrait of the Woman with the African Shawl, *Edwin A. Harleston, oil on canvas, 1930. This painting was one of the last Teddy Harleston made in the year before he died.*

Rev. Caesar S. Ledbetter, *Edwin A. Harleston, oil on canvas. One art historian has pointed out that Teddy's portrait of Reverend Caesar Ledbetter, pastor of Plymouth Congregational Church in Charleston, resembles the young nobleman in Il Bronzino's* Portrait of a Young Man, *made in Florence between 1535 and 1540. In the Il Bronzino painting, a princely youth wearing a smug expression holds his hand on his hip and props his other on a book, precisely like Teddy's commanding, seemingly egotistical minister. The Renaissance painting had been shown in April 1915 at the M. Knoedler and Company gallery in New York and was later hung in the Metropolitan Museum of Art, which Teddy visited on frequent trips to the North.*

Pierre S. Du Pont, *Edwin A. Harleston, oil on canvas, 1924. When Teddy painted the rich industrialist Pierre S. Du Pont, the sitter was a household name. At fifty-four, Du Pont was chairman of E. I. du Pont de Nemours & Company, an explosives maker that cornered the munitions market during World War I, and president of General Motors, of which he owned a one-third stake. Du Pont had peeled off $3 million to build eighty-three "colored" public schools in Delaware, an act for which a teachers' group thanked him by commissioning this painting.*

The Nurse, *Edwin A. Harleston, oil on canvas. Teddy Harleston made many genre paintings depicting nameless people in common acts, and the best of them were like portraits. The attentive, sad, startled look of the model makes* The Nurse *a personality study, while the tray is a look at materials—the glint on the teapot, the texture of an orange, the drape of the napkin.*

Teddy Harleston was in the last year of his life when his wife photographed him in the studio they shared in 1930. The long curtain was a backdrop for Elise's photography and the wooden riser on the floor a place for subjects to sit up high and pose. Teddy probably didn't paint while wearing a necktie, and the picture on the easel has since been lost.

Teddy Harleston died in 1931, at age forty-nine. The day after the funeral Elise took a camera to the Unity and Friendship Cemetery to photograph the flowers spilling over the sandy hump of her husband's grave (center). *Edwin "Captain" Harleston, Teddy's father, had died three weeks earlier, and his fresh grave can be seen to the left. Teddy's grave remained unmarked for sixty years, until 2001, when his niece Edwina Whitlock placed a stone in the Harleston family plot.*

Reverend Daniel Jenkins, founder of the Jenkins Orphanage and husband of Teddy's sister Ella Harleston, around 1930. Still handsome at nearly seventy years old, Jenkins wears the contented face of a patriarch, a role he filled maximally during almost fifty years as a surrogate father to thousands of parentless black children.

Ella Harleston Jenkins helped run the Jenkins Orphanage with her husband, beginning in her mid-twenties and continuing into her own old age.

The Jenkins Orphanage, at 20 Franklin Street in Charleston, about 1918. The orphanage attracted a procession of vivid characters, including "holy rollers" like traveling evangelist "Black Billy Sunday," whose appearances Reverend and Ella Jenkins advertised on billboards. Parked on the street is the bus that carried the Jenkins Orphanage Band up the East Coast on annual fund-raising tours.

SAVE YOUR MONEY AND GET READY!

A Great Time is Coming.

THE FIRST ANNUAL PRIZE FAIR,

—OF THE—

COLORED ORPHANAGE,

WILL BE GIVEN IN THEIR BUILDING,

20 FRANKLIN STREET, CHARLSTON, S. C.

AND WILL BEGIN

The first Monday Night in Feb., 1894.

And will continue Each Night until all Prizes are taken up.

All Schools in Charleston are invited to perform during the fair. There will be five Prizes given to the five best displays among Schools. Speaking and Singing, also Essays, (not exceeding over sixty words.) Address all subjects to the Colored Orphanage. We ask all friends to the Colored Orphanage of Charleston, far and near, to send us something, with their names attached to the article, to be exhibited in the Fair. There will be great attention paid to such. The President is determined to make it the Grandest Exhibition of the kind ever known in the history of Charleston, provided the friends of the poor and helpless will lend us a helping hand.

N. B.—229, B. B., the Phillip Division, will be quite a feature in the Fair. The boys will parade and play their Brass Band, which contains fourteen pieces. There will be a Prize Drill, Art Neadle Work, Painting, Pottery, Wood Articles and Comforts, and something in every line in Works of Art.

The boys and girls will perform Fifty Different Pieces during the Fair. Speaches by divines and other able men. We are determined to make it interesting. Selections which cannot fail to satisfy the intellectual tasts of our patrens. Purchase a Pamphlet at the door, at Ten Cents a Copy, containing a description of all to be shown, and which will be on sale February, 1894, at the Orphanage, 20 Franklin Street. Rev. D. J. Jenkins, President.

Only Tickets Purchased Before February 1st, 1894, is Good for a Prize.

All persons desiring to take part in the Fair will please notify the President, ten days before the Fair.

An early fund-raising poster for the Jenkins Orphanage announces (in 1894) that "the boys will parade and play their Brass Band, which contains fourteen pieces."

In 1914, a New York talent agent arranged to send the Jenkins Orphanage Band (with Reverend and Ella Jenkins) to a world's fair in London, where they were billed as the "Famous Piccaninny Band," a misspelling of the racist nickname for black children in America, "pickaninnies." Though the agent asked for the youngest children, the Jenkins couple brought the oldest instead and put a baton in the hand of an impish "conductor," who led the group for laughs.

The fame brought by their international travel made the Jenkins Orphanage Band a concert draw, and the few instrumentalists multiplied into bands #1, #2, and #3 (ranked by talent and fund-raising skill). Each was outfitted in expensive uniforms and publicized with mailings, as in this postcard from around 1920.

Beginning in the 1890s and continuing for nearly fifty years, the Jenkins bands were a familiar sight on the streets of Charleston. (The picture dates from about 1925.) The bands traipsed through town accompanied by an adult, usually their music instructor (far right, *playing the baritone), who paused periodically to ask for money from onlookers.*

It's possible that the Jenkins bands originated "the Charleston," the most popular dance of the 1920s Jazz Age. The orphans were famed for tossing their bodies in jumpy steps to get a laugh, so much so that white women (in front of the orphanage about 1925) made their way to the Charleston children's home to learn the signature dance of Prohibition.

Only a few of the hundreds of children in residence at the Jenkins Orphanage played music or sang. Most were down-and-out boys who might look forward to a career as a newsboy or housepainter.

Although boys staffed the brass bands, Reverend and Ella Jenkins organized girls from the orphanage into choirs, four of whose members salute the camera about 1935.

Some children in the Jenkins bands went on to big careers in jazz, including William Alonzo "Cat" Anderson (seated in front, *with trombone), who was possibly the best high-note trumpet player of his generation. He and several other Jenkins veterans formed a touring and recording act called the Carolina Cotton Pickers. Cat Anderson gained lasting renown after 1944, when he joined the Duke Ellington Orchestra. His wild solo playing, which rarely dropped below the upper octave of the trumpet, included notes so high other musicians said "only a dog can hear them."*

Rhythm guitarist Frederick William "Freddie" Green (right, *with guitar) was born in 1911, started on banjo at the Jenkins Orphanage, and left Charleston for New York as a teenager; he eventually landed a job with the Count Basie Orchestra, with whom he played for fifty years. When Billie Holiday (at the microphone around 1940) began recording with Count Basie, she and Freddie Green started a love affair. Years passed, Billie Holiday married someone else, but the affair continued. Eventually, Holiday was arrested for heroin use, neatly halting her career and ending her relationship with Green.*

Trumpet player and former Jenkins orphan Jabbo Smith (right) *at a music festival in New York City, with trumpet player Wynton Marsalis, about 1981. Jabbo Smith, born in 1908, was a prodigy who left the Jenkins Orphanage Band at age seventeen to join the house band at Smalls' Paradise, one of the flashiest nightclubs in Harlem. In the late 1920s, Jabbo briefly became a rival of Louis Armstrong.*

Marie Forrest (right), *the sister of Elise Forrest, Teddy Harleston's wife, married Robert Harleston, Teddy's younger brother, in 1915. (Two brothers married two sisters.) Marie gave birth to Gussie Harleston* (left), *but in 1918, when her husband returned from wartime training camp with a case of tuberculosis, she contracted "the white plague" and shriveled to ninety-one pounds. Here she sits in a tuberculosis sanitarium in Asheville, North Carolina.*

Gussie Harleston (seventh from the right) *and her class at Dart's Kindergarten, a school for the children of Charleston's black middle class, around 1920, the year her mother died. Gussie is above and slightly to the left of the fishbowl, wearing a big bow.*

Gussie Harleston, born 1916, photographed by her aunt and adoptive mother Elise F. Harleston, about 1924.

When her adoptive father, Edwin A. "Teddy" Harleston, died in 1931, Gussie changed her name to honor him, becoming Edwina Harleston. As a high school senior, Edwina was president of her class and had the lead role in the school play, Smilin' Through, *a Broadway hit about a young British woman who falls in love with an American soldier.*

Graduating from Talladega College in Alabama in 1939, Edwina went on to graduate school in journalism at Northwestern University in Evanston, during the 1940s. She paused to have this picture taken, and the photographer applied Vaseline to the lens, a glamour technique of the period.

Edwina Harleston, Chicago, 1940s.

In 1945, Edwina married Henry O. Whitlock of Gary, Indiana, publisher of a weekly newspaper for black readers, the Gary American. *She moved to her husband's hometown an hour from Chicago and became a young homemaker (perched here in her reading chair), as well as a columnist for the* American, *writing weekly dispatches on black social life in Chicago and its suburbs.*

The Christmas card sent in 1949 by Edwina (right) *and her husband, Henry Whitlock, after the birth of their son Henry, the first of four children.*

Edwina H. Whitlock (right) *talks with friend Margaret Bush Wilson* (center) *and an unidentified woman at Fox Lake, Indiana, about 1958.*

Edwina H. Whitlock in Hawaii in 1999, at age eighty-three, at the wedding of one of her sons.

made him famous. Langston at that time made his living as a sailor in the merchant marine. He had come in on some boat on the way to the Caribbean and stopped to see Teddy, whom he knew about because of Teddy's art. Anyhow, in the entrance to our apartment there was a sofa settee, and behind that, on a table—the first thing you saw—there was a Victrola. The Victrola was a good-size box with a great big horn coming out of the top. Teddy loved classical music, and he knew it, and he had a collection of records with the music of different composers. Langston Hughes was wearing a sailor-boy outfit, the kind with a cape collar and bell-bottomed trousers. He was a sizable man with a nice manner, and he came in unannounced. I remember he sat in a cane chair that was in the front room and he broke right through it because of his weight. After that, I thought I had to entertain him, so I turned on the Victrola. I remember putting on a classical record, a famous opera solo, the one my friends and I sang that sounded like 'Oh, my, my, mommy-oh.'"

(The piece was the well-known aria for tenor, "La donna è mobile," from act 3 of Giuseppe Verdi's 1851 *Rigoletto,* an opera about a hunchbacked jester at the court of Mantua who suffers a curse when his daughter is seduced by a duke.)

"I stepped back and looked up at him, and waited for some approving remark, but it didn't come. Hughes was a modern artist, and he evidently didn't approve of European music. He probably liked jazz. He listened for a minute to the aria, and then turned down his mouth and said, 'Do you have anything else?'"

Other modern artists traipsed to Teddy's door looking for comradeship. Another visitor to the Harleston Studio was a white novelist, Julia Peterkin. In the 1920s, Peterkin, the wife of a cotton plantation owner, became an unlikely new voice of Afro-America with novels about black cotton hands fighting a losing battle with their chaotic lives. Her books, including *Green Thursday* and *Black April,* were, like *Porgy,* written partly in Gullah dialect and brought the writer national acclaim. When Peterkin's third novel, *Scarlet Sister Mary,* won a Pulitzer Prize, the attention emboldened her to cross the color line and pay a visit to a pair of black artists, Teddy and Elise, whom she had never met.

"We have had interesting visitors, famous visitors, who were sorry to have missed you," Elise wrote Teddy, after Peterkin showed up unexpectedly. "This morning I saw by the *News and Courier* that this very famous person was in town and at 4:30 the Author came up, Julia Peterkin, made herself at home and fell in love with your home and wife, and asked to come again."

Within two or three years of his most public commissions, Teddy's successes stopped coming. Before it could be finalized, the portrait of President Coolidge fell through. William M. Butler, chairman of the Republican National Committee, wrote to say that although the painting was a good idea, Coolidge had no time to sit for it. Next, the fee for the Du Pont picture failed to materialize. Teddy wrote Elise nervously when he realized the money might not be coming: "I am still on the edge over payment for the job as a reorganization of the interested persons caused a delay and postponement of collections." Teddy never wrote to Pierre Du Pont about his payment troubles, because the portrait wasn't the rich man's commission. According to family tradition, the head of the charitable group that had hired Teddy embezzled the funds. He wrote repeatedly to the picture's sponsors, but after several years received only a portion of his fee.

The job in Washington, D.C., also didn't pan out, even though, as he had been asked, Teddy had studied landscape painting and sent his work to the hiring authorities. He had gone back to the Art Institute of Chicago for a second summer, in 1925. But prospects eventually faded, and no formal job offer appeared, which meant that Teddy's plan to escape from Charleston withered away.

Through his various setbacks, Teddy tried to keep a sense of humor. "Well cheer up, my love," he wrote Elise, "we aren't dead yet nor so terribly old, and the dynasty may yet be kept in succession."

He was helped along by a few successes as well. W. E. B. Du Bois sent Teddy a telegram with the news that he had won the NAACP's Spingarn Prize for his drawing *A Colored Grand Army Man*. The NAACP had established the Spingarn Prize in 1914, naming it after the group's national chairman, Joel E. Spingarn, one of the founders of the publishing firm Harcourt, Brace and

Company. The prize came with a sizable check, which alleviated the humiliation of the Du Pont episode. After the prize, Teddy felt cocky and wrote his wife: "Well, well! It does come a little, eh my partner? A little at a time. Remember, it is not mine, it is ours."

But as the decade wore on, Teddy faced a series of disappointments. Sometimes the blows came from other artists, who leaped ahead while Teddy stayed behind. One rival in these years was a black painter named William Edouard Scott. Three years younger than Teddy, William Scott had studied at the Art Institute of Chicago and in Paris, and had had several shows in France before returning to America to make his career—and undermine several of Teddy's projects. In 1925, Teddy proposed to the black-owned insurance company Supreme Liberty Life Insurance the idea of commissioning a mural for its office building. T. K. Gibson, Teddy's old friend from Atlanta University, was a high-ranking executive in the firm and in a position to help. Supreme Liberty took Teddy's suggestion and advertised a competition, but then hired William Scott to paint the mural. The same year, Teddy suggested to a black-owned cosmetics company that a portrait be offered as a prize in a national contest. The company followed his advice, but the portrait commission went to Scott.

By the second half of the decade, the jobs slowed to a trickle, and Teddy grew accustomed to being let down. In New York, attention drifted toward not only the young Aaron Douglas but also other newcomers, such as Archibald Motley and Hale Woodruff. In Charleston, things were somewhat easier, because Teddy was used to being ignored; but on the national scene, it was too late for him to catch up.

In the spring of 1926, Teddy was relieved when things took an unexpectedly good turn. In mid-March, the mayor of Charleston, Thomas Stoney, visited him in his studio. Stoney had long known the Harleston family and was an old friend of Teddy's father, the Captain. Mayor Stoney arrived at the Harleston Studio with two other whites on what appeared to be a social call. Teddy showed his guests some paintings, and after a look around the studio, the mayor left.

On March 31, Thomas Stoney wrote to Laura Bragg, the director of the Charleston Museum, to describe his reaction.

> I visited the studio of Charleston's colored artist. I was agreeably impressed that a member of the colored race had gone out and perfected himself in the art of painting. If you could arrange it so as to exhibit some of his work, I think you would be rendering a real service to a worthy individual of the opposite race.

Bragg took the suggestion and contacted Teddy with the news. The idea that he might have a show in his hometown astonished Teddy. Charleston was a place where he was barred from things whites took for granted, including the freedom to even enter the museum. Warily, Teddy arranged to see Laura Bragg to discuss the exhibit. The two met, and Bragg explained that although the Charleston Museum had never shown a piece of art made by a Negro, the time had now arrived.

The deal was finalized. By agreement, Teddy would make a selection of paintings and charcoals and share with the museum some news clippings and exhibition catalogs from previous shows. He would also send a list of black ministers to whom the museum could mail publicity material, so that word of the show could travel more easily in black society. Teddy began to crate some of his paintings for the trip over to the galleries. A courier was to pick them up.

About this time, John Bennett, the writer and friend of DuBose Heyward, took a sudden interest in Teddy Harleston. Bennett was a member of the Charleston Etchers' Club, the printmakers' collective at the museum, and it seems the group had been discussing Teddy's breakthrough. Elise told her husband of the unusual attention Bennett had shown during a week that Teddy was away: "Captain was stopped on the street this week by Mr. John Bennett, who asked if it is his (Capt's) son who is the artist. 'Yes? Where is he? Does he have a studio? Where? Can I visit it? Yes? I shall do so in the near future.'"

It's difficult to say exactly what happened next, but it seems the mayor's office began to hear from white voters who were unhappy at the prospect of a "Negro exhibition." At least some supporters of the museum let Laura Bragg know that they did not want throngs of black people in the lobby, let alone in the galleries. Pressure mounted on Laura Bragg to change her mind. In the end, after the museum director met with the mayor, the museum peremptorily withdrew its offer for a Harleston show. Laura Bragg wrote Teddy a letter informing him that his show had been "postponed," but the letter was delivered to the wrong address. A couple of days later, Bragg received a package from Teddy containing the exhibition catalogs she had requested. Bragg realized her letter had not been delivered, and she wrote a curt note putting him in his place.

> Unforeseen circumstances made it necessary for me to have a conference with the Mayor in regard to your exhibit and we decided that, in consequence of this, the exhibit would hurt and not help you at the present time.

Teddy was devastated. He could hardly bring himself to work in the funeral office, much less return to his studio. Two weeks passed before he answered.

> I should have been a little better prepared for the disappointment caused by the postponement of the exhibit. This delay in replying to your kind letter has been due to the disconcerting statement of the postponement without my knowing the cause of it. It might not make much difference whether I know or not, but you will understand that I have been very much embarrassed.

Teddy knew the "postponement" was permanent, that no explanation would be forthcoming, and that the museum would never make good on its offer.

"Uncle Teddy had crated the paintings for the exhibition, in nailed wooden cases," Gussie Harleston says. "After the museum canceled its offer, Teddy had to uncrate the art. I remember hearing the sound of his wrenching open the wooden crates below our apartment. He was banging things around downstairs, and the nails made squealing noises as they came out of the wood."

Sometime later, Elise wrote a letter that shed some light on the episode. In the story Elise tells, a friend of Laura Bragg, the museum director, paid a visit to the Harleston Studio. "She said Miss Bragg told her to say that a very eminent critic praised your work so highly that quite a spirit of jealousy was aroused in those who 'aspire to paint' in Charleston, which forced her and others responsible for the idea of your exhibit to abandon it." At the time, the Charleston Museum was host to the Charleston Etchers' Club, a group of nine white artists who "aspired to paint." John Bennett, the writer who had bumped into Captain Harleston on the street, was a member.

Teddy kept angling for work and recognition. In 1926, a real estate developer named William E. Harmon gave five hundred thousand dollars to endow a new group of annual awards for achievement by blacks in science, education, religion, and art. In June, Teddy sent photographs of several of his paintings to New York to be considered for a Harmon Award. In a few days, much to his surprise, he received a visit from the philanthropist himself, William Harmon. It seems Harmon had bought a plantation in South Carolina and was restoring it; he and his wife, Helen, entered the Harleston Studio without ringing the doorbell. Harmon was enthusiastic about Teddy's paintings and encouraged him to send five of the original artworks to New York for the judges to have a look. The philanthropist flattered Teddy, saying the technique of other painters couldn't compare with his technique and that he deserved first prize.

First prize came with a five-hundred-dollar honorarium. Buoyed by the attention, which came on the heels of the museum fiasco, Teddy shipped three drawings and four paintings to New York. In the fall, three judges convened to study the entries, with William Harmon joining them at the conference table. But as Harmon watched in shock, the judges mocked Teddy's work, dismissing it as old-fashioned. "The judges were, in my opinion, saturated with modernism," Harmon wrote Teddy by way of apology. A painter named Francis Jones, who had been asked to judge when others weren't available, was especially cruel. Jones told the Harmon Foundation that he thought Teddy's paintings were "not more than the work of a photographer," and added, "as Mr. Harleston is a man of forty-six years, he will probably not do any better work than what he has already exhibited." (Teddy was actually forty-three at the time.) Another judge, William Boring, an architect at Columbia University, concurred that Teddy, "while he could paint well, did not have the poetic soul of the real artist." When William Harmon suggested that Teddy at least should receive an honorable mention, Boring replied, "It would be best not to make the honorable mention unless the man would be so instructed as to disabuse him of the idea that what he is doing is of a superior quality." The first prize went to Palmer Hayden, a former letter carrier and janitor who had struggled his way into painting after a stint in the army.

Teddy fell into a depression and painted nothing for a year. Photographs from this period show that he began rapidly to age. Within a period of twenty-four months, his hair turned white and he began going bald. An increased sadness appeared in his eyes, which replaced the more familiar melancholy that his family knew he carried.

Elise did what she could to console him. "Through it all, I love you," she wrote when Teddy was out of town and feeling alone. "I repeat, I love you. It is all that keeps me, it is all that constrains me. When the cruel iron turns in my heart and my thoughts are blacked, I think I feel your arms, hear your voice. I remember sublime moments, and I hold on."

While her adoptive parents struggled, Gussie Harleston was growing up. She turned twelve in 1928.

"I spent my time after school at our apartment, but also at the funeral home. The Captain was there and it seemed like there were things going on. I got into trouble, too, like the one time I peeked behind Captain Harleston's office and into the embalming room.

"One of the boys from the orphan band, by the name of Sammie Walker, had just come home dead. The band was playing in Maine in the summer, and Sammie drowned up there. His body wasn't well prepared for the trip down from Maine, and it arrived at the funeral home in awful shape. I wanted to see him, because I knew him, so I went in the embalming room. Sammie was lying out on the cooling board, with most of his clothes on. But the thing I remember is that his body was oozing water. His eyes and his mouth were all dribbling water, or some fluid, I don't know what, down onto the table. I got out of there quickly, and that was the last time I went in the embalming room.

"It seems like a lot of things went on back there behind the Captain's office. In the Captain's bathroom was the first place I saw a boy's penis. I had a cousin who was nearly the same age as me. We were playing around the back and he said, 'Come here, Gussie, I want to show you something.' So we went in the bathroom and he closed the door, and he said, 'You ever see one of these?' Then he opened his pants and pulled out his penis, and let it sort of sit there. I remember thinking it was kind of an unimpressive thing. He didn't make me touch it, fortunately. Anyhow, I told him I wasn't impressed and to put it away, and I left."

Searching for distractions, Teddy turned to the family business. But where he had once been able to console funeral clients with genuine empathy, his attitude about things now became more detached and ironic. "No one has passed on here since I wrote last, but they tell me there are several 'prospects,'" he told Elise. There were also new management problems. After decades of relying on horses, Captain Harleston had decided to put his money in motorcars. A white funeral director in town, A. C. Connelley, was replacing his automobile fleet and offered to sell the Harlestons his limousines and hearses, which were made by a firm called Cunningham. When the Captain took the offer, suddenly the horse stables had to be converted into automobile garages, new tools had to be bought, and mechanics had to be hired to keep the Cunninghams rolling. If this weren't enough, Teddy received a notice in the mail from the South Carolina State Board of Embalmers that his embalming license was about to expire. He had to decide whether to renew it, a choice that seemed to carry all the weight of his discouragement. Finally, Teddy resigned himself, took the test, and got a new license. He was an embalmer, after all, and only incidentally an artist. Later, he allowed himself to be elected president of the state Colored Funeral Directors' and Embalmers' Association, a group of twenty-five undertakers. As the months passed, he sank more deeply into the funeral business and let his artwork dwindle.

"Teddy had many disappointments," Gussie remembers. "He was sad, but he never stopped caring for others. I remember the day my father died, on July 25, 1929. He had finally succumbed to tuberculosis at age forty. I was standing at the top of the stairs looking down at the entrance to our apartment, and I saw Uncle Teddy walking slowly up toward me. He had a sweet, sad expression on his face. When he got to the top and told me the news, he said, 'Gussie, don't let anybody tell you that you're an orphan. Because you'll always have your Uncle Teddy.'"

The last years of the 1920s were desultory ones, as Teddy's painting career slowed to that of an amateur. There was occasional work, but it wasn't always paid. In 1928, Teddy painted Yolande Du Bois, daughter of W. E. B. Du Bois, as a wedding gift when Yolande married the poet Countee Cullen. Adding to Teddy's injury, every fall came the news from New York about the Harmon Awards. The prizes became the signal that other black painters were being catapulted over Teddy's head. Archibald Motley, a painter from Louisiana, won first prize in art in 1928, and in 1929, it was twenty-eight-year-old William H. Johnson, an artist who, in addition to his youth, had the added cheek of having been born in South Carolina. Johnson, who imitated the blocky brush strokes of Paul Cézanne, succeeded at being a modernist where Teddy had failed.

In 1929, Teddy suffered the strange humiliation of being spoofed as a character in a novel. That year the author of *Porgy,* DuBose Heyward, published his second story of black life, *Mamba's Daughters,* about the tribulations of a passionate but poor woman in Charleston. To ornament the story, Heyward invented the character of Frank North, a light-skinned Negro artist who had graduated from Avery Institute and gone to art school on the East Coast. The fictitious painter exemplified what Heyward called "the sharp cleavage between full-blooded Negroes and mulattoes." Frank North, says the narrator of *Mamba's Daughters,* practiced a kind of "copybook gentility" that was modeled after whites. He and his circle "seemed to spend all their time saying how glad they are to be negroes while all the time they're trying their damnedest to be white." In one scene in the novel, a pretentious group of black professionals pays a visit to North's studio to inspect his paintings.

> The studio was a large airy second-story room, and a number of portraits were already hung, while many more were stacked against the walls. . . . The likenesses were good [but] in spite of the fact that the drawing was well done and the features characteristically negro, they gave an effect of not being negroes at all, but white people painted in darker shades—some subtle racial element was lacking.

Teddy could have done without the mockery from a white novelist, especially after his recent string of rejections. But locked behind the fences of Jim Crow, he could hardly respond to it.

In February 1929, Claflin College, a black school in Orangeburg, South Carolina, invited Teddy to give a talk about his work. The president of the school, J. B. Randolph, was an admirer. In need of the lecture fee, Teddy accepted, but decided to give a demonstration in lieu of a talk. At home, he worked up notes for his presentation, which he called "The Building of a Picture."

The show that Teddy put on in Orangeburg was part art history, part magic trick. At the lectern, the artist began by describing the components of an old-fashioned painting—pose, light, color, and form. Then he set up his easel and asked for a volunteer from the audience, and a student named Richard Brown presented himself. Posing his subject, Teddy rapidly made a painting of Brown, drawing the crowd into rapt silence. To his surprise, when he was finished, the demonstration elicited waves of applause. A week later, the president of the school wrote to say that he had the study of Richard Brown on the wall of his office and that he was still in disbelief: "Now and again I look up from my work and I am reminded of that marvelous evening last week when the painting of that picture actually happened before my eyes. To me it is amazing." Rereading the letter, Teddy realized he had a new source of income.

About this time, a young woman named Sue Bailey, then an administrator for the YWCA in New York, came to Charleston to see about setting up a new branch of the association. Teddy asked if he could paint her picture, and during sittings, it emerged that Bailey was organizing a meeting of regional YWCA chapters, which was to be held in a school at King's Mountain, a retreat near Asheville, North Carolina. Bailey listened to Teddy's description of his painting demonstration and asked him to do it at her event. "We have much to offer you," she said in a letter, "one hundred brown college girls, adorable and adoring, and an interesting group of conference leaders splendid to play with and famous and comely enough to paint."

King's Mountain sits in an especially beautiful fold of the Appalachian range. On June 6, 1930, Teddy arrived at the mountain enclave to find the meeting as advertised, that is, peopled by beautiful women. He wrote home to Elise, "All the fixin's Miss Bailey promised are here—the mountains, the moon, and the brown girls—and white—an inspiring lot." Family memory states that speakers at the conference included theologian Reinhold Niebuhr, who taught at Union Theological Seminary, activist and educator Benjamin Mays, and union leader A. Philip Randolph, of the Brotherhood of Sleeping Car Porters. Teddy gave his talk and was rejuvenated by the response. He found that all he had to do to get more speaking invitations was to ask for them. Later that year, Teddy found venues for his painting demonstration at Howard University in Washington, D.C., Talladega College in Alabama, a school in Fort Valley, Georgia, and a half dozen other places. He charged seventy-five dollars for a talk, and another twenty-five dollars to bring a group of paintings for a one-day exhibition.

Boosted by this new kind of acclaim, Teddy decided to make one last push to lift himself into the national spotlight. The Harlem Renaissance was nearly ten years old, and at age forty-eight, he was very much on its margins. But Teddy felt that one painter, Aaron Douglas, seemed sympathetic enough to approach for help. He had never met Douglas, who lived in New York, but the younger artist had gotten a great deal of attention for his work. From the grapevine, Teddy knew Douglas had been awarded two mural commissions, one for the Sherman Hotel in Chicago, and the other for the walls of a new library at Fisk University in Nashville. Teddy realized he could no longer ignore the current trends and that he would have to do more than paint beautiful likenesses. Mural painting was in the midst of a revival, so Teddy wrote Douglas to ask whether he would like some help.

In July 1930, Douglas answered:

> Dear Mr. Harleston:
>
> Your desire to come over here [to Nashville] is interesting. I have an idea that you could help me considerably with the actual execution of this work. Can you stay and assist me until about the 15th of September?

Knowing that he was gambling, Teddy asked Elise to look after the funeral office one last time and headed to Nashville.

Aaron Douglas, thirty-one, was in the enviable position of having too many commissions. A new library building going up at Fisk was scheduled to open in fall 1930. When Teddy arrived in Nashville, Douglas was working simultaneously on panels for the Sherman Hotel and the murals for the library.

Douglas's plan for the library murals was to try to show a five-hundred-year span of African American history, what he described as "the pageant of the Negro from Central Africa to contemporary America." Although this ambition was far from Teddy's idea of an art subject, he forced himself to learn. Teddy and Douglas faced a cruel deadline and had to finish more than 650 feet of murals in eight weeks, a giant task. There were to be panels in two reading rooms, more panels in the catalog room, and still others in the periodicals section. Teddy described it to Elise as a "scheme of many figures silhouetted in a kind of smoky blue-gray." With its spare lines and flat figures, Douglas's design resembled Egyptian funereal art as it might be painted on the side of a sarcophagus.

Teddy told Elise that he was "to play second fiddle for a change," but that his collaborator was pleasant enough. "I found Mr. Douglas a very cordial and affable fellow and I think we shall get along all right. You may not like his things as pieces but on the whole they will go very well."

The two painters created a schedule that called on Teddy to do most of the work. Aaron Douglas would arrive at the library at 8:00 A.M. and give the day's instructions to Teddy; then he would disappear into his studio and work on the hotel mural. At 5:00 P.M., he and Teddy broke for dinner. At night, from 7:00 until about 9:30, Teddy usually helped Douglas with his panels for the Sherman Hotel. After several weeks of this routine, Douglas put aside the hotel murals and began to appear in the library, where he and Teddy started working side by side. The two began a seven-day-a-week schedule, furiously trying to finish the job for the October deadline, working twelve-hour days for nearly six weeks. Teddy wrote Elise that he was painting part of the section of the mural about America and all of the section that dealt with West Africa. The African theme was a complete departure for Teddy. Until this moment, he had been the picture of the "Negrotarian"—cautious, even prim, with his emotions always in check. But on October 4, 1930, Teddy wrote Elise with a little cheer for black nationalism: "We are done! Finished! Completed!! 670 linear feet of mural decoration of a unique type done by Us, of Us and for Us."

Teddy returned home with high hopes that he had restored his career, even if it was within the dubious new arena of modernism. Aaron Douglas wrote to Teddy from New York to thank him. "I consider your decision to come over and give me a hand a little short of cosmic. I don't know how I could have made it through without your skill, your sound judgment and your helpful friendly attitude. I only hope that you are a free man when I get another commission."

In a burst of self-promotion, Teddy got in touch with Atlanta University, which was planning a mural for one of its buildings, and offered his new skills as a painter of panels. He contacted the Harmon Foundation, which had previously passed him over, and shipped four paintings to enter in the annual awards. This time the gambit worked. In mid-February 1931, Teddy received a letter saying he had won a Harmon Award for his painting *The Old Servant*. His spirits lifted, Teddy began touring the South, giving his painting demonstrations and describing the murals he and Douglas had done. Things looked promising, money was coming in, and he was back in the game. And it even looked as if he and Elise might manage to leave Charleston, after all. They could go to Washington, D.C., or even New York, where they would face the real competition.

In April 1931, Teddy was away from Charleston when he received word that his father was ill with pneumonia. The Captain was seventy-six now, and he had kept the funeral home running throughout Teddy's peripatetic years. The business he had founded was the Harleston family's cash generator, and the Captain knew that one day his children would have it for their own. By the time Teddy arrived back in Charleston, his father was in bed, and near death. Ever dutiful, he kept a vigil by the old man's bedside, staying close even though the pneumonia could be spread. Captain Harleston died on April 21. A few days later, he was buried at the Unity and Friendship Cemetery, where he had overseen the interment of hundreds of people in more than thirty years of burials.

Teddy was unprepared for his father's death. The old man had once made it possible for him to go to art school, and although he wanted to leave the South, he had always been grateful for this education. But Captain Harleston's death was a blow to his son's art career, because the funeral home had now fallen into Teddy's hands. Teddy's two sisters were married and living elsewhere; his brother Robert was dead, and his brother Moultrie was rarely seen. He and Elise had dreamed of moving away, but now they were trapped, and Teddy again sank into a depression.

Within a few days of his father's funeral, however, Teddy was himself ill. The family physician, Dr. Seabrook, came to call and found Teddy in bed. Seabrook's examination determined that during the vigil with his father, Teddy had evidently picked up a mild case of pneumonia. Dr. Seabrook must have noticed Teddy's listlessness, because according to family tradition, the physician said Teddy had a good chance at recovery, but that he didn't seem to want to get well. Gussie says:

"I remember when Uncle Teddy got sick with pneumonia, which he had gotten from his father. As he lay in bed, his face was kind of flat, and he spoke very little. Years later, I had a conversation with Miriam Seabrook, the wife of Teddy's doctor. She remembered her husband saying that Uncle Teddy could have kicked the infection, but that his spirit was gone. He had lost his desire to live and seemed to have given up. The family had a vigil, and Elise ministered to him. She brought some food, which he didn't eat, and he got weaker and weaker. Before anybody could believe what was happening, he was dying. Uncle Teddy died on a Sunday, on the afternoon of May 10, 1931."

The funeral was held on May 13, at 2:00 P.M., a little more than two weeks after Captain Harleston's funeral, at the same cemetery. Teddy was forty-nine.

"I was fourteen when Uncle Teddy died. My Uncle Teddy was against expensive funerals, strangely enough. He thought about all those poor people putting their money in the ground. So Tantie got him a pine box. At the interment, the Reverend Ledbetter offered last rites. Everybody was standing around the hole in the ground weeping. Tantie was in another world, just staring. They started to put the coffin in the ground with a lowering device. But right at that moment a carload of mourners arrived at the cemetery. They had driven in from Atlanta, and they were in terrible distress about Teddy. Uncle Eddie Mickey was in charge of the arrangements. He knew all these people from Atlanta, who were some of Teddy's friends from college. Those people wanted to see his face one last time. Uncle Eddie made the gravediggers bring the coffin back up out of the hole. Tantie was pretty unhappy about that. In front of everybody, they pried open the coffin, and we could see Uncle Teddy in there. You can imagine the reaction of the people all around, his body lying out in the open air, everybody looking at that chiseled face of his."

The letters and telegrams of condolence came in bundles, addressed to Elise.

"I am tempted to question," wrote Ada Hinton from Massachusetts. "It does seem that the world has so few 'Teddys' that he might have been spared."

J. B. Randolph, the college president in Orangeburg, said, "I admired your gentle husband for the latent power that was so capable, for his poise, his culture, and for his unusual power of mind and greatness of soul. He was one precious jewel in the crown of our race. There are not many."

Helen Baker, a friend of Elise's from the Bronx, wrote, "It does seem so strange for one so creditable to his race and his profession to have to pass on when we think of so many worthless lives that are spared."

W. E. B. Du Bois, who was not known for his graceful manners, neglected Elise and sent a telegram to Eddie Mickey instead: "WE ARE SENDING OUT NOTE OF HARLESTON THROUGH OUR PRESS SERVICE I AM VERY SORRY TO HEAR OF HIS DEATH GIVE MY CONDOLENCES TO HIS FAMILY—W E B DU BOIS."

Elise telegrammed several newspapers: "REGRET TO ANNOUNCE DEATH OF EDWIN A HARLESTON, SUNDAY FOUR THIRTY PM—MRS E F HARLESTON." A dozen papers ran obituaries. The *Atlanta Independent* carried an item on its front page. The lead story that day, May 14, concerned Franklin D. Roosevelt, the governor of New York, who was being considered a likely presidential candidate for the following year.

Teddy was buried in the family plot next to his father, but despite the attention, no headstone was placed at Teddy's grave. The Harlestons were distraught over the two almost simultaneous deaths, so Teddy occupied an unmarked place in the family plot at the Unity and Friendship Cemetery, one of the graveyards of the old colored elite.

The day after Teddy's funeral, Elise returned to the grave and brought her camera with her. Although twenty-four hours earlier hundreds of mourners had milled about, now the graveyard was empty, except for a fantastic display of cut flowers. From one end of Teddy's plot to the other were strewn piles of blossoms, wreaths, and arrangements. Elise set up her tripod and put the black cloak over her head. Then, as though capturing a landscape, she carefully photographed the mounds of flowers on her husband's grave, taking care to include every petal in the frame.

Chapter Eighteen

ELISE Harleston sent out the appropriate black-bordered mourning cards, one of which has survived: "The family of Edwin A. Harleston desire to express their deep appreciation of your sympathy."

The colored elite followed old rituals of mourning. A black crepe or scarf was often affixed to the door, and the room last occupied by the dead was draped in black or gray. Family members of the deceased wore fabrics associated with death: black crepe, black and blue-black bombazines, black alpacas, black silks, black kid, and black cotton. At the burial, a widow might wear a black veil with a white lining that appeared on the inside of the bonnet.

The expectations were especially high for Teddy's widow when it came to mourning him. During the first year after death, the spouse of the deceased was expected to wear only dull black. After eight months, flat tones could give way to black silk or crepe de chine; after twelve months, violet and purple were permitted; and after eighteen, white. Ordinary clothes returned at the end of two years. A widow was also expected to use special stationery throughout her mourning. The first kind was white or gray with a quarter-inch black border. After one year, the border shrank to one-eighth of an inch, and at eighteen months, to one-sixteenth. Perfumed pages and postcards were proscribed.

But while Elise followed the code of her class, her niece Gussie chose another way of expressing her grief.

"I was fourteen when Uncle Teddy died, and I changed my name in his honor. I told everybody I was no longer Gussie Harleston. Uncle Teddy was Edwin Augustus Harleston, so I became Edwina Augusta Harleston. I didn't do it officially with the state registrar, but I just did it. My friends kind of made fun of me, but that was a small price."

Elise was devoured in sadness. Teddy had died in May 1931, and Elise wanted to be alone, so she arranged to send her nieces, Gussie Harleston and Doris Forrest, away to friends in a nearby town for the summer. She then took a trip to New York and through New England to be with friends and to try to alter her mind. But Elise wasn't able to get Teddy so easily out of her system. Back in Charleston and alone again in the family apartment, she brooded about her husband. The setting was too much, and Elise began to act impulsively. On a whim, she sent several of Teddy's paintings off to friends as gifts, prompting them to write to ask whether she was trying to sell them. She returned obsessively to the studio where she and Teddy had made their art together.

Teddy did not always sign his paintings, because he thought he should do so only when he was sure they were finished, so not surprisingly, Elise discovered dozens of canvases with no inscription. In a distressed state, she went around the studio and signed paintings in her own hand, "E A HARLESTON," on the theory that if they weren't signed, she couldn't prove they had been his. Elise couldn't imitate her husband's careful script, so she used block letters instead. When she was done, a large number of Teddy's paintings had a signature that in no way resembled his own.

Teddy had bought a life insurance policy from his friend T. K. Gibson of Supreme Life and Casualty Company, and for a time at least, Elise was free from worry about money. The Harleston Funeral Home, however, no longer had a manager. There was some talk that Eddie Mickey might acquire the business, but the Mickeys were having trouble keeping their own business alive and were in no position to acquire a new one. Teddy's sister Ella Jenkins consulted with their other sister, Kitty Fleming, and the two of them decided to hand the daily operation of the business to their only surviving brother, Moultrie.

Moultrie Harleston, fifty-two, was the unpredictable one of the siblings. Elise had once described Moultrie in a letter as "the laziest man I have ever met." He had never married, but he had two children by a sometime mistress. Until the funeral home was thrust upon him in the summer of 1931, he had had nothing to do with it.

Arriving on the scene, Moultrie let it be known that he was in need of money to keep the business afloat until he could learn the essentials. Although Elise had no other resources, she reluctantly agreed to loan Moultrie the majority of Teddy's insurance money. This arrangement worked for a short time; as he collected from funeral clients, Moultrie repaid Elise with small installments. But it turned out money was not the only thing Moultrie had on his mind. Moultrie had always admired his brother Teddy, and Teddy's wife, and six weeks after the funeral, Moultrie began to call on Elise.

Edwina Harleston, the former Gussie, recalls:

"Uncle Moultrie did Tantie terribly. He borrowed the insurance money, and then he asked her to marry him. She told me about this later. She said no, because he was uncouth. After Uncle Teddy, what man would measure up? So Moultrie turned on her. He was paying back the insurance loan, and then he got mad and sent Tantie a bill for Teddy's funeral. Uncle Eddie Mickey had run the funeral, but the Harleston business had provided the services. Uncle Moultrie owed Tantie a lot of money, and he just sent her an invoice for that amount. He called it Teddy's burial expenses, and marked it 'Paid in Full.' He never gave her the rest of the insurance money. He probably had spent it anyway. That's when Tantie decided to get out of Charleston."

Other members of the Harleston family also cooled toward Elise. Ella Jenkins treated her shabbily, gossiping behind her back that she was unreliable, and Eddie Mickey, blinded by his jealousy of Teddy, offered no help.

Isolated, with her inheritance pilfered by her brother-in-law, Elise became desperate. To raise money, she tried to sell some of

Teddy's paintings to the Tuskegee Institute and to Atlanta University, as well as to friends she thought could afford them. But her timing was poor. The crash on Wall Street had seemed a distant event when it occurred in October 1929, but the stock market failure had cut into the demand for art. Letters came back asking Elise to wait until the financial danger had passed.

"I am sad but must return the pictures to you," wrote her friend Helen Chisholm from Connecticut. "This quarter I lost a very considerable dividend and I can not do things which I had hoped to do. One of them was purchase that lovely Harleston picture. This is a bad time with banks toppling on top of each other and investments not worth the paper they are written on. Even the man in the ditch is feeling the waves of uncertainty."

Elise realized that at the minimum she would have to go to work. She knew she couldn't make a living as a photographer, and the funeral home was closed to her, but she and Teddy had an old friend who offered to help. In Georgia, Forrest B. Washington was the director of the Atlanta School of Social Work, and he suggested that Elise come to Georgia to study at the school. Friends encouraged the move. "Social work is about the only field that colored folks like us can enter now since we are about barred from teaching," one wrote. With mass layoffs being reported and the Great Depression on the horizon, another friend said, "Never have social workers been so busy as this year, when all the world is being ministered to."

In a single fitful week in January 1932, Elise solved her outstanding problems. She packed up her niece Doris Forrest and sent her to live with another member of the family. Then she told Gussie, now known as Edwina, that she had to take up residence with her aunt Ella Jenkins, because "Tantie" was moving to Atlanta. Edwina summarizes the fateful day:

"Tantie did not like Aunt Lou, and Aunt Lou didn't like her. Maybe Tantie was jealous because Reverend Jenkins had servants and all kinds of help. Tantie called over to the Jenkins Orphanage and requested that they send the wagon, but she didn't tell Aunt Lou she was sending me over to live with her. The wagon was a kind of mule cart, dusty and dirty. Tantie and I loaded up my mother's silver and other heirlooms, and we piled the things high on the cart.

The wagon driver was scared and didn't know what to do, because he knew when we got to Aunt Lou's house he would get in trouble for bringing me over without any warning. So the driver just kept riding around and around town, and we must have clip-clopped for two or three hours before we landed at the Jenkins home."

With that, Elise left behind her two adopted children and moved to Atlanta.

The Jenkins mansion stood at 34 Magazine Street, up the block and within shouting distance of the orphanage. Everyday life at the house was an uncomfortable blend of luxury and penury, as the many comforts were thrown into relief by the deprivation of the orphans, who seemed to be everywhere, some of them scrubbing the floors and serving dinner. Olive and Mildred, Ella's daughter and stepdaughter, had married and moved away, but Edwina's twelve-year-old sister, Sylvia, had been living with the Parson and Ella Jenkins from infancy. When Edwina arrived, Ella had two Harleston girls to pamper.

The girls attended the Avery Institute, the only rigorous high school in Charleston for colored children, and Edwina became a superior student.

"When I was a first-year student at Avery, the older girls didn't want to walk to school with me. So every night I would go home and learn a poem. The next morning I would walk behind these three or four downtown girls, who were seniors who wanted nothing to do with me, a little freshman girl. I would recite two or three lines of the poem at their backs, and I would taunt them, 'Bet you can't finish it!' I had the feistiness of the Harlestons, but I didn't have the tact. A classmate took me down a peg when she accused me of being stuck-up. She said, 'You think you're so much 'cause your name is Gussie Harleston, but your hands look like they belong to your grandmother and your hair smells like fox piss!'"

The income of the Jenkins Orphanage was largely dependent on the generosity of ordinary people. In 1932, before the Depression cut into collections, the orphanage had one of its last good fund-raising seasons. That year, Ella Jenkins, a well-dressed matron who had a liberal way with money, asked the Harleston sisters to choose between two luxurious rewards. The city of Chicago was planning a World's Fair, scheduled to open in 1933, that would be similar to the one Ella had once attended in England. Ella offered the girls a trip to Chicago—or, if they preferred, they could have a tennis court in the backyard.

"They've almost finished the tennis court," said a letter from one of the girls in August 1932. "It may be finished next week if they work steadily on it."

The justification for the tennis court was that black people were not allowed to use the tennis facilities in public parks. After the court was ready, the Jenkins household became a social hub for Avery students, and even their teachers, who made a direct line from school to the Jenkins house. The Harleston girls' new court was made of yellow clay, which required upkeep. Mr. Capers, the caretaker from the orphanage, was given grooming equipment along with instructions on how to keep the surface smooth and to maintain the fault lines. Edwina remembers:

"Sylvia and I decided there would be another World's Fair, but tennis courts didn't come around often. Mr. Capers was ignorant, but he was strong, and he loved us. He used a big metal roller to make the clay even. He had been in the World War, and since then, President Herbert Hoover had given the veterans a pension. Mr. Capers called him 'Mr. Hooger.' He would say, 'Mr. Hooger made me free from worry with that money.' Mr. Capers took his pension money and had all his teeth taken out, and all gold teeth put in. He walked around with the court roller, and you could see the teeth kind of glinting in the sun.

"Once Uncle Joseph and Aunt Lou were late coming home from church, and one of my male cousins dared me to drive the orphanage truck around the tennis court. So I got in the truck and drove around and around, though I didn't know how to drive. I chewed up the clay and I couldn't stop the truck, and I kept looking out the window for Uncle Joseph. Finally I said to my cousin, 'We own the house next door,' and I just drove into the wall of that house. The hood popped open and the radiator cracked, and water sprayed up everywhere. Uncle Joseph and Aunt Lou came home, but they didn't find out who did it, and I didn't tell them. I think Mr. Capers cleaned up the court."

The Jenkins household attracted an ever-changing procession of vivid characters. As a Baptist minister, the Reverend Mr. Jenkins had friends who were "Holy Rollers," traveling evangelists who earned their living hopscotching from town to town to conduct revival meetings. One of the most flamboyant of this group was a man known as "Black Billy Sunday." A large, loud, demanding, and sensual black preacher, Billy Sunday made Charleston a highlight of his annual revival tour, and he regularly arrived with his entourage for weeklong stays at the Jenkins household. The evangelist and his companions took over the third floor of the mansion, pushing Edwina and Sylvia out of their rooms. The Parson's staff helped the minister by advertising his appearances with billboards put up next to the orphanage.

Black Billy Sunday spoke the language of the New Testament and carried himself like a country prophet, but according to one of the grandchildren of Parson Jenkins, the big preacher was "fresh." He would often coax the orphan girls into his rooms at the Jenkins house, where he would try to kiss or grab them. When the young women complained, the Parson was philosophical. "God sends the good with the bad," he said, "and Billy Sunday does so much good that he's not bad."

Although Edwina was settling into a comfortable new life, some of her cousins across town were struggling. About this time, the Mickey family, rivals of the Harlestons in the funeral business, were running into financial trouble. After years of competing head-to-head with the Harlestons' burial franchise, often unsuccessfully, Eddie Mickey looked for ways he could finally siphon off some of the Harlestons' loyal clients. Hoping that a show of luxury and promises of better service would draw new business, Mickey decided to build a big, new, and expensive funeral parlor on nearby Cannon Street. It was a risky expansion made more precarious by the deepening Depression, and it turned out to be a failure. Eddie Mickey couldn't get enough business to pay off the loans on the new facility. Business records of the Mickey Funeral Home from this period show a stream of collection notices sent to the gleaming new office by casket manufacturers and florists. Unable to pay the bills, Eddie Mickey had to shut down the family firm, after nearly fifty years in business. At this point, with their source of income gone and facing social embarrassment, the Mickeys made another dramatic decision. Eddie Mickey, his mother, Hannah, and several other members of the family decided to move to New York, where they opened a new funeral parlor in Harlem, called the Carolina Funeral Home, on 124th Street. After that, the family was seldom heard from in Charleston.

Edwina remembers the Mickeys:

"My uncle Eddie Mickey was one of five children. His mother was the one who was the Harleston. It's sad, but none of those five children had any kids of their own, and that was the end of the Mickeys. I think the problem with the Mickeys was not that they were infertile, but that they were gay. Uncle Eddie was a homosexual. In today's argot, he was closeted. He had been married for the sake of appearances, briefly, but his wife left him—I think because she was disappointed with his dormant tools."

In August 1932, a little more than a year after Teddy's death, the news came that Elise Harleston was to be married again. She had short-circuited the obligatory two-year mourning, and her husband-to-be was a physics teacher named John Jones "Jack" Wheeler. Uncannily, Jack Wheeler's first wife had died on the same day as Teddy Harleston, a fact that provided the couple nothing if not a shared interest. Elise had met the thin, formal science teacher by chance on a visit to Baltimore, where he lived. Wheeler and Elise saw each other only ten days before Wheeler asked her to marry him. Elise said no—that she had come to Baltimore to see a friend of Teddy's about the disposition of some paintings. Wheeler asked who that person might be. Elise told him it was W. E. B. Du Bois, who was in town at the local branch of the NAACP. As it turned out, Du Bois and Jack Wheeler were fraternity brothers. When Elise went to see Du Bois, she emerged from the office to find Jack Wheeler waiting for her. He asked her to marry him again, and this time she said yes.

Jack Wheeler was fifteen years older than Elise, but he had a good income, was predictable, and he promised to support his new wife in a trouble-free style. The wedding took place in New York, and after signing the marriage papers, Elise summarily dropped out of the Atlanta School of Social Work and moved to Baltimore.

"I was mad at Tantie because she married Jack Wheeler, and I didn't write her for a long time," says Edwina.

Elise came back to Charleston long enough to pack her belongings and the contents of the studio, including dozens of Teddy's paintings. Then she said good-bye to the chilly Harlestons and vanished from Charleston for good. After moving in with Jack Wheeler, Elise invited her eleven-year-old niece, Doris Forrest, to live with them as the couple's adopted daughter. Doris went to Baltimore to be raised by Elise, and Edwina eventually resumed her relationship with them by mail.

Now and then, Edwina Harleston's pampered life was punctuated by duty.

"The Harlestons reared their children according to the W. E. B. Du Bois theory of the 'talented tenth.' That was the tiny group of Negroes that was supposed to lead the rest. Uncle Joseph, the Reverend Jenkins, was influenced by the Booker T. Washington theory that everybody had to work. 'Throw down your buckets where you are,' Washington said. He meant start with what you have, use your hands, and don't rely on abstractions. Uncle Joseph did that with the orphanage kids, and he did it with us. I had to work, but he gave me a choice among jobs. There was a farm the orphanage ran outside of town where a lot of the children were sent to dig sweet potatoes. I didn't want to be digging potatoes, so I took the second choice, which was to work on the newspaper, the *Messenger*. You had several things you could do on the *Messenger*. One of them was folding the paper and using liquid glue on strips of yellow paper to hold it shut for delivery. But the glue smelled to high heaven, and I couldn't stand it. So the alternative, you could call up people—different churches, community leaders—and ask them questions, and write little items they called 'locals.' It was just like being a reporter, and I liked it. I'd write up the church news, or that somebody was in town, or there was a fair going to happen. I became a reporter because I didn't want to dig those sweet potatoes."

Edwina became president of her high school class, sang in the Avery glee club, and during her last year, was chosen to play the lead role in the school play, *Smilin' Through*. The play was an old Broadway hit about a young British woman—named Kathleen, played by Edwina—who falls in love with an American soldier and then discovers that his father murdered a member of her family. Edwina was especially happy to wear the glamorous costumes the role required. Eventually, she mended fences with her aunt Elise, and the two began to correspond.

> Tantie, you mean more to me than anyone else in this world. When I wrote the last time, I said I was quite busy, but if I was then, there is no word to describe the situation now. *Smilin' Through* is to be presented next week. I don't think, by the tickets we've sold, there will be enough room to seat the people. We've been practicing every afternoon from 3:00 to 6:00 for two weeks. I don't have to worry about my dresses for the play: Aunt Lou just wouldn't let me make them. Miss Sarah Chisolm is making them. One of them is a peach, flat crepe trimmed with green. It's a little afternoon dress and I'm going to wear it for my piano recital. The other is a very pretty, blue georgette dinner dress. It's trimmed with lace. I'm going to wear it to the formal reception.

At the end of her first letter to Elise in some time, Edwina referred to a single ominous cloud that darkened her otherwise bright world:

> Sylvia hasn't been well at all. She attends school very irregularly, and is sick and cranky most of the time. Aunt Lou has taken her to several doctors, and I think is going to take her out of school.

In 1935, fifteen-year-old Sylvia Harleston had just been elected "Miss Avery" by her high school class for the second year running. But then, inexplicably, she had started to feel weak and to lose weight. Sometime in early 1936, her condition suddenly worsened.

"Sylvia and I were driving to the finals of the South Carolina high school basketball tournament," Edwina wrote in a letter to Elise. "We were driving over to the gym when Sylvia had a hemorrhage. They sent for the college doctor, and there in the hospital he thumped her chest, and shook his head. Our family physician called it 'galloping consumption' and arranged for her transfer to the T.B. sanitarium."

Like both of her parents, Sylvia had been taken with tuberculosis. An old euphemism for tuberculosis used to be "the white plague," though the irony of the light-skinned Harlestons falling to a pale affliction was probably lost on its victims. While waiting to be admitted to the sanitarium, Sylvia spit up blood and rarely left the house, but the teenager tried to keep a sense of humor about her predicament. When tuberculosis was epidemic, institutions fighting the disease used a symbol or logo, a red cross with two horizontal bars rather than one. Everyone knew the emblem, which was printed on stationery and mounted on buildings. One Sunday, the principal of the Avery Institute and his wife came to visit. Sylvia came downstairs in her pajamas, and the couple, knowing her diagnosis, and that her chances of living were slim, was sad and unsure how to act. But on the lapel of her blouse Sylvia had pinned the tubercular cross, a morbid joke that lightened the funereal mood.

The city of Charleston had built its first sanitarium, known as Pinehaven, in the early 1920s. At the time, the death rate for tuberculosis in South Carolina was nearly double the national average. Pinehaven consisted of two long, low buildings (one for white patients, one for colored) on fifty acres of land, eight miles north of the city. Edwina recalls what her sister went through to get there.

"The sanitarium was in the middle of a pine forest, with good air. Our doctor applied for a place in the ward, but it was full, I mean the colored section was full. They might have had beds in the white ward, I don't know. But the colored ward was smaller than the white, and there was no room at the inn, so to speak. We waited for months before a bed opened up, because you had to wait till another Negro died.

"I remember the day we brought Sylvia to Pinehaven. She was failing rapidly. We all got in one of Uncle Joseph's cars and made the half-hour ride. We had brought Sylvia's bedroom things, little knickknacks to put around. She was in her pajamas, smiling faintly, but every once in a while she would pass out. We got her in the bed and sat around it for an hour or more. The nurse told the young people to go on home, but Aunt Lou stayed. Then the nurse told Aunt Lou to leave the room, because Sylvia would not die as long as she was there. That night—it was the Sunday before Easter, Palm Sunday—I got a phone call saying that she had died three hours after being checked into the sanitarium. She was sixteen."

Early in the morning of March 17, 1933, a fire swept through the second floor of the Jenkins Orphanage. One hundred forty-two boys and thirty-five girls were evacuated at 4:00 A.M.; eight children were injured. The *Charleston News and Courier* seemed to take special interest in the danger to the orphanage music program, reporting that when the fire started, "The sextet of negro youths, including the youngster with the 'bullfrog' voice, who sings on the streets, were sleeping in the dormitories." Parts of the walls of the upper floor collapsed and several of the rooms were gutted. Overnight, the old orphanage building was made uninhabitable.

Fortunately, the previous year Parson Jenkins had acquired a three-story, ten-thousand-square-foot building on the corner, the former South Carolina Medical College, by assuming its mortgage of seven thousand dollars. The Parson and Ella directed that the nearly two hundred children be housed there pending repairs to the burned building.

For decades, Parson Jenkins had relied for his success on the neglect of white authorities. As long as the city government ignored the welfare of black children, he had an opportunity to help hundreds of families, and in turn to help his own family financially. But the fire called unwanted attention to the little empire in the Parson's hands. After the blaze, a series of newspaper articles appeared that criticized Jenkins for his neglect of fire safety. White families whose houses bordered on the orphan compound had never been happy about the loud music, traffic, and nighttime high jinks of runaway delinquents. The fire gave them a chance to act.

A public meeting was called of the neighborhood's white residents. Charleston fire chief John H. Wohlers came and denounced the orphanage for its inadequate fire escapes and old electrical wiring. Property owners complained about the noise, and the police offered stories of crime. Someone made a call for eviction. A second public meeting was called and this one turned into a rally. Black citizens, who had the most at stake, were barred from both meetings. A white-owned newspaper quoted the Reverend Mr. Jenkins as saying that no one at the orphanage had received an invitation to the meetings, and therefore none from his office had attended. "We do not want to embarrass anybody," he said, by sending colored citizens to a white rally. At the second meeting, according to the paper, the "embattled citizenry of Ward 4 voted with only a few dissenting to 'condemn the orphanage'" and force it to move out of town. Coming to Jenkins's defense, a committee of black citizens organized a counter-demonstration, reported as a "mass meeting of Negroes," which passed a resolution demanding the orphanage be allowed to stay. An inquiry was launched. There had been rumors of malfeasance for years, and someone would have to get to the bottom of them. Relatives of the orphans had complained of beatings and other abuse of children at the hands of orphanage staff.

Rev. Daniel Jenkins was seventy years old and had lost some of his vigor; he could see the end of his life. Influential white board members, who had long protected him, had grown old and were dying; and since the financial turndown, the trips that the band took to the North were no longer as lucrative as they once were. But Ella and the Orphanage Man still had influence. Facing the groundswell of resentment, the two made a decision to fight until the last hour, to give their creation a chance to live.

Edwina Harleston grew into a beautiful young woman. She was slender, her eyes were alight, and she was sly, with a charming laugh and seductive smile. During four years in the Jenkins mansion, despite working at the *Messenger* and walking past the dormitories every day, Edwina had spent little time with the young men from the orphanage. The Parson and Ella Jenkins had kept her away, especially from the do-anything boys who played in the bands. Ella made sure Edwina had supervision, aware that the orphans came from another social class. But during her senior year, Edwina had a flirtation with at least one appropriate boy, a fellow student at the Avery Institute. His last name was Hamilton, and his first name was Gussie, Edwina's old name.

Gussie Hamilton was a slight, handsome teenager. Though he wasn't as accomplished at schoolwork as Edwina, he had a sweet disposition and was attentive to her needs. Gussie took Edwina to chaperoned dances, choir concerts, and parties at the homes of friends, and sometimes they kissed. But Gussie was thoughtful; when they were in the midst of a kiss, he never reached for her body. In any case, the habits of the colored elite made such a thing impossible, because every moment of Edwina's courtship with Gussie Hamilton was overseen. Even when the couple was alone, a door was left ajar and an adult sat in the next room.

Though only a handful of black women had the opportunity to attend college, Edwina spent her senior year in high school contemplating a choice of schools.

"I wanted to go to Howard University, in Washington, D.C., because my mother had gone there. But Aunt Lou wouldn't let me go to Howard, because she was superstitious about tuberculosis. Reverend Jenkins had a daughter who was a graduate of Howard, and she had died from tuberculosis. My mother had graduated, and then died of tuberculosis. Howard was on a hill in Washington. Aunt Lou said, 'Everybody that ever went to Howard, and had to climb up and down that hill, they all died of tuberculosis.' And my sister, Sylvia, had died of it. I said, 'But Sylvia didn't go to Howard.' And Aunt Lou said, 'Doesn't matter.' When Aunt Lou wanted to give me a talking-to, her foot would kind of twist around at the end of her leg, because when people are nervous, they jiggle their feet, and when Howard came up, her foot would be going strong.

"My second choice was Fisk University, in Nashville, where Uncle Teddy had done the Aaron Douglas murals. But Aunt Lou said I couldn't go there either, because the president of the school was white, and Uncle Teddy had told her that when he was on campus working with Aaron Douglas, the president didn't tip his hat to any of the black faculty. So Aunt Lou said, 'You're going to Talladega.'"

Talladega College, founded in 1867, was a small, liberal arts school in the pineland Alabama town of the same name, a hundred miles west of Atlanta. It was operated by the Congregational Church and had a thirty-year-old white president, a minister named Buell Gallagher; many of the teachers were white. Enrollment stood at about five hundred, and nearly all of the students came from light-skinned African American families in urban centers—New Orleans, Washington, Birmingham, Memphis, Charleston.

To Edwina's pleasure, Gussie Hamilton had also chosen Talladega, and they were to enroll at the same time. But before she left for college, Edwina's aunt, Ella Jenkins, dealt her a blow to the heart. Ella, who had been seduced as a young woman, seems to have thought that the danger of sex was too great to allow her niece to go off to school with a boyfriend. Acting as Edwina's substitute mother, she told her niece that she had to "widen her horizons" and give up Gussie Hamilton. After some argument, Edwina obediently broke off with her boyfriend. The thin, good-looking eighteen-year-old was stunned and hurt, and to make matters worse, Edwina and Gussie had to share the same train to Alabama, a long ride they spent in extended silence.

Arriving at college in September 1934, Edwina registered for a slate of required courses. She took English composition and public speaking, natural science, American history, and a language elective, French. Edwina thought she might want to be a writer. At eighteen, she was already a veteran contributor to the orphanage *Messenger*, where her pieces were distinguished by gossipy storytelling and unexpected twists. Dispensing with her course requirements, Edwina decided to concentrate on English literature. In rapid succession, she took classes in Chaucer, Shakespeare, the Romantic poets, and the history of the English novel. Turning to fiction, she enrolled in a course on the short story.

Edwina came from a sheltered world, and during the Depression, few Americans, let alone black Southerners, had the desire or leisure to study Chaucer. But in another respect, Edwina was like everyone else. Once she got away from home, and away from Gussie Hamilton, she began to think about men. She remembers:

"I was a prim Harleston girl, but these things came under pressure when I got to college. The women outnumbered the men at Talladega. I was popular and happy, but I learned there were rules in getting dates with young men. They used to have what we called 'front door girls' and 'back door girls.' The front door girls were the ones the young men took to dances and fraternity affairs. The back door girls were from equally prominent families, but they didn't get to go out of the bedroom for their dates. They usually didn't marry, either. In my freshman year I had a boyfriend, named Jacko, who was really after me."

Within a month of settling into her freshman dorm room, Edwina wrote to her aunt Elise, "I'm in love with the most popular, smartest, most well-thought-of man on the campus, Jacko." Despite being an orphan, Edwina had long been supervised by straitlaced adults; at eighteen, she was now on her own for the first time. The courtship with Jacko pushed Edwina to face a choice that she described in a further letter to her aunt.

> Dear Tantie,
>
> I may as well be perfectly frank with you. Most of the girls here on the campus, or anywhere else, for that matter, of my age and my standing think nothing of having sexual intercourses with their boyfriends. They laugh and disbelieve me when I insist that I haven't, but I suppose I thought too much of you and Uncle Teddy's teachings to do a thing like that. However, it's the thing that fellows here expect their girls to do. I'm afraid for myself, or at least I was. I suppose that's why I'm going to quit Jacko. He gave me the impression that that was expected of me. That's where I draw the line.

"I spelled it 'intercourses' because I didn't know nothing about it," Edwina says.

Chapter Nineteen

IN 1933, after the fire in the orphanage, Ella and the Reverend Mr. Jenkins stood accused of negligence. The city opened an inquiry, and the Charleston Bureau of Social Welfare appointed a "Committee to Investigate the Conditions in the Jenkins Orphanage." Hearings were called and reports drawn up on "The Revenue of the Institution" and "Sums Received for the Care of the Children from their Relatives and Friends." Local writer John Bennett summarized events in his weekly diary:

> The Jenkins Orphanage investigation seems a futile thing, saying little drastic, recommending less. Rev. Jenkins intimates he cures tuberculosis by giving the patient a horn in the band and making him blow it. Nothing at all appears to have come from the loud accusations of self-enrichment by Jenkins of which we heard so much, from the usual Negro sources mostly.

Ella and the Parson managed to hold on under scrutiny, but in 1935, two years after the blaze, it didn't help the Jenkinses' case when *Time* magazine fed a story to its national readership that described the family's huge profits from the music program. A reporter had gone up to Harlem to question the Parson during his working summer tour and found the old man an easy target:

> Grizzled, black-garbed and ailing at seventy-four, Daniel Jenkins is Charleston's No. 1 Negro citizen. . . . The Jenkins Band once had five units simultaneously on tour. Today its 125 players, aged 10 to 18, earn from $75,000 to $100,000 a year for the Orphanage. The hard-working youngsters play spirituals, sweet ballads, and hot arrangements of tunes like *Dinah* and *Sweet Sue* on their rusty cornets, trombones, French horns, drums. Rich old Rev. Daniel Joseph Jenkins sits in his institution's Northern headquarters in New York's Harlem scrutinizing detailed weekly reports of his bands' doings.

As it eventually turned out, the Jenkinses' critics in Charleston were interested in more than Ella's bookkeeping: they wanted the Jenkinses' turf. With Franklin Roosevelt's election to the White House, new sources of federal housing money had begun to become available, and the orphanage stood in the middle of a quarter-mile of old buildings. A plan was gaining favor with wealthier homeowners to evict the citizens of the poor and black neighborhood that was the waifs' domain. If Jenkins could be pushed out, then his fief could be razed and new housing put up.

"Slum clearance" was the name being given to the idea of a bulldozing sweep. "They have a housing idea in view," Edwina wrote her aunt Elise in Baltimore, "a plan to tear down all the houses in Cromwell Alley next to the orphanage and build a government housing project for the whites."

Around this time, the Reverend Mr. Jenkins became sick. According to letters, he went to bed for a month with a virus of some kind, and even when finally he got up, he seemed diminished. But the aging minister still managed to defend his work. He wrote letters to the newspapers refuting the allegations made by *Time* magazine, stating that the bands brought in only three or four thousand dollars each year. (It's likely the truth lay somewhere in the middle: the *Time* reporter exaggerated to improve his story, and Jenkins understated the bands' income to improve his.) When the city council suggested that Jenkins surrender the orphanage to public control and leave, he refused; in reply, the city canceled a two-thousand-dollar appropriation for the care of black children. The Parson showed his resolve and sent back a defiant letter:

> I plead guilty to failing to consider the recommendations made to me to turn over the Jenkins Orphanage and all of the property to the city. I have been a member of the National Conference of Social Workers for over twenty years. I am no stranger to every move made for the welfare of children. But God is not dead. I must prove true first to God, secondly to my fellow men, and thirdly to my heart and conscience—no more, no less.

At Talladega, Edwina Harleston rebounded from her troubles with her boyfriend. By the spring of 1935, she had gotten rid of Jacko and acquired a new, older companion—Nelson Palmer of Newport News, Virginia.

"Nelson is a junior and an Alpha Phi Alpha man," Edwina wrote Elise. "He's about a half head taller than I am, and about your color. His dad's the principal of the high school in Newport News. I still like my other friend, and know he likes me, but Nelson is slowly creeping up in my affections. We had our first altercation the other day, and I liked him far better when we made up than I did before. I can overlook the fact that Nelson drinks occasionally."

Nelson Palmer was tall, well spoken, and fiercely intelligent. His charisma had landed him the lead role in a production of Oscar Wilde's *The Importance of Being Earnest,* put on by the Talladega theater department. He was unusually ambitious and planned to become a sociologist. But he did have a problem with his drinking.

National prohibition had ended in 1933, but many rural parts of the South elected to stay dry. Talladega lay within a dry county, where liquor was illegal but bootleggers operated freely. Edwina remembers: "I didn't drink at Talladega. Well, maybe one time I had a taste. Nelson drank sometimes. We used to call the liquor 'fee-co,' but I don't know why. There was an undertaker in Talladega named Strickland, who was the source of the fee-co, which he sold in bottles out of his office. Every weekend some of the fraternity men would go over to Strickland's at night and buy a bottle of corn liquor. We used to say, 'They gonna die because Strickland's selling them embalming fluid.' "

Edwina joined a sorority and the debating team; she kept in shape with sports.

"I was president of my sorority, Delta Sigma Theta. All the girls from South Carolina were Deltas. The captain of our debating team was a boy named Louis Goodwin, very handsome, and also on the team was Jacko, my former sweetheart. I once had to give a talk in an oratorical contest on the subject of 'The Abdication of King Edward VIII.' That's when the king had given up his crown to marry that divorcée from Baltimore, Wallis Simpson. I thought it was so romantic he would do that. I remember his speech on the radio: 'I cannot fulfill the responsibilities of my office without the love and support of the woman I love. This I do with all my heart.' And I made the basketball team, where my position was guard."

Edwina moved in the thick of club life, supplementing one club with another, and trying to keep a sense of humor about them. With a group of young women, she formed a secret society called the Sacred Order of Ancient Pigs. Members of SOAP got together on slow nights with tuna fish salad, crackers, and gossip. Fraternity boys and "fee-co" were barred. "There are fifteen girls in this room," Edwina reported to Elise, "and I'm sitting on my bed trying to write letters. They are all looking out the window admiring two pledges."

When the Sacred Order of Ancient Pigs grew too big, Edwina set up a two-person club with a friend from Indianapolis, A'Lelia Ransom.

"A'Lelia Ransom and I were tight friends, I mean real ace boon coons. We formed an offshoot of the SOAP's called the Ain't-Got-Nothing Club. A'Lelia and I were in a psychology class where the professor announced that there were two categories of mental defectives, 'idiots' and 'morons.' That gave us our club names. In the Ain't-Got-Nothings, A'Lelia was 'Idiotia' and I was 'Moronia.' Our ritual was small. Every week, A'Lelia's father sent her a clipping from her hometown paper in Indianapolis, and he would enclose a dollar or two. We would lock the door and read aloud from the clipping, and look to see how much money had come, which was never much. Then we would chant a kind of invocation:

We're the Ain't-Got-Nothin' club!
Ain't got no pretty face

Ain't got no gorgeous hair
Ain't got no new winter coat,
Ain't got nothing!

"That was our prayer. Afterwards, we went out and spent A'Lelia's money."

A pretty and gregarious nineteen-year-old, Edwina spent a great deal of time thinking about her looks. She filled her letters with descriptions of outfits she wore to the parties, concerts, and dates with Nelson, and lobbied Elise to help her build an appealing wardrobe.

> I've been doing sewing. I made a blue lace dress that I think is very pretty, and a white corduroy shirt. Today I'm going to work on a black linen suit that Aunt Lou sent me. I want to ask you a huge favor. On Saturday night, there is a formal banquet. I'd like to have a tan and blue flowered chiffon dress, with a jacket to it, and some tan and blonde T-strap sandals to match. I take a 6½-A in a sandal. I'd like the dress to come to my ankles, and the jacket with long sleeves gathered at the wrist. You know the kind of thing appropriate for an afternoon tea. Tantie, if you don't have the money for the dress, I can get it here and you can send the money later.

Whenever one of the fraternities gave a dance, Edwina's letters omitted all mention of her course work.

> Well, all of last week I sewed. I made my dress for the Alpha dance. I finished it 7:35 p.m. the night of the dance. It's tan starched lace trimmed with brown linen, with a little brown linen coat that's used as an evening wrap.

Things got to the point that for one Christmas break, Edwina planned her vacation around what she would wear. She wrote Elise:

> I'm coming home for Christmas. I've decided at last that I'd like to have a dressy, green woolen dress. If you want to select it, I'd be delighted. If not, I'd be just as glad if you sent me the money, and I'll

> get it in Charleston. This weekend I'm trying to get a frock together. It's to be white piqué trimmed with big black appliquéd flowers at the hemline—no back—little black straps coming from underneath a collar in the back. Tantie, I'd like to have some lace stockings (flesh colored) with black toes and heels, size 9½, to wear with my dress. Do you think you can get me a pair here by Friday?

In 1935, Moultrie Harleston, who had been running the Harleston Funeral Home, died after a short illness. This time there were no obvious heirs to take over the company, so Ella Jenkins and Kitty Fleming had to keep the doors open while Kitty's college-age son could train to become an undertaker. At the end of her first year at Talladega, Edwina Harleston volunteered to get a job to ease the burden on the family. Ella accepted the offer, but the work Edwina found took her up to the state of Maine. The Reverend Mr. Jenkins's daughter Mildred was living there, in the coastal town of Ogunquit, and had gotten Edwina a summer position as an assistant pastry chef, working for a middle-aged Bostonian named Miss Tingley. "I heard from Miss Tingley and she wants me to come to Maine as soon after June 1st as is possible," Edwina told Elise.

The job in Ogunquit was straightforward enough: a prosperous white woman operated a bakery that specialized in decorated cakes, and Edwina was to help make and garnish them. But the unexpected thing about spending the summer in Maine was that Edwina's job brought her into the midst of an unusual colony of highly cultured lesbians.

Mildred Jenkins, then in her late twenties, was trying to break into the concert circuit as a soprano. Preparing for her debut at Jordan Hall in Boston, Mildred had retreated to the summer home of her voice teacher, Miss Tingley. At the time, Ogunquit, Maine, was a retreat for middle- and upper-class lesbians from Boston and elsewhere, and Miss Tingley summered there with her partner, Miss Foor.

"These women, Miss Tingley and Miss Foor, lived in a glass-enclosed house right on the ocean. The waves used to crash on the rocks just outside the door. Miss Foor was a librarian who worked at Boston University, and Miss Tingley was Mildred's music teacher. Miss Tingley was plump and wore dresses, and she was creative, not only as a musician, but the way she baked those cakes. The two women had adopted a daughter, whom they named Judith Tingley-Foor. Miss Tingley ran the cake oven, and my job was to wash dishes and clean up and occasionally help ornament the cakes. In this summer colony, all the chauffeurs and maids were black, and all the guests and tenants of cottages were white women. Once I got a job as a waitress at a big social function, a lawn party in somebody's backyard, and all the guests were women. At the time I thought it was strange. One member of the lesbian group was Ethel Barrymore, the actress. My letters home are about decorating cakes and going swimming in the ice-cold water, and nothing about the other goings-on in Ogunquit. None of the women ever made an advance toward me, but the black servants, of which I was one, lived apart from the white women and rarely spoke to them, except to take instructions."

After this peculiar summer, in September 1935, Edwina went back to Talladega and the embrace of Nelson Palmer. Nelson made Edwina feel wanted. "Nelson sends me ice cream and fruit," she told Elise. "Tantie, I feel this is the real thing, this time." But the couple always stopped short of "intercourses." Edwina and her boyfriend told each other they would get married at some future date and would wait until then. Though they didn't exchange engagement rings, Edwina gave Nelson her sorority ring, and he gave her a crucifix that came from his mother.

In January 1936, Edwina left school with a back injury. She had been playing tennis and had fallen hard on the court. Physicians couldn't rid her of the pain, so she left school until the trouble could be treated. While Edwina was back home in Charleston, Nelson graduated from Talladega and enrolled in a Ph.D. program at the University of Michigan. Edwina underwent surgery on her lower spine and could not visit her boyfriend at his home in Newport News before he left for the Midwest. In the fall of 1936, Nelson went off to Michigan and Edwina stayed in Charleston, where she was recovering from the surgery and working as a secretary in the orphanage's office.

Edwina missed her boyfriend terribly, and the couple exchanged letters, though none has survived. Nelson Palmer hoped to become a scholar at one of the black schools, possibly Fisk, Tuskegee, or Howard University; he would be preparing at the University of Michigan for several years. They had not been separated a year, however, when the bloom began to fade from their romance. From friends of friends, Edwina started to hear disturbing stories about her fiancé's new life in the Midwest. In fact, the rumors drifted down from halfway across the country that her boyfriend was seeing another woman. Edwina didn't believe it, but the stories continued to filter back; and in the spring of 1937, they got worse.

"Aunt Lou said she was sending me to Talladega to widen my horizons," Edwina says. "I was engaged to Nelson, which was a big leap in my experience. But at the University of Michigan, he impregnated a fellow student."

In the 1930s, unless a woman knew an abortionist, pregnancy was usually followed by a quickly arranged marriage. Edwina was frightened and jealous. Not only did she suspect her boyfriend had betrayed her, but Nelson might be snatched away for good. Edwina went back to Talladega, not sure whether she would ever have Nelson for herself. In the late spring of 1937, Edwina's fears were confirmed. Nelson had finished his year at Michigan and had taken a summer job at Fisk University in Nashville. On a trip back to the South, he stopped at Talladega to give Edwina the news.

"Nelson spent the day here last Saturday," Edwina wrote Elise, "on the way to Fisk where he will be working all summer. He plans to be married in September and he and his wife will go back to school in Michigan in October. Viola Goins of Connecticut. I hate her. Funny, eh? I have gotten my ring back. Tantie, please look again for Nelson's mother's cross. I would like to relieve myself of that obligation."

Twenty-year-old Edwina had been jilted. She went home to Charleston for the summer, vulnerable and distraught. Trying to be a good niece, she started back to work at the orphanage's office, but her mind was elsewhere.

On weeknights, Edwina stayed at the Jenkins mansion, but on weekends she moved to the upstairs of her old home at 118 Calhoun Street, across from the funeral home. There was minimal supervision in the old house. Thinking much about her lost love, Edwina began to see an old high school boyfriend, Gussie Hamilton. Like Edwina, Gussie Hamilton (who also attended Talladega) was home for the summer and glad to renew the affair. Edwina remembers what happened with Gussie on a day she would have preferred to spend with Nelson:

"Like a fool, I had fallen out with Gussie Hamilton before we both went to Talladega. And we were kind of spending time together again. This was after I had gotten the news about Nelson having gotten that girl pregnant. Gussie used to call me 'homey,' because I was the only girl from Charleston my freshman year, when we were both at Talladega. Anyhow, Gussie Hamilton was real handsome, and I think it was a Saturday afternoon that he was walking me home to Calhoun Street. I don't remember what I was wearing. We got to the building at 118 and into the hallway and on the stairs. The hallway was wide, and Gussie Hamilton and I were climbing up. I knew nobody was at the house, and I probably told him that. He turned to me at the top of the stairs, kissed me, and then grabbed me. Things happened very suddenly. It felt like a wave crashing over me. I was a virgin, but he must not have been. It was my first time, but I don't remember the sex in particular, only the shock of how it happened. We did it standing up, right there in the hall, on the landing, against the wall."

The Jenkins Orphanage—under siege, its famous music program wounded by the Depression—seemed close to demise. In the North, the boys in uniforms who came back to the same street corners and parks year after year were no longer looked on as an entertaining act. At a time when soup lines provided the backdrop, the children seemed more like beggars. Donations were off, the music program had been cut down to two bands, and the quality of the music was suffering.

Still, boys who had gotten through in earlier years were going on to careers. Jazz, once a fugitive music played for underground fans, had swept the commercial mainstream, and the 1930s were its golden years. Radio and a well-oiled circuit of hundreds of ballrooms fueled the national hunger for danceable rhythm. By 1935, jazz reached a mass audience that the musicians from its early days had only dreamed about in the years after World War I. The small ensembles of Louis Armstrong and Jelly Roll Morton had given way to the big bands, fifteen or twenty musicians who could fill a stage and swing a dance floor with the jitterbug. As jazz grew in popularity, white bands edged out the black bands and took over the profits, with acts like Benny Goodman, Tommy Dorsey, and Artie Shaw dominating record sales. But black bandleaders like Duke Ellington and Count Basie did almost as well on a parallel circuit of nonwhite nightclubs and record labels. The upturn was good for a few orphaned teenagers from Charleston, some of whom hit the big time almost as soon as they ran away to New York.

Rhythm guitarist Frederick William "Freddie" Green was born on March 31, 1911, in Charleston. Freddie Green wasn't an inmate at the orphanage: an organizer for one of the bands befriended the boy and taught him to read music. The bandleader saw Green's talent, so he agreed to let Green in the orphanage to play, starting him out on banjo. Freddie Green was a slight, handsome man with a quiet manner and low-key presence. In an interview late in life, he remembered the scene on the Charleston streets:

> I used to stop whatever I was doing and follow the Jenkins band all over the city. I figured if I could get away—I knew I wouldn't be able to make it in Charleston, not playing music. I had to get away. And because of the fact I had been to New York—I had been there, and I had seen, and I had heard, and I knew whatever it was, it had to come from New York. And I went with the Jenkins band, we stopped in New York, and I just stayed in New York and I didn't go back.

Green lived with an aunt in New York, finished high school, and taught himself to play rhythm guitar. At first he played rent parties, but beginning about 1933, a bandleader named Lonnie Simmons gave Green a real job. Simmons and Green played at a nightclub called the Yeah Man, which they soon left for a competing gig at the Exclusive Club. From there they leaped to a jazz club in Greenwich Village called the Black Cat. There, a music producer named John Hammond heard Green on the guitar and introduced him to Count

Basie, a bandleader Hammond had been working with. In March 1937, Freddie Green joined the Basie band.

William "Count" Basie was short and sturdy, a native of the town of Red Bank, New Jersey, who had gravitated to Harlem in his teens during the early 1920s. There he encountered the stride pianist Fats Waller, who gave him the fundamentals of the later Basie style. William had picked up the label of "Count" in 1935 during a radio broadcast. Signature numbers of his thirteen-piece band included "One O'Clock Jump" and "Jumpin' at the Woodside."

Freddie Green became Basie's guitarist. In 1930s jazz, guitarists were not soloists, but rhythm specialists who kept the band swinging four beats to the bar, strumming reliable chord changes on their unamplified instruments. The stars with solos were the people in the horn and reed sections. Count Basie was known for his smooth but relentless beat, which was held together by just four musicians. In the jazz press they were called the "All-American Rhythm Section": Count Basie on piano, Walter Page on bass, Jo Jones on drums, and Freddie Green on guitar. Green's rhythm guitar was like the metronome of the band, a sound whose flavor gave him the nickname "Pepper."

In the late 1930s, the amplified electric guitar was invented, and jazz guitar began to break away from its traditional place at the back of the bandstand. The pioneer of the guitar solo was Charlie Christian, a twenty-three-year-old black guitarist with Benny Goodman, who plugged in his instrument and started playing light and showy solos like those usually heard from a clarinet. In January 1938, when Count Basie recorded a song called "On the Sentimental Side," Freddie Green allowed himself a few solo phrases, an unusual move. According to one of Freddie Green's band mates, Green could have been a soloist like Charlie Christian, but the All-American Rhythm Section fell apart whenever Green took a solo.

There is a story connected to Green's short-lived experiment with solo guitar. At some point, Charlie Christian gave Freddie Green an amplifier, which he plugged in and used for several rehearsals. Others in the Basie band didn't want a louder guitar, so one night somebody removed a tube from the amplifier. Green had it repaired. A second night, the plug disappeared. Green replaced it. Finally, other players took out all the insides of the amplifier. When Green arrived at the session, he plugged his guitar into an empty box. Green was angry, but he agreed not to take any more solo breaks.

Solo or none, the job with Count Basie was nearly perfect for Green. A year after he joined, the guitarist appeared with the Basie band in a Hollywood film called *Policy Man*. The Basie contract paid well, though it didn't make him rich. For the year 1940, Freddie Green's tax returns put his income from the Basie orchestra at $3,277, a middle-class salary for that time.

More than money and recognition, the Count Basie job allowed Green to surround himself with unusual people. When they were on the road, the Basie band traveled in a bus that musicians nicknamed the "Blue Goose." At the end of the 1930s, a sometime singer with the Basie orchestra was a sad-eyed twenty-five-year-old from Baltimore named Billie Holiday. Holiday was the only woman on the Blue Goose. The men had given her the sobriquet "Lady Day," meaning First Lady (with Day being a shortened form of her last name). After some weeks of crisscrossing the highways together in the bus, Billie Holiday and Freddie Green began an affair.

The lovers stayed together off and on for years. When they were both with Basie, Freddie Green would check into hotels with Billie Holiday. When Lady Day went off to sing with other acts, as she did with the white Artie Shaw band, she would make a point of finding Green when he was in the same city. Freddie Green's papers and memorabilia contain little evidence that would explain what he had that appealed to the great jazz vocalist. In fact, other musicians in the Basie band complained that Holiday and Green fought all over the Blue Goose. But whatever his charm, Freddie Green is said to have been the only musician in the band with whom Holiday ever got involved.

In 1941, Billie Holiday got married (to someone other than Green) and the lovers stopped their affair, though it resumed soon enough. According to one jazz historian, Holiday's new husband was an addict, and Holiday started using hard drugs in 1942. That year, she had a run-in with police in Chicago. Five years later, she was arrested in Philadelphia for heroin possession, and sentenced to a year and a day at the Alderson Reformatory in West Virginia, neatly halting both her jazz career and her love affair with Freddie Green.

The list of Jenkins musicians who made a substantial mark on jazz in the 1930s includes William "Geechie" Harper, who played clarinet and alto saxophone. Born about 1895, Harper entered the orphanage at the beginning of the century and became a music instructor after 1912. In the 1920s, he played with LeRoy Smith's Society Orchestra in New York; for a decade, he was with a variety of radio orchestras and big bands until finally he joined arranger Fletcher Henderson in 1944. Another Jenkins success was trumpet player Herbert "Peanuts" Holland. Coming to Charleston from Norfolk, Virginia, Holland played with the orphanage band in the 1920s before going on to work with bandleader Jimmie Lunceford in the 1930s, and later, with Coleman Hawkins.

But it was a trumpet player who walked away with the brightest spotlight, beginning in the middle of World War II, and made the orphans in Charleston cheer with envy. William Alonzo "Cat" Anderson, born on September 12, 1916, in Greenville, South Carolina, was a child when both of his parents died and he was placed at Jenkins. Other inmates gave him the nickname "Cat" for his slashing fighting style in the yard of the orphanage. Cat Anderson started on the trombone and baritone, but later moved to cornet, and eventually trumpet.

Cat Anderson became the best high-note trumpet player of his generation. His professional path started in 1932, when Anderson was sixteen, and he and several other Jenkins veterans formed the Carolina Cotton Pickers. The group toured and recorded until 1935, when Anderson left for a string of acts, including the Sunset Royals Orchestra, Lucky Millinder's band, Lionel Hampton, and Erskine Hawkins. But he gained his lasting renown after 1944, the year he joined the Duke Ellington Orchestra.

Concert film of the Duke Ellington band shows Cat Anderson to have been a stocky man with a cool presence and sedate smile. With Ellington, Cat Anderson became famous for his high-note style, playing solos that rarely dropped below the highest octave it was possible to reach on the trumpet. Duke Ellington usually had Anderson come out from the horn section to stand at the front of the stage, where he gasped and strained to reach notes so high, other musicians said that "only a dog can hear them." The trumpet player's restraint was usually in evidence until the solo, which transformed his face into a knotted mass of muscles and contorted cheeks that looked as though they might burst.

On July 19, 1937, Rev. Daniel Jenkins was resting in the heat of a Charleston summer when he suffered a cerebral hemorrhage. He managed to hold on to life for some days, but finally died at 11:15 P.M. on July 30, with his wife, Ella, by his side. According to the examining physician, the cause of Jenkins's death was the hemorrhage, complicated by "chronic interstitial nephritis." Jenkins was seventy-five and had been a minister for forty-six years.

When the Parson died, the two Jenkins brass bands and one choir were away on tour. (One band was in Saratoga, New York, playing at horse tracks; another was in Boston, while the girls' choir was in New York City.) Although Ella wanted the bands for her husband's funeral, they could not get back to Charleston in time. As the day of interment approached, the flood of condolences and press attention surprised Ella. The *New York Times* ran an obituary, and Mrs. Jenkins received a cavalcade of well-wishers. Ella realized there wouldn't be enough room for the mourners at the New Tabernacle Fourth Baptist Church on Palmetto Street, where her husband had preached, and she arranged to have the service moved to the larger Morris Street Baptist Church to accommodate the expected crowds.

On the day of the funeral, colored citizens were amused to read in the white-owned newspapers, "A section for white people will be reserved in the church." Some two thousand people came to the service. A photograph shows the crowd pushing its way into the sanctuary and streaming out onto the sidewalk. The cortege accompanying Jenkins's body from the church to the graveyard wrapped around the block.

Rev. Daniel Jenkins died without a will, so on August 7, 1937, the probate court appointed Eloise C. Jenkins administrator of the estate, placing her in charge of distributing her husband's assets. Much of the property the Jenkins family used had been held in the name of the Orphan Aid Society, a nonprofit organization whose board, at least on paper, operated the philanthropy. For various reasons, Parson Jenkins kept few holdings in his own name; thus, an inventory of his personal estate contained only an abbreviated list of his assets: several houses, pieces of real estate, a life insurance policy, automobiles. As executor, Ella Jenkins had the discretion to divide the inheritance as she saw fit. In the end, Ella chose four heirs to the estate: herself, her husband's daughter Mildred Jenkins of Boston, and two of the Parson's grandchildren by his first wife, Livingston Howard of Charleston and Corrine White of Brooklyn. Left out of the division was Olive Jenkins, Ella's love child, conceived when Ella was a twenty-two-year-old secretary.

It may have been that Ella had grown so distant from her daughter that as far as she was concerned, Olive Jenkins did not exist. Or perhaps Ella wanted to include Olive in the Parson's inheritance, but she worried that such an action recorded in a public document might invite scrutiny. The most important thing was that the nature of Ella's relationship with Olive be kept secret, and so once again, as a sacrifice to appearances, Olive was stung by her mother's neglect.

When Edwina Harleston graduated from Talladega on June 5, 1939, she was at a loss for what to do. Elise, who knew of her niece's vivid writing from a stream of letters, suggested she study journalism. Elise gently forced Edwina to apply to the School of Journalism at Northwestern University in Evanston, Illinois, north of Chicago. She was accepted and moved to Chicago in the fall of 1940.

According to a transcript, Edwina registered for night classes, first in "Newspaper Law" and then in fiction writing. She dropped out for a while, but came back in the fall of 1943. The United States had entered World War II in December 1941, and by this time, all journalists focused their attention on the war. Edwina took a course called "Interpreting War News" and one in "Typography and Production."

"Strangely, my move to Chicago put me in contact with raw racism for the first time," Edwina says. "In Charleston, I had been sheltered from it, because the white world and the black world were parallel, never touching. Then I got to Northwestern, the so-called great Methodist institution. Two things happened that surprised me. The star football player, who was black, was meeting the requirements of his major, but he was not allowed to swim in the university pool. His coach had to bring him to the black YMCA for separate lessons. There was also the policy of this supposedly religious university that prevented black students from living in the dormitories on campus. Eventually, I found a room in the home of a minister's family, across the tracks in Evanston. Once I was studying for finals with a friend who wasn't black. I was invited to her dorm room, but at midnight was told by the matron I had to leave because I was colored. I was frightened and furious, and had to stumble back across the railroad tracks to my room at the minister's house."

The years in Chicago put Edwina in touch with other trends, including political changes seeping through black society.

"One of my boyfriends in those years was a theology student named Moran Weston. It was after the 1930s, and the Communist Party was making inroads into parts of the black community. Moran was a Communist, and I loved him. He went on to study at Union Theological Seminary in New York. After a while, he had me out on the street, making speeches about the workers. I would have done anything for him, and I did. But Moran wrote his thesis in seminary on a comparison of Christianity with Communism. In his paper, he rejected Christianity, and the seminary rejected him. He had a hard time working as a minister after that.

"Eventually at Northwestern I met the man I would marry. His name was Henry Oliver Whitlock, and we were in a class on typography together. He was quiet and shy, but he was also quiet because it turned out he knew all this stuff about newspapers already. Henry's father owned a newspaper in Gary, Indiana, just a little drive east from Chicago. I remember seeing this young man in class, who was so pale that he looked white. In fact I thought he was white, except he kept looking at me during the lecture. So I asked one of the only other black students there, my friend Johnny Johnson, to do an experiment. After class, all of us were getting ready to take the elevated train, the 'El,' back home. Johnny, Henry, and I walked up to the platform not saying anything. Johnny and I stood behind Henry and I whispered to Johnny, 'Do you think he's one of us?' Johnny looked Henry over surreptitiously, and then we walked a little bit away. He nodded his head and said, 'Yeah, he's a spook.'"

After the Parson died, and Ella Jenkins took over the orphanage, the city continued to apply pressure to push the old welfare institution out. Repairs had never been finished after the fire, with the result that a half-burnt hulk served as the focus of operations for day programs, while the children's dormitories were spread out in several other buildings.

Ella Jenkins wanted to carry on her husband's work, and she was helped by two longtime aides, a married couple named John and Sarah Dowling. It wasn't long, however, before Ella found that she was not as skilled at fund-raising as her husband. In the late 1930s, the orphanage spent money faster than Ella could bring it in. According to a family member, Ella became distraught and started behaving irrationally with money. Hoping to raise enough to finish the repairs, she turned to dubious sources.

Once, during a particularly stressful period, Ella let superstition overrun her. At the orphanage farm north of the city, she met a charming middle-aged man who promised to solve her money worries. It turned out he was a con man with an outrageous scheme. Edwina tells how Ella reacted.

"Aunt Lou had driven to the farm and had met this con man. He had persuaded Aunt Lou that if she buried bags of twenty-dollar bills around the roots of a certain oak tree on the farm, the bags would grow. My Aunt Lou didn't have a college education. I think she had gone to the Hampton Institute, in Virginia, for one summer, and she had a superstitious frame of mind. Anyhow, she and this man buried paper bags of bills Aunt Lou had taken out of the bank. The man told her they would pick them up after a certain number of weeks or days and find the money had just multiplied. Some time passed, and Aunt Lou went up there with this man and dug up the bags. There seemed to be more of something in them, and Aunt Lou brought them home. The man warned her only to open the bags at night, when she was alone, otherwise the money would vanish. The man went home, and Aunt Lou took the bags in her office in one of the parlors of the house. I was in the house with her at the time. Night came, and I heard her in the office, opening up these things, and she was just moaning. Nothing but newspapers, because the con man had taken the money."

Within a year of her husband's death, Ella had struck a deal that surrendered the orphanage to the city. She gave away the main building and several surrounding lots, but kept the Jenkins mansion. In return, the city promised to build a new facility for the Orphan Aid Society north of town, which Ella could continue to operate.

Completed in 1938 at a cost of thirty-three thousand dollars, the replacement compound, eight miles out in the farmland, resembled a work camp more than a home for children. A high wire fence surrounded it, which, in the words of one newspaper, was meant "to discourage runaways." Children were confined to four long barracks, and the buildings began leaking as soon as they were occupied. Also, the site was too far to link to the city sewers, so before many months passed, the plumbing didn't work.

In Charleston around the old orphanage, two blocks of houses built in the nineteenth century were torn down, leaving an empty lot the size of a soccer field. The bulldozing "has been particularly desired," said a white newspaper. The old city jail, the Jenkins mansion, and the former orphanage at 20 Franklin Street were three of the only buildings left standing in the swath. A brick housing development for low-income tenants, the first in the United States financed with federal money, went up on the open tract.

When the children moved to the new facilities, the Jenkins bands no longer paraded through town, a tradition that went back forty years. The loud, dancing children were once a familiar sight, but now the distance to town was too great to travel, and besides, the music program was a faded image of the old one. In 1939, a newspaper reported a band's appearance on the street as though it were an event. "Yesterday there was a parade of the orphans in the city for the first time in many months," said the *News and Courier*.

A crowd of two hundred black children followed the boys down the alleys and up the old boulevards.

* * *

In Chicago, Edwina Harleston and Henry Oliver Whitlock got engaged. The wedding was to take place on April 29, 1945, at the borrowed home of a friend, and Edwina's guest list was short. Elise Harleston Wheeler was invited to come from Baltimore, along with Elise's niece, Doris Forrest; Ella Jenkins came from Charleston. A photograph of the wedding party shows the handsome couple flanked by their attendants. Edwina is beaming, but her new husband looks nervous and wary. "Aunt Lou said that Henry looked like he knew what this family was all about," says Edwina. "She said he looked like a lamb being led to slaughter."

Part VI

~

A Trunk in the Grass

Chapter Twenty

THE Harleston family was dying out. Of the original eight children of William Harleston, the white founder of the family, and his mistress, Kate Wilson, only one, Susan, or Sister Sue, was still living. She had buried her brother, Captain Harleston, her last surviving sibling, and at seventy-four lived alone, widowed, in the same house that William had bought for Kate. The original address of the house, where the first colored generation of Harlestons had grown up, had been 28 Laurel Street; but the street had been renamed and renumbered, giving the same building a new address, 50 Ashe Street. In April 1941, Sister Sue sold 50 Ashe for twenty-five hundred dollars.

Around this time, Ella Jenkins, fifty-eight, was forced to sell the Jenkins mansion to the city of Charleston because she could no longer afford the upkeep. To curtail expenses, Ella and Sister Sue, her aunt, moved in together, setting up housekeeping in the little cottage behind the funeral home that had once been occupied by Mamie Randall, Captain Harleston's mistress. In front of their new home, the funeral home was also drifting a bit from its origins. It was now in the hands of Sister Sue's grandnephew, Maithlun Fleming, the son of her niece Kitty Harleston Fleming.

Edwina Harleston Whitlock had moved away to the North, and all the Harlestons were fading from her mind.

"After Tantie moved to Baltimore, and Aunt Lou sold the Jenkins house, there was very little for me to come home to in Charleston," Edwina says. "I had to get along with my life."

In the summer of 1945, freshly married, Edwina arrived in her husband's hometown of Gary, Indiana. Forty-five minutes by car southeast of Chicago, the city of Gary was a classic company town. It had first arisen at the start of the century, when the United States Steel Company built a plant on the sandy shore of Lake Michigan. Elbert H. Gary, founder of U.S. Steel, made sure all the major streets of his new town led right to the mill entrance. Gary had grown from 2,000 residents in 1907, to 55,000 in 1920, to about 150,000 by the time Edwina arrived.

To its black citizens, Indiana was far from a racial utopia. In fact, the state had a rich tradition of Ku Klux Klan activity. The KKK had surged into daily life after World War I, and Indiana Klansmen (and –women) soon composed a statewide movement with some three hundred thousand adherents. The Indiana Klan denounced African Americans, but they were also against Catholics, Jews, sexual freedom, and drinking. In 1924, when Indiana voters had elected Edward Jackson as governor, it was with KKK support. But two years later, the Klan lost its momentum in a scandal involving the Grand Dragon of Indiana, David Stephenson. Stephenson kidnapped, beat, and raped an Indianapolis woman named Madge Oberholtzer; the victim tried to kill herself by swallowing poison and later died. The Klansman was convicted of second-degree murder and sentenced to thirty years in prison, leading many Indianans to look elsewhere for inspiration. Nevertheless, an old center of Klan activity was the city of Valparaiso, a short drive from Edwina's new hometown of Gary.

Blacks in Indiana made up only 8 percent of the population, but the city of Gary was a rare colored enclave. With the steel mill and dependent businesses, Gary had filled up with black migrants. At the end of World War II, just as they had during and after World War I, blacks left the Southern states to find well-paid work they couldn't get at home. By 1960, when the new wave of migration would begin to fade, some 40 percent of the nation's African Americans would live in the Northern and Western states.

Edwina's marriage and move to Gary, Indiana, was part of this second exodus.

Henry Whitlock was a gentle man, thirty years old when he married Edwina, soft-spoken and mild, with pale skin and a small mustache. The Whitlock family ran the *Gary American*, a weekly newspaper founded by Henry's father in 1937. Readership consisted of colored subscribers who wanted to read news about local black life. Black families in Gary had more money than Southern Negroes, and they also were more literate, leading the *American* to have newsstand sales and advertising revenue that kept the Whitlock family in comfortable circumstances.

In the 1940s, black-run newspapers numbered in the hundreds. Most were weeklies, but three daily papers cast a shadow over all the others: the *Chicago Defender*, with circulation of about 250,000, the *Pittsburgh Courier* (350,000 readers), and the somewhat smaller *Baltimore Afro-American*. In 1946, the *Courier* put out more than ten editions and operated news bureaus in twelve cities. But compared to these, the *Gary American* was a small operation that occupied an office on the second floor of a modest brick building. Henry Whitlock ran the paper with two or three helpers, his secretary, and now his wife. Edwina remembers her husband on the job:

"We owned the paper, but he ran it. I helped somewhat. Henry was the only person I met who could sit down and write an editorial at the linotype machine, where you couldn't correct your mistakes. I would write some of the front-page headlines and captions. A young fellow who worked for us, Admiral Murphy, would get the stuff and make a copy of it. Henry would take it and enter it on a linotype machine, which produced the type, and then he would make copies and run a galley for me to proofread. The printing press, which Henry operated, was in a room adjacent to our office."

When Henry and Edwina Whitlock set up house, in the summer of 1945, World War II was just ending, following the bombing of Hiroshima and Nagasaki and the surrender of Japan. These were boom years for international news, but still the *American* stayed firmly local.

"My husband was very much of a fighter," Edwina says. "The *American* was a local paper, and we fought to get black bus drivers in Gary, when there were none. We fought the electric utility to hire black women, because they didn't have any. Henry's father, who started the paper, was on the board of the Urban League, and tried to get certain jobs in the steel mills opened to Negroes, because not all of them were. All our circle and all our friends belonged to the NAACP, and attended the annual meetings."

Edwina Whitlock was twenty-nine, and her memories of Charleston were mainly of her childhood. Like many black Southerners, she had cut herself off from her family's roots by moving away. Edwina wanted home life, interesting work, children, and to be like any white American. She got all these things in Gary. Trying out her prose, she began a column in the newspaper called "First Person Singular," which offered social news and gossip to colored readers starved for both.

"I used to get my hair done at a parlor, because the best place to go to for news was the beauty parlor. I sometimes got my hair straightened, and they used to curl the front above the brow, but by that time I wasn't that meticulous. I mainly used to go because those women who owned the place knew everything, and I could get material at the same time as having my hair done."

Edwina wanted children, and she miscarried three times before her first child, Henry Whitlock, Jr., was born in 1949. After that, the rest came in rapid succession: Sylvia Elise Whitlock in 1950, Mae Eloise Whitlock in 1951, and Edwin Harleston Whitlock in 1953. When the children were all at home, Edwina wrote "Tantie" to report: "Speaking of Mae, she is quite fair. (I call her 'white folks.') Is as dainty as a Dresden china doll. And I'm not prejudiced."

Home life for the Whitlocks was comfortable most of the time. Although Edwina didn't have a housekeeper when she first got married, when the kids came, she and her husband hired a live-in servant named Pauline. About twenty-five and dark black, Pauline prepared meals and cleaned the house daily.

One year, in addition to their place in Gary, the Whitlocks bought a summer home in Michigan.

"There used to be four places in America where the black bourgeoisie went on summer holiday: Sag Harbor on Long Island, Martha's Vineyard in Massachusetts, Fox Lake in Indiana, and

Idlewild in Michigan. My husband gave me a summerhouse for one of my birthdays. It was two stories, with seven bedrooms, and it stood on ten acres near the town of South Haven, Michigan, which was south of Idlewild and on the other side of Lake Michigan from Chicago. We called it 'the Farm,' even though there was no one doing any farming, and we took the kids there during summers for eight years."

Edwina's life was often fulfilling, but any trip could be a reminder of the things she had left behind.

"I took my kids to see Aunt Lou in Charleston when my oldest son was three and my daughters were two years and six weeks old. We went on the train, and we had to change to the Jim Crow car in Cincinnati, because that was the edge of the South, just on the border with Kentucky. Black folks couldn't sleep on the train when you got below Ohio. In Cincinnati, all the colored people started getting on with their chickens and practically their goats. The colored car didn't have sleeping cabins or pull-down sleepers, so when we got to Charleston, we were exhausted and must have looked pretty bad. Aunt Lou met us at the station. She was getting old by then, must have been in her seventies, but she was still pretty wry. She said, 'When I saw you coming with all those babies, I thought, My Lord, here come the immigrants.'"

A center of Edwina's social life was a group of twelve couples that called themselves the Curl Club. Six times a year, the Curl Club assembled at a member's house to cook chitterlings, or chitlins, an old fixture in the diet of black Southerners. Chitlins were the cut-up intestines of hogs, a slice of flesh that was straight when it went into the pot, but which curled as it cooked. While growing up in Charleston, Edwina thought chitlins tasted awful and were even beneath her, but living in Gary, she missed the South, and pig intestines became a nostalgic symbol of what she had left behind. She remembers:

"The Curl Club met six or seven times a year, but not in the summer, because chitlins were seasonal and couldn't be bought in the hot months. The men in the group seemed to know all about chitlins, such as where to get them and how to cook them, but the women, for the most part, hadn't been exposed to chitlins before joining the club. First we had to clean them by turning them inside out and getting rid of all the corn that adhered to the lining in the intestines, and then we'd cook them for hours in a kind of stew. We used to invite guests, and if they didn't eat the chitlins, they weren't invited to the next meeting. After we ate, we would do some made-up preaching about chitlins, I mean, testify about them like in church. Somebody would stand up and imitate a preacher, and say, 'Now sisters and brethren, we is gathered here to talk about our chitlins! And first we will hear from Dr. Professor Smith.' Then one of the men would give a short sermon about the chitlins and how good they were. Each of us in the Curl Club would try to outdo the other with these sermons. Somebody would say, 'Praise God! These chitlins is the bes' I had since my grandmamma been take her chariot to heaven. The texture of them is like the silk that wrapped Jesus' body! The flavor is like da bread from da Las' Supper. The way they flavored is as big as Jesus feeding the multitude with the loaves and two fishes!' Somebody else would say, 'Can I have a witness! Now hear the man! Amen!'"

The Curl Club, however, came to an end after several years, because the men who belonged to it started dying. Two or three members were buried within a year, and those who were left disbanded, joking that the chitlin club was the reason, and that they didn't want to die.

But life in Gary, even among the black bourgeoisie, was far from ideal. A relatively high standard of living disguised plenty of social scars.

"A lot of black men who were professionals, like dentists, had a hand in the rackets to make the big money. . . . There were disproportionate numbers of black businesspeople with double lives. A pharmacist could also be a silent partner in the numbers, because our folks, colored folks, played the lottery, which was illegal then. The rank-and-file people made dime bets, and it added up big.

"And there was also lots of fooling around outside people's marriages. I had a secretary at my office who later became a state senator. She found out her husband was having an affair with a white gal in Valparaiso, which was forty miles across the top of Indiana from Gary. Long-distance calls started appearing on their phone bill, and she realized what was happening. My secretary, whose name was Carolyn, told me that she started to call the other woman. When the white gal picked up the phone, Carolyn said, 'Nigger lover,' and then she hung up. Carolyn thought she had a fool for a husband and would tell me this laughing.

"But I didn't play around. I really and truly loved my husband, and I wasn't going to be cheating on him, because with my luck I would have got caught."

Henry Whitlock was hardworking, but he had heart trouble. Early in their marriage, the newspaper editor suffered a mild heart attack, and then another. Edwina worried what might happen if her husband's health worsened, but she persevered with her full calendar of family and work. With four children, she spent the mornings as a mother and the afternoons as a journalist.

"We got a new housekeeper, Geneva," Edwina wrote Elise in 1955. "She comes each day at noon and stays until 6:00 p.m., and I don't pay her but $15 a week. In that way I spend the mornings with Harleston [the younger son], get the girls out to school by 9 a.m., and Hank [the older son] off to kindergarten at 10:30 a.m. I take Harleston shopping with me from 10:30 until the girls get home at noon. Then most of the afternoons I spend in the office."

Office work was draining and brought unpredictable troubles.

"We had a new office girl," Edwina told her aunt. "For a couple of weeks I tried paying her $30 a week, but couldn't see that it was being spent for the maximum good to the paper, so I let her come three days a week. However, she had three epileptic seizures in my office one day, fell out on the floor and was quite ill, so naturally I had to let her go. She was sort of dumb, too, and I had to explain things to her over and over again."

In 1960, Edwina's husband decided to run for the local office of justice of the peace. He had name recognition in the black neighborhoods, and the *American* was a free advertising medium for the campaign. Henry Whitlock's opponent was an Italian-American named Rocco Schiralli, who won the election after a vote count that went on for two days after the ballot closed.

"They stole the election," says Edwina. "Gary was crooked. Two days after the election, my husband and I were mulling it over in the living room. The stress of the campaign had been pretty high, and Henry was probably feeling strung out. We were both sitting down, and I was reading the paper, and I was tired. Henry wanted to go to a party we'd been invited to, so he could relax. There was a dance at the time called the Madison, and Henry got up out of his chair to show me something about the party and the dance. He said, 'All they're gonna do there is the Madison.' I said, 'You go to the party if you want to do the Madison. I don't want to go.' Henry said, 'One of them dancers is going to be like Wilt the Stilt Chamberlain,' meaning the basketball player, 'and I'm gonna be Roy Campanella,' who was a baseball catcher. 'All Campanella had to do was squat.' Henry got up to show me the squat and fell over, splayed out, with a heart attack.

"I jumped up and called our friend Kenneth Washington, a doctor who lived ten minutes from our house. While I waited for Ken to arrive, I could hear the death rattle in my husband's throat, and by the time he got there, the rattle had subsided, and Henry was dead. It was about eleven at night. The kids were upstairs asleep. Then the people from the church came. They had to scrub the floor.

"Henry died May 5, 1960. We had been married fifteen years, and when he died, he left me knee-deep in debt. I ran the newspaper by myself for about a year, but it was too much for me. On top of that, my husband's friends were trying to console me, I mean sleep with me, and their wives thought I was letting them. It was a mess."

In some ways, Edwina's situation after her husband's death paralleled that of Elise Harleston when she became a widow following the death of Teddy Harleston. Some of Edwina's in-laws, apparently stunned by Henry Whitlock's death, cooled toward Edwina. Practically alone, and preoccupied with running the newspaper, Edwina saw little of the Whitlocks after her husband was buried. A few relatives made insinuations that Edwina would not be able to run the *Gary American* and raise her four children. Because the Whitlock family had been in Gary for generations, whereas Edwina was from far-off South Carolina, there was even talk that they might try to take away Edwina's children so they could be reared in the right churches and clubs.

"Several of my in-laws thought I was crazy—I couldn't get along with some of them—and I knew I had to get out of there. I was forty-four years old. I had my four children, who were like little stair steps, each one a little taller than the last, up to the oldest one, who was just eleven. I was in a fix and I didn't know what to do. After Henry died, it was six months before I was able to train a young woman to take over the printing equipment. I got her trained, then I decided to get some advice from people I trusted, so I went to New York to see my oldest friend from Talladega College to ask her what I should do. That was A'Lelia Ransom Nelson. A'Lelia was my former sister in the Sacred Order of Ancient Pigs, and my partner in the Ain't-Got-Nothing Club. She lived at 5 West Ninety-first Street, a few steps from Central Park, in New York City.

"I arrived in the city and made my way to uptown Manhattan. A'Lelia and I talked and talked, and she bolstered me for several days. I brooded and walked around the streets of New York. A'Lelia suggested I move to Los Angeles, because Tantie, my aunt Elise, had just moved there from Baltimore with her husband, Jack Wheeler. I thought it was a good idea. But the day before I was to go home to Gary, I went to see another person in the city to get another opinion. That was the visit that changed me."

Thirty blocks north of A'Lelia Nelson's apartment was an undertaking firm, the Carolina Funeral Home, which belonged to Eddie Mickey, the former Charleston mortician who had moved to New York after the Mickey Funeral Home failed.

"I walked uptown and into the office of the Carolina Funeral Home to see the owner, Eddie Mickey. I knew Uncle Eddie from my childhood growing up in Charleston, and I loved him and worshiped him. Uncle Eddie had closed the Mickey Funeral Home in Charleston about twenty-five years earlier and moved up to New York. When I went to see him he was in his late seventies, but he was gentle and elegant. I hadn't seen him much for decades since I moved away from home.

"When I stopped in at the Carolina Funeral Home, I was feeling kind of vulnerable. Uncle Eddie's business was in a brownstone, one in a row of them, I think, and his offices were on the first floor. His apartment was up the stairs, above the casket parlor, so I entered the hall and climbed the steps to find him in his bedroom. The room was decorated in a kind of Victorian style, with lace curtains and dark furniture, because the Mickeys had all this old furniture. There was Uncle Eddie, lying in a four-poster bed with a canopy on it. He said he was on his deathbed. I had heard this before, because Uncle Eddie had always been a hypochondriac. We talked about my husband and how he'd died, and Uncle Eddie said he didn't think he was going to live long, either.

"'I'm going to die soon,' Uncle Eddie said. 'But before I do, you need to know something.'

"'What is it?' I asked him.

"'You need to know who you are.'

"On the bedside table I saw a red ledger book, lying there closed. Uncle Eddie picked up the book and opened it.

"'This is your legacy—the Harleston family legacy,' he said.

"I opened it and saw the pages were full of notes and charts, and all sorts of dates. The little red book was the result of Uncle Eddie's research into the family. He had listed all the Caucasian as well as the black connections of our family. I don't know how he found some of that stuff, because he wasn't light enough to pass for white, and we black people couldn't go anywhere to do research in the South. We couldn't go to the library or the museum, because we would have been turned away. But Uncle Eddie was determined, and had done his digging. I guess because Uncle Eddie lacked a family of his own, he had poured his energies into genealogical research.

"Anyhow, I knew that the Harlestons were an unusual group, but I did not know the whole story. Uncle Eddie talked about William Harleston and Kate Wilson, the white patriarch of our branch and his black consort, who started the family so many years before. He talked about the white Harlestons, and how one of them had stolen Kate's inheritance from her. He talked about Kate herself. I didn't know any of this, you see. I didn't know about our connection to white people, because my elders didn't talk about it when I was young. I knew we were different, lighter than other people, but there was no story behind it. Maybe it was too painful, I don't know. That day in Harlem was the first time I thought about how our family might be related to whites. And after he showed me the red book, Uncle Eddie, lying there in his bed under that canopy, gave me an injunction.

"'You are the writer in the family,' he said, 'and I'm entrusting this book to you. I want you to promise me that if I give you my findings, you will finish what I've started.'

"It was really supposed to be a social call, and the little red book was unexpected, but I told him I would try to follow his wishes. I took Uncle Eddie's book, put it in my pocketbook, and made my way into the streets. When I got home to Indiana, I put it on my shelf. I didn't have time to study or write, because I had to go to work and raise my kids.

"But sometimes I took Uncle Eddie's book down and looked at it. It had snippets of letters, handwritten copies of wills, genealogical charts. Uncle Eddie had written a copy of the will of his grandfather, the Caucasian, William Harleston. He had some notes he had made about Benjamin Huger, the nephew of William Harleston who had gone to Kate Wilson's house to get her to sign over her inheritance. There were lists of names, and diagrams that connected people to one another. Uncle Eddie had written down the names of my uncles and aunts, grandparents and great-aunts, all the others. In between, he had written up some family stories I'd heard, and many I hadn't.

"Fortunately, Uncle Eddie wasn't actually dying. He got better and lived another five years. That wasn't the first time he thought he was going to die, but he was never embarrassed by these episodes. And he didn't ask me to return the red book, because he had given me his injunction, and I had given him my word."

In 1962, Edwina Whitlock moved to Los Angeles with her four children. The family lived in several rented apartments before finding a home in south Los Angeles, at 124th Street and Figueroa, on the edge of the predominantly black neighborhood of Watts. Edwina got a job working in public relations at a local bank called Watts

Savings & Loan and sent her children to public schools. Elise Harleston Wheeler, now seventy-one years old, lived nearby, which meant Edwina's children were able to grow up around "Tantie" just as Edwina herself had done in Charleston. Elise helped with the children when she could, and gave support and comfort to Edwina, who struggled as a single parent.

In August 1965, three years after Edwina and her family had settled in Los Angeles, a racial rebellion erupted in Watts. For nearly a week, buildings burned and gangs of teenagers stormed the streets, looting and smashing windows, burning and upturning cars. Thirty-four people were killed and some one thousand injured in the riots, and Governor Edmund G. "Pat" Brown called out sixteen thousand National Guardsmen to cordon the neighborhood and restore order. An estimated thirty-five thousand African Americans took part in the uprising.

"The riots were terrifying," Edwina remembers. "When they started, we had friends visiting from out of town, and they were supposed to come over to our house, but they couldn't get through because the National Guard had roped off Figueroa. We could walk down to Figueroa from our house and see the fires. You could see the soldiers in their uniforms, with their guns across their shoulders. They walked around with bayonets on their rifles, and they wouldn't let you cross the lines. My office was close to where the fires were, and I couldn't even go to work."

The riots subsided, but Edwina knew they were a signal of some kind. She had been thinking about her uncle Teddy Harleston, who had founded a chapter of the NAACP in Charleston so many years before. Although Edwina was no activist, she knew doing public relations for a bank would mean less in the long run than another kind of career. Quitting her job at the bank, Edwina decided to train as a social worker.

"I studied for the 'War on Poverty,' which is what the Lyndon Johnson administration called it. I guess I was one of the advance soldiers of that war. I mean, these people were smart, the training staff for the social workers—they were idealists, and we all believed in what President Johnson promised about finding good jobs for blacks."

After passing a civil service exam, Edwina got a position with Los Angeles County as a "Social Caseworker I." The work took her throughout the city and into the homes of poor families, both black and white. For the first time, nearing the age of fifty, she began to spend every day with people less privileged than herself, who happened to be white.

"I wasn't around white people that much till I got into social work, because before that I was a journalist. The first assignment I had was 'general relief,' in Torrance, which is south of Los Angeles. General relief was the scum of the earth—low-class blacks and whites, but mostly whites—and you had to go in and interview them and do an assessment of their situation. It was the lowest form of welfare, merely subsistence, maybe ninety dollars a month, and they got food stamps. A whole lot of the whites were drunks, and I got a good look at white families for the first time. Some of them I made come out on the porch, because I couldn't even go in the houses, since I'm sensitive to odors.

"One of the differences between white people and blacks is that white people smell different from us. Everybody has an odor if you're really sensitive. It's not scientific, but it's my own deduction that whites have a special smell. You notice this when you get near them, that white people smell like an old copper penny that you've been holding in your hand. Pennies have a peculiar odor, I don't care how clean they are, especially in the summer when it's hot, and you hold them in your hand. The smell of a copper penny is not a pleasing odor, or a displeasing odor, but it's an odor.

"In Charleston there was a family, the MacBeths, who lived on Calhoun Street. John Macbeth was my rival in school. They were really pale colored people, and they finally left and started living as white folks, crossed over. Well, the Macbeth family, when you went into their house, it smelled like copper pennies. In Los Angeles, when I was a social worker, I was in white people's houses for the first time, and there was that copper penny odor again."

While she worked for Los Angeles County and reared her children, Edwina forgot about Eddie Mickey's injunction to study the Harleston family story. Several years passed. In 1970, her daughter

Sylvia, who was studying at Talladega College, her mother's alma mater, sent a letter to Edwina saying that she wanted to write a term paper about her great-uncle, the artist Teddy Harleston. Edwina's interest in the family was rekindled.

"After Uncle Teddy died, Tantie almost never talked about him or about her own work as a photographer. She closed that door behind her and went on to life with her second husband. I knew that Tantie had some papers that belonged to my uncle Teddy, but I didn't know how much she had, and I went over to her house in Los Angeles to ask. Tantie's husband, my uncle Jack, had died, and she had moved in with a grandnephew. She was about eighty, and failing. I didn't get all the material I needed for my Sylvia, and in fact I didn't have any idea Tantie had all the stuff I later found out she did. But I told Tantie's grandnephew, a young man named Albert Hinton, about my curiosity, and then I left.

"I put together a couple of the clippings I had and some photographs, and I sent them to my daughter at Talladega. Sylvia wrote her paper. That episode put my mind back on the subject. I kept thinking there was something more to it, because I had Uncle Eddie's little red book and his stories. Well, a little later I went on a cross-country trip by car, which took several weeks. When I came back, I found that somebody had left a steamer trunk in the grass in our backyard. I thought at first the trunk had automobile parts in it, because my son Hank was working on cars at the time. I got someone to help me bring it inside, and when I opened it, it was full of dusty papers and old pictures. Turned out this young cousin of mine, Albert Hinton, had brought the trunk over from Tantie's house and left it in the grass. It was Tantie's whole collection. The box had the smell of old paper and old houses, like the bottled fragrance of Charleston released in the living room. It was providential, because it hadn't rained in the weeks while I was away, which could have destroyed everything, and all of it was perfectly preserved. There were letters between Uncle Teddy and Tantie, which the two of them had written to each other before they were married. There was their whole correspondence when they were husband and wife. And there were Uncle Teddy's papers and photographs, and funeral home things belonging to the Harlestons.

"Ten years passed again. I tried but I wasn't able to write Uncle Teddy's story, even with these papers. My friends started complaining to me that it was all I ever talked about. I was sixty-five, and my kids were gone off and had their lives, and I was living alone in Los Angeles. Then one summer, something brought me to Detroit, and there I encountered a woman named Josephine Harreld Love, who operated a children's museum. Josephine said she wanted to organize an exhibition of Uncle Teddy's paintings and asked if I would help her. When I got back to California, I took an early retirement so I could work on the retrospective. I had worked for Los Angeles County as a social worker for seventeen years. I started to travel, looking for Teddy's paintings in private collections around the country. The exhibition opened in Detroit in 1983, and then it went to other museums. I wrote an essay for the catalog that was published with the show. I thought I had done at least some of my job, getting more notice for Uncle Teddy's paintings."

In 1983, Edwina Harleston Whitlock moved back to Charleston, hoping to continue researching the family. She had lived away from the city for fifty years, ever since she left the Jenkins mansion to go off to college, but her hometown had never left her. Now there was the added draw of trying to fill out the Harleston story. In Charleston, Edwina renewed ancient friendships and became active in several clubs. She sifted her memories, paged through the collection of papers that had once belonged to Elise. And, above all, she thought of her uncle Teddy.

Chapter Twenty-one

Atlanta, Georgia—2001

IN southwest Atlanta there is a section of rolling pine forest shaped into a subdivision, where Edwina Harleston Whitlock lives in a one-story brick house nestled in a dip below the road. She is eighty-four and moved to Georgia in 1994 to be near her two daughters, who have houses in the neighborhood.

Edwina Whitlock starts the day at 9:00 A.M., when she pricks herself with a needle to draw blood. For many years she has suffered from diabetes, and every morning she measures the level of glucose in her system. A red drop oozes onto the tip of Edwina's finger, and she presses it into the sensor of an electronic device that beeps and flickers out a number on a screen. Using a preloaded syringe, Edwina then injects herself with insulin. Although on several occasions I've observed her drawing blood herself, I have not seen Edwina make the injection, which she usually does in her stomach, under her bathrobe. Edwina then takes some prescription pills, eats breakfast, and sits down at a card table in the living room in front of a stack of old photographs. Most of the pictures she lingers over show long-deceased members of the Harleston family, but there are a disproportionate number of photographs of Teddy Harleston, Edwina's uncle, and these photos take the most time.

She is the last member of her family who knew personally the generation she calls "Kate's eight," the children of William Harleston and his slave concubine, Kate Wilson. Kate's eight were old when Edwina was a girl; nevertheless, this part of the Harleston story will be diminished when Edwina dies, because no one who actually spoke with Captain Harleston and his siblings will be left to describe them.

Edwina was once a profligate letter writer who maintained a handwritten correspondence with lists of friends, but now it takes her five minutes just to write her own signature. It is difficult for her to read, and she moves carefully. She has lost some of her hair, which sometimes leads her to wear hats indoors. And yet Edwina's mind bristles with incident.

We sit in the dining room, talking about the past. On the wall are several paintings by Teddy Harleston. Gesturing a little with her hands, Edwina tosses out impertinent remarks as though she were feeding a pigeon.

"Do you know the story of the black airplane passengers?" she asks.

"Well," Edwina says, "this mother was on a flight with her little boy. I should say they were black. And the pilot comes into the cabin before takeoff to make an announcement. 'I'm sorry, but we have too many people on this flight, and we have to bring down the weight. We need two people who will volunteer to get off, but we're going to do this alphabetically.' The captain looks around and says, 'Are there any African Americans aboard the plane?' Nobody answers. The little boy looks at his mother. The captain says, 'Are there any blacks who will volunteer to get off?' Silence. The little boy looks at his mother again. The captain is frustrated, and says, 'How about colored people getting off the plane, now!' The little boy leans over to his mother and says, 'Momma, aren't we one of those?' The woman whispers, 'Shut up, son! Today, we's niggers!'"

Teddy Harleston, who died in 1931 at forty-nine, came to a cheerless end. His work later appeared in a few exhibitions—at the National

Gallery of Art in Washington, D.C., in 1931, and at the "American Negro Exhibition" in Chicago in 1940—but within thirty years, many Harleston paintings were lost to the winds. A 1983 catalog included a partial list of his work, but a greater number of canvases have vanished. When alive, Teddy was locked out of exhibitions by Jim Crow and pushed to the margins of the art market by modernism. In death, he has been introduced to indifference.

Edwina tells a parable about one of Teddy's paintings, *The Bible Student,* which he finished in 1924. The picture shows an old man intently reading.

"Everybody who owned *The Bible Student* seemed to have had bad luck. That banker, Jesse Binga of Chicago, bought it after he saw it on the cover of a magazine. A couple of years later, Binga was under investigation for fraud. He was arrested for some kind of shenanigans and went to prison. His bank failed, his life was destroyed, and when he got out of prison, he sold *The Bible Student*. Years later, my husband and I were spending the night in a hotel in Chicago, and there was *The Bible Student* hanging over the fireplace. We asked about it, and the owner of the hotel came out. He had bought it and was pretty proud of the thing. Next I heard about this man, a year or two later, was that he had gone out of business and lost his hotel. He had to sell the painting to a collector, a woman named Frankie Singleton. I didn't get to see it when she had it. But this Frankie Singleton was having an affair with a married man, and a short time after she bought Uncle Teddy's painting she died an early death. Bad luck again. Then her sister inherited it. I went to visit the sister, and she took me to Frankie's old apartment around the corner. There was *The Bible Student,* and the sister said, 'This should be in a museum.' That's the last I heard, and I'm afraid to inquire about the sister."

Teddy Harleston's wife, Elise Forrest, had a somewhat kinder fate. After Teddy's death, she made a clean break from photography and remarried. Edwina says that during her second marriage (to Jack Wheeler, the physics teacher), Elise didn't talk about Teddy or her former career as a pioneering photographer, a period in her life that ended painfully. She let her archive of pictures become lost and dispersed, and little remains of it. Elise was evidently happier having left the Harleston family in South Carolina, and keeping her personal memories of Teddy locked away. Jack and Elise Wheeler raised Doris Forrest, Elise's niece, and the couple lived in Baltimore and later in Los Angeles, where they moved in the 1940s. Jack Wheeler died in the 1960s, and Elise died in California in March 1971, at eighty.

The Rev. Daniel Joseph Jenkins fared better than his in-laws, the Harlestons. When Jenkins died, in 1937, he was a majestic figure to thousands of blacks and possibly an equal number of whites. In 1985, a painting of Jenkins was commissioned and hung at city hall in Charleston, the only portrait of a nonwhite citizen in the building, even at that time. During the first fifty years of its operation, the Jenkins Orphanage gave shelter to upwards of five thousand children. Jenkins's work with young people, although strict and sometimes self-aggrandizing, improved the lives of uncountable families. The Jenkins Orphanage has remained in operation, although its name has changed and the music program has been suspended.

Edmund T. Jenkins, the musician who died in Paris in 1926 at age thirty-two, was a promising composer who might have done more but for his early death and has since been largely forgotten. His sheet music appeared to have been lost for decades, until it surfaced in the 1980s in Boston. But none of it was performed in the United States until 1996, when several of his pieces appeared on a concert program in Charleston.

Eloise Harleston Jenkins lived in Charleston after her husband's death, where she ran the orphanage for several years before moving to an advisory role. In 1949, *Ebony* magazine published a feature story entitled "Wealthy Widows," and Ella gave her permission to be photographed for the story, which described her as follows: "Mrs. Eloise Jenkins inherited an orphanage and valuable real estate in Charleston, South Carolina, when her businessman husband died in 1937. Widow disposed of holding in orphanage in 1941, but now has regular income from funeral business grossing $25,000 yearly in addition to rentals from fifteen pieces of property."

During her long widowhood, which lasted nearly forty years, Ella never reconciled with her daughter, Olive, the "foundling." Ella kept

Olive at a distance, always refusing to acknowledge her as her own child. Ella Jenkins died in Charleston in May 1975, at ninety-three.

"Aunt Lou was blind by the time she died," Edwina remembers. "The funeral took place at New Tabernacle Fourth Baptist Church, where Reverend Jenkins had preached, and Aunt Lou was disgraced at her own funeral. The service bulletin was prepared by one of her nieces, and it announced that Olive was Aunt Lou's surviving daughter. I've never seen a funeral where people were whispering and pointing like they did, and Aunt Lou's memory was pretty trampled. The protocol of a funeral is firmly established, and in the procession, Olive was first in line behind the body, laying claim to her mother in the role of chief mourner, although her mother had rejected her throughout life."

Olive Jenkins, born in 1906 in Wigan, England, was married twice, first to Carroll Frederick, a South Carolina attorney, and then to Chester W. Campbell, of Yonkers, New York; she had children by both husbands. Olive made many attempts over the decades to please Ella Jenkins, and win her acceptance. In 1974, a short time before Ella's death, sixty-eight-year-old Olive wrote a letter to a relative, which read in part:

> What can I gain from my mother's death? It is what she could have and should have given to me in life: a Name. The minute she married my father, she then removed the stigma of illegitimacy from my name, but she did not do one thing about it. How do you think I felt when I saw her in June, and she said Yes, she knew and remembered me, her "adopted daughter." Can you understand what this did to me? The fact that Pilate, and not his wife, washed his hands of the matter of the Crucifixion may be the solution. As for my mother, she is even now paying the price for her heartlessness.

On the night of August 12, 1987, in Yonkers, Olive Jenkins Campbell was returning home from a meeting of a club to which she belonged. As she often did, she had been playing the piano that night. While crossing a dark street, Olive was struck by a car and killed. She was eighty years old.

* * *

Fate has dealt unevenly with other figures in the Harleston and Jenkins world. Doris Forrest, whom Elise and Teddy Harleston adopted in 1924, grew up in Charleston and Baltimore. She married twice, moved to Los Angeles late in life, and had several children.

Mamie Randall, the mistress of Captain Harleston who once lived behind the funeral home, fared well until her lover's death, and then her life collapsed. Edwina remembers that the police were at the door before the Captain's body was cool:

"My grandfather's lady friend, the one we called 'Mama Sister' that the Captain built the house for in the back, she didn't think that house was good enough. She wanted a brick house, instead, so in the 1920s Captain Harleston built one for her, and she moved out with her children and lived in it until he died. Mama Sister had at least two, perhaps three, children with the Captain, and they attended his funeral. At the service, Mama Sister put on a real performance. It was acceptable for colored folks to carry on at a funeral, and people wept openly. Mama Sister was screaming, 'Oh, Cap'n so good! He's sooo good!' Everybody at the service was kind of nudging each other with their elbows, saying, 'What was she talking about? Was he good, or was it good in bed?'

"When he was alive, the Captain protected her, because of the abortion business. But when he died, there was no one to keep the police away, so they came and arrested her. Mama Sister spent time in jail for performing abortions, but I don't know how long."

At some point in the 1930s, Mamie Randall was released from prison, because ten years after Captain Harleston's death, she appears in a city directory as the owner of a lunch counter at 68 Calhoun Street in Charleston, the Ritz Sandwich Shop. Mamie Randall died in the 1940s.

The Mickey family, Charleston undertakers who once competed with the Harlestons, came to disparate ends. Some of them went quietly, while at least one disappeared in scandal. Eddie Mickey, proprietor of the Mickey Funeral Home and keeper of the "little red book" of Harleston genealogy, died in New York in 1965, at eighty-two. He had no children and was buried in Charleston in the Mickey plot at Unity and Friendship Cemetery. But another family member, Ellen, ran into trouble. Edwina and I sit on her sofa as she recites the last facts of the Mickey story.

"My uncle Eddie had a sister, Ellen Mickey, who, like her brother, was also gay, which led to a notorious incident with one of her lovers. Ellen was the one who drove the Mickey hearse and stole all those crystal glasses from Uncle Teddy. She could be very angry, and her anger was dangerous, and finally caught up with her. Ellen had become involved with a woman who was married. She got jealous of the woman's husband, and so she went out and got a gun. Ellen waited for the unsuspecting fellow, her rival, and then just shot him. I don't know if he died, or what happened, but this time the Harlestons pulled together and protected one of their women. There is a letter written by Uncle Teddy that describes the final blowup. In the letter, he said, 'The episode has shocked everyone, and we've had to spirit Ellen out of town for her own safety.'"

With that, Ellen Mickey vanishes from the Harleston scene.

The children of the Jenkins Orphanage did better or worse, depending on whether they participated in the music program. Nearly all of the Jenkins inmates grew up to become blue-collar workers, from domestic servants to shoe repairmen, housepainters, and seamstresses. Most of the hundreds of musicians did likewise, becoming taxi drivers or construction workers when their talent wouldn't propel them further. But a lucky few overcame the odds and made lives in jazz.

Tom Delaney, the first of the orphans to have a music career, went on to fifteen years of songwriting after his first success, "Jazz-Me Blues." In 1935, Delaney settled in Baltimore, where he played an occasional show, but he worked mostly outside of music. He died at age seventy-four, in December 1963.

Drummer Tommy Benford played and recorded with dozens of musicians in a working life that spanned sixty years, from 1920 (including nine years in Europe, from 1932 to 1941), until he was well into his eighties, by which time he played only weekend gigs. Benford died in Mount Vernon, New York, in 1994, at eighty-eight.

Guitarist Freddie Green had what music critics have called "the longest job in jazz history." He was part of the Count Basie band, with only a short interruption, for fifty years, from March 1937 until March 1987. Green played with Basie at President John F. Kennedy's inaugural ball in 1961, and in 1981, President Ronald Reagan honored Green (as a member of the Basie band) for achievement in the performing arts. Freddie Green died in Las Vegas in 1987, after walking off a stage where the Basie orchestra was finishing its second show of the night. He was seventy-five years old.

Jabbo Smith, the trumpet player who jousted with Louis Armstrong, reached the peak of his playing in 1929, when, at age twenty, he recorded twenty songs for the Brunswick label. The Jenkins prodigy became erratic after that; his work suffered from his drinking. In the 1930s, Jabbo played with at least a dozen acts, and in 1939, he had a small band at the New York World's Fair. For several years in the 1940s he played in Newark, New Jersey, and then moved to Milwaukee to be with a girlfriend. He got fewer jobs and eventually quit playing, finding work at a car rental company. In 1959, he got a job on trumpet with a rock-and-roll band, while his wife drove a cab. But in 1969, his career revived when he was featured in a book on early jazz. At sixty-one, Jabbo traveled to Europe for the first time, playing in Holland and West Germany. In the late 1970s, a theater producer heard him play and asked Jabbo to join the cast of a musical he was putting together called *One Mo' Time*. The show succeeded and Jabbo became a regular act again. He appeared frequently at the Village Vanguard in New York, where, one night in 1979, Edwina Whitlock walked through the door. The two hadn't seen each other in more than fifty years.

"I was visiting my old college friend A'Lelia in New York, who had consoled me after my husband died, and she and a couple of friends went down with me to Greenwich Village, to the Village Vanguard. There was a man playing this trumpet like he owned it. I said, 'Seems like to me he's from the Jenkins Orphanage.' My friend said, 'You think anybody of any good at all is either from Charleston or from the orphanage. How much you want to bet?' The fellow who owned the club was a Jewish guy, but a Chinese woman ran it. I asked the Chinese woman if Jabbo Smith was in the Jenkins Orphanage, and she said she would ask him. I got back to the table and sat down. Jabbo came out from the back and sat down, too.

First thing he said was, 'You Sylvia Harleston's sister? Oh, how sweet.' He had known my sister when she lived with Aunt Lou at Jenkins's house, when Jabbo was an orphan in the band. My friend A'Lelia was pretty chagrined at that."

The trumpet player lived for a short time at the Village Nursing Home in Greenwich Village before his death in 1991, at age eighty-two.

In the twilight of her life, Edwina Whitlock misses her uncle Teddy. He was the crux of her youth and the person whose memory she venerates. For many years, Edwina has wanted to place a marker on Teddy's grave. Although the Harleston plot at the Unity and Friendship Cemetery is large, with a brick and stucco border around it and the name HARLESTON chiseled on a marble slab, the place where Teddy lies buried is bereft. (His grave went without a tombstone after he died in 1931: the family had just buried his father, Captain Harleston, and in the difficulties that ensued with Teddy's widow, Elise, no stone was ever placed.)

Seventy years after her uncle's death, Edwina ordered a stone to mark his grave. To see it properly placed, she made plans to visit Charleston one last time from her home in Georgia.

Teddy's new stone is white marble with gray veins; it stands sixteen inches tall and twenty-five inches wide, with a surface for inscriptions that tilts away from the viewer at a forty-five-degree angle. It looks like a lectern from which a reader could speak.

Edwina and I make our way to her family's cemetery north of Charleston. On one side of the road are the graveyards full of the colored elite, and on the other side are the white cemeteries. We arrive at the site to find the Harleston plot has been trimmed and cleaned. In the middle, flanked by freshly planted flowers, is the new stone, which reads EDWIN A. HARLESTON—ARTIST—MARCH 14, 1882–MAY 10, 1931.

Edwina is wearing black trousers and a lavender blouse and carries a large bouquet. She moves slowly so as not to stumble on the lumpy grass.

"I have recurrent dreams about Uncle Teddy," she says. "The dream takes place on the stairs of our house. Those stairs were wide, and I used to count the steps. There were twenty-three steps to the first landing and twenty-five to the second landing. Those were the stairs where Uncle Teddy came up and he told me that I wasn't an orphan, and that I would always have him. In the dream, I'm floating down from the top floor all the way to the front door."

The day is warm and moist, and the air is full of bugs that buzz around our heads. Edwina looks around the graveyard at the thousand tombstones stretching out in every direction. She waves away the gnats, adjusts her hat, and we move over to the place where the stonecutters have put Teddy's marker.

"There's a stereotype that black people like to spend money on funerals," says Edwina. "Well, we do, we do. Black people like extravagant funerals, and I guess I should be grateful that they do, because of Uncle Teddy's business."

I ask Edwina if she thinks there are any differences between white people and colored people, and she answers:

"You know, statistics show colored people do not commit suicide as often as white people. That's because we are accustomed to the slings and arrows of outrageous fortune and can stand things more than you-all. White people expect more to be given to them, so they have more things to be unhappy about. We can console ourselves, and you see it with our songs, in our dance, in our spirituals. We can get together and sing, and make jokes about our condition. But whites, they're unable to withstand the kinds of things we've withstood for generations. They have trouble translating suffering into art, or pleasure, for instance."

Edwina places her bouquet on the ground in front of her uncle's stone.

"Also, white people, they're conscienceless. I'm sorry, but your set of ethics stinks. Black people are far less materialistic than white people, and we wouldn't, as a rule, do the things that white people do to get money—not unless we were brought up around white people. And another thing, we're not nearly as materialistic as you-all. White people are far ahead of us in the love for things. You do all these ugly things to other people to get money, and then on top of it you-all don't tell the truth, you know, about real historical facts. You don't face them. That's my theory, don't take offense."

Early in her life, Edwina apparently didn't hold such strong opinions about white people, but she developed them over the years. She once told me she didn't have any white friends, and I had noticed a kind of wariness in her dealings with whites. Like anyone would, she probably kept her worst experiences to herself, and was cautious after having been stung.

Teddy's tombstone is the newest marker in the cemetery, which has not seen a funeral in some time. Here and there in front of other stones are vases full of plastic flowers, which offer their faded red or yellow color against the elements.

"I have never attended a white funeral," Edwina says, "so I don't know how you people behave around death. I've never been close enough to a white family that had a death, but I've been to a lot of black funerals. I'm used to people dying, so when I hear the news of somebody's death, I'm never shocked. I was four when my mother died, and my father died when I was twelve. When I was fourteen, my grandfather died, and my adoptive father, Teddy, both in the same month. My sixteen-year-old sister died after that. I got married and my husband dropped dead at my feet. I fully expect that somebody's going to die, and I'm inured to death. But I ain't a weeper."

Edwina kicks her flowers to push them into the right spot.

"I'm ready to die—at any rate, I'm ready to be relieved of my anxiety. And I'm also curious about the other side, what comes next. I don't know what's on the other side, and I have not fortified myself or stuck to any belief. But I'm not afraid."

Family history is history in miniature. If the skin of one person is scratched, the world dribbles out from inside and assumes recognizable shapes. Maybe tragedy appears, or a utopia. The Harlestons aren't different from other people in that way, not finer because of the scale of their world. But they do have one gift, which is that they can see something of where they've been.

Notes

Sources

This book attempts to give a "true" account of one American family and some of the people associated with them from the mid-1800s until 2000. It grew from family papers, interviews, and secondary material. But the Harlestons met the fate of all historical figures: their experiences combined with a writer's projections and they became characters in a plot. I tried to make the personalities reflect those in the family, remembering that different portraits could have been made from the same material.

I relied to an extent on the Harleston Family Papers, housed at the South Carolina Historical Society in Charleston. Edwina Harleston Whitlock loaned me duplicates of those papers—which she had kept after she placed the originals in that archive in the 1980s—as well as many manuscripts she had gathered later that were not yet in public hands. This combined set, the family papers plus the private collection, I've called the Edwina H. Whitlock Papers. Half are in public domain, and all will eventually land there.

My researcher, Nichole Green, uncovered other primary material, especially from the Jenkins Orphanage. Many of the records of the orphanage were destroyed in a fire in 1933. The music, manuscripts, photographs, and interviews with former inmates of the orphanage came from scattered archives and individuals, noted below. Though most material is documentary, I interviewed members of the Harleston family and people who had known them in the 1900s. A framing signal, such as "according to" or "family tradition says," announces a source in oral testimony.

I've taken one or two liberties with primary sources. Occasionally words or phrases were removed from letters without inserting ellipses, and sometimes two letters closely spaced in time were compressed into one. Interviews with Edwina Whitlock, which amounted to some seventy-five hours, were edited and statements from different interviews combined. Even with these alterations, the story survives scrutiny under the standard of preponderance of evidence, as the following sources attest.

Chapter 1

3 Harleston family: The Hut: Henry A. M. Smith, "Landgrave Ketelby's Barony," *South Carolina Historical and Genealogical Magazine,* 15:3 (1914), 146–47; Harleston genealogy (white): Theodore Jervey, "The Harlestons," *South Carolina Historical and Genealogical Magazine,* 3:3 (1902), 151–73.

3–4 Harleston family during the Revolutionary War: Fort Sullivan: *American Revolution Roster, Fort Sullivan, 1776–1780* (Charleston, SC: Daughters of the American Revolution, 1976), 172; Edward McCrady, *History of South Carolina in the Revolution, 1775–1780* (New York: Russell & Russell, 1901), 143; Bobby Gilmer Moss, *Roster of South Carolina Patriots in the American Revolution* (Baltimore: Genealogical Publishing Co., 1983), 415; loans and slaves: Isaac Harleston to Commissioner of the Treasury of South Carolina, Mar. 22, 1780; State of South Carolina to Isaac Harleston, 1779 and 1780, in accounts audited of claims growing out of the Revolution in South Carolina, South Carolina Department of Archives and History, Columbia, SC.

4 Isaac, son of William Harleston Sr.: Edward C. Mickey, "Red book" of Harleston genealogy, ledger book ca. 1950, papers of Edwina Harleston Whitlock (hereafter EHW); Harleston family oral tradition.

5 Black majority: Fourth Federal Census of the United States, 1820; Inter-University Consortium for Political and Social Research, Study 00003, *Historical Demographic, Economic, and Social Data of the United States, 1790–1970* (Ann Arbor, MI: ICPSR).

6 Thomas Jefferson and Sally Hemings: Annette Gordon-Reed, *Thomas Jefferson and Sally Hemings: An American Controversy* (Charlottesville, VA: University Press of Virginia, 1998); James Akin, "A Philosophic Cock," Newburyport, MA, ca. 1804, hand-colored aquatint, American Antiquarian Society, Worcester, MA.

6–7 Guardianship: Will of Sarah Harleston, July 20, 1821, Charleston Probate Court, Will Book F, 348.

7–8 Kate Wilson: Death Certificate of Kate Harleston, Nov. 14, 1886, Charleston County Probate Court; presence at Elwood: Charleston County, South Carolina, Register of Mesne Conveyance, 19–77 & S10–248, Schedule and appraisement of the plantations Mepshew, Pimlico & Kecklico, also of the Negro slaves . . . July 20, 1830, Ball family papers, Duke University, Durham, NC; Anthony Wilson: Mickey, "Red book" of Harleston genealogy.

Chapter 2

9–10 Harleston family: Birth dates: Charleston Death Card file, 1821–1926, Charleston County Library; Sibby: Mickey, "Red book" of Harleston genealogy, misc. notes.

12 "A white man laid a nigger gal": Former slave Rose Maddox, quoted in Thelma Jennings, "'Us Colored Women Had to Go Through a Plenty': Sexual Exploitation of African-American Slave Women," *Journal of Women's History,* 1:3 (Winter 1990), 61.

13 Education: 1834 law: David J. McCord and Thomas Cooper, eds., *The Statutes at Large of South Carolina,* 10 vols. (Columbia, SC, 1836–1840), 7:468–70; reading his mail: Edwin A. Harleston, Boston, to Elise Forrest, Charleston, Nov. 15 & 27, 1923, EHW. (All letters hereafter from papers of EHW unless otherwise noted.)

13–15 "Mulattoes": Christine B. Hickman, "The Devil and the One Drop Rule: Racial Categories, African Americans, and the U.S. Census," *Michigan Law Review,* 95:5 (March 1997), 1161–265; distribution of "mulattoes": Joel Williamson, *New People: Miscegenation and Mulattoes in the United States* (New York: Free Press, 1980), 24–27; manumission laws: McCord and Cooper, eds., *The Statutes at Large of South Carolina,* 7:440–43, 459–60; 11:154.

17–18 Civil War: 1860 enslavement crisis: Michael P. Johnson and James L Roark, eds., *No Chariot Let Down: Charleston's Free People of Color on the Eve of the Civil War* (Chapel Hill and London: University of North Carolina Press, 1984), 8 passim; Williamson, *New People,* 66–67; Harleston petition: Petition, James Rose et al. to the Senate of South Carolina, 1860, Manuscript Division, Library of Congress, accession #11386004, copied from Records of the General Assembly, South Carolina Department of Archives and History, Columbia, SC; Etiwan Rangers: Index to Records of Confederate Soldiers Who Served in Organizations from South Carolina, Charleston County Library; John Harleston Read: Johnson and Roark, eds., *No Chariot Let Down,* 133 n, 136 n.

18–19 Reconstruction era: Education: Thomas R. McDaniel, ed., *Public Education in South Carolina: Historical, Political, and Legal Perspectives* (Spartanburg, SC: Converse College, 1984), 16; Ku Klux Klan: Lou Falkner Williams, *The Great Ku Klux Klan Trials, 1871–72* (Athens, GA: University of Georgia Press, 1996), 19, 35; Walter Edgar, *South Carolina: A History* (Columbia, SC: University of South Carolina Press, 1998), 398–401.

19–20 Kate Wilson and family in Charleston: 28 Laurel Street: Thomas Nestor to William Harleston, conveyance, May 17, 1870, Register of Mesne Conveyance, Charleston County; later, Richard Harleston appears to have been living in a multigeneration household, with his daughter, Fannie, and her family: Charleston Death Card file,

1821–1926—"Infant daught. of Fannie Harleston and Joseph Robertson, 20 Laurel Street, d. Oct 4, 1879"; neighbors: Census of the City of Charleston, 1861 (Charleston: Evans & Cogswell, 1861), 17; population: Ninth Federal Census of the United States, 1870.

21–25 Death of William Harleston: Inheritance law: *Statutes at Large of SouthCarolina, 1861–1877*, 13:62–63, quoted in Joel Williamson, *After Slavery: The Negro in South Carolina During Reconstruction, 1861–1877* (Hanover, NH: University Press of New England, 1965/1990), 297; Will of William Harleston, proved Jan. 11, 1875, and Inventory and Appraisement, Estate of William Harleston, Mar. 30, 1876, Charleston County Probate Court; "Died at 'The Hut,' in 1874, unmarried": Jervey, "The Harlestons"; Kate Wilson's inheritance: Recorded Account of the Estate of William Harleston, May 18, 1876, Charleston County Probate Court, Book S, 84; Huger family: "Descendants of John Huger of South Carolina," anon. printed circular, Charleston, ca. 1910; Kate Wilson and Benjamin Huger: Harleston family oral tradition.

25 Aftermath of Reconstruction: Ariela J. Gross, "Litigating Whiteness: Trials of Racial Determination in the Nineteenth-Century South," *Yale Law Journal* (Oct. 1998), n. 300; 1879 law: "An Act to Prevent and Punish the Intermarrying of Races" (1879), *Acts and Joint Resolutions of the South Carolina Assembly* (Columbia, SC, 1897), 3; codicil: Will of John Harleston, Nov. 15, 1877, Charleston County Probate Court, Will book P, 386; death of John Harleston: Charleston Death Card file, 1821–1926, Charleston County Library; end of Reconstruction: Edgar, *South Carolina*, 408–15.

26–28 Kate Wilson's last years: 28 Laurel Street, in trust: Thomas Nestor to William Harleston, conveyance, May 17, 1870, Register of Mesne Conveyance, Charleston County; Charleston Death Card file, 1821–1926, Charleston County Library.

Chapter 3

31–33 Edwin G. Harleston's early adulthood: His wife, Louisa Moultrie: Tenth Federal Census of the United States, 1880; Moultries as free people of color: James Lee to Guillaume Antsin, Bill of Sale of Negro Woman Lucy and child Flora, Nov. 15, 1804, and Lucy Wilkinson to Flora Wilkinson, Deed of Emancipation, Dec. 5, 1820, EHW; Edwin Harleston at 28 Laurel Street: *Charleston City Directory, 1881* (Charleston: Sholes & Co.), 276; *Beulah Benton: Charleston City Directory, 1901* (Charleston: W. H. Walsh), 484.

33–34 Years after Reconstruction: Edgar, *South Carolina*, chapter 18; Jack Temple Kirby, *Darkness at the Dawning: Race and Reform in the Progressive South* (New York: J. B. Lippincott Company, 1972); Williamson, *After Slavery;* lynchings: I. A. Newby, *Black Carolinians:*

A History of Blacks in South Carolina from 1895 to 1968 (Columbia, SC: University of South Carolina Press, 1973), 59–61.

34–37 Edwin Harleston's siblings: Charleston Death Card file, 1821–1926; health and marital status: interviews with EHW; Mickey, "Red book" of Harleston genealogy, misc. notes, papers of EHW; jobs: Charleston City directories, 1880–1910; Edward H. Mickey Sr.: Register of Signatures of Depositors in Branches of the Freedmen's Savings and Trust Company, 1865–74, Bureau of Refugees, Freedmen and Abandoned Lands, National Archives, Washington, DC; William Hine, "Black Politicians in Reconstruction South Carolina," *Journal of Southern History*, 49:4 (Nov. 1983), 579.

37–38 Census figures for 1890: Inter-University Consortium for Political and Social Research, Study 00003; varieties of "colored": Williamson, *New People*, 112; E. Franklin Frazier, *Black Bourgeoisie: The Rise of a New Middle Class in the United States* (New York: Collier, 1957), passim.

38–39 Jim Crow: Kirby, *Darkness at the Dawning*, 10–15; Benjamin "Pitchfork Ben" Tillman: Edgar, *South Carolina*, 437–38; "unlawful and void": article III, section 33, South Carolina Constitution of 1895 (Columbia, SC: Bureau of Publications, University of South Carolina, 1927).

41 "Harleston Brothers": "The Harleston Funeral Establishment," brochure, ca. 1925.

Chapter 4

43–44 Blanchard family: F. R. Blanchard, stable master, charges for horse medicine and horseshoes, in Account book, Oct. 4, 1913–Dec. 18, 1934, Mickey Funeral Home Papers, Avery Research Center, Charleston.

44–46 African American funeral practices: Elaine Nichols, ed., *The Last Miles of the Way: African American Homegoing Traditions, 1890–Present*, exhibition catalog, June 4–Dec. 1, 1989 (Columbia, SC: South Carolina State Museum, 1989); Cynthia Connor, "'Sleep on and Take Your Rest': Black Mortuary Behavior on the East Branch of the Cooper River, South Carolina," M.A. thesis, Anthropology, University of South Carolina, 1989; burial societies: Michael A. Plater, *African American Entrepreneurship in Richmond, 1890–1940: The Story of R. C. Scott* (New York: Garland, 1996), 101 ff.

46 "Undertaker": *Oxford English Dictionary* (Oxford: Clarendon Press; New York: Oxford University Press, 1989).

46–47 Mickey and Harleston profits: Mickey and Harleston Account book, 1909–1913, and Account book, Oct. 4, 1913–Dec. 18, 1934, Mickey Funeral Home Papers, Avery Research Center, Charleston.

47–53 Coffins, history of embalming, Captain Harleston as embalmer: Robert W. Habenstein, *The History of American Funeral Directing*, 4th ed., rev. and ed. Howard C. Raether (Milwaukee, WI: National

Funeral Directors Association, 1955/1996), 165–80, 204–20 passim; Champion fluid: Account book, Oct. 4, 1913–Dec. 18, 1934, Mickey Funeral Home Papers, Avery Research Center, Charleston.

54–55 Demise of funeral business: Dissolution and manner of disposition of certain assets and real estate, May 5, 1913, and Statement of personal property, n.d., Mickey Funeral Home Papers, Avery Research Center, Charleston.

Chapter 5

56–60 Daniel Jenkins and Jenkins Orphanage: John Chilton, *A Jazz Nursery: The Story of the Jenkins' Orphanage Bands* (London: Bloomsbury Book Shop, 1980); Jeffrey P. Green, *Edmund Thornton Jenkins: The Life and Times of an American Black Composer, 1894–1926* (Westport, CT: Greenwood Press, 1982), chapter 1; "God Dealing with Rev. D. J. Jenkins," pamphlet, ca. 1913, Charleston County Library, Jenkins Orphanage vertical file; Certificate of Death, Daniel Joseph Jenkins, South Carolina Bureau of Vital Statistics, July 30, 1937, in Jenkins Orphanage Papers, Avery Research Center, Charleston; U.S. Census, 1900, Charleston, SC; interview, John Dowling/Sarah Dowling/Mary Stiles, Feb. 7, 1979, in Edmund T. Jenkins Papers, Schomburg Center for Research in Black Culture, New York Public Library.

60–63 Captain Harleston's children: oral tradition of Harleston family; deceased siblings: twins named Willie and Arthur (b. Nov. 30, 1885, d. age one month, and 2 or 3 yrs., respectively), and Ethelyn (Mar. 16–Apr. 25, 1888), Charleston Death Card file, 1821–1926, Charleston County Library.

64 "Corn bread and molasses": Daniel Jenkins, draft of letter to the editor of the *Charleston Evening Post,* Feb. 3, 1936, Jenkins Orphanage vertical file, South Carolina Historical Society, Charleston.

67 Editorial: *London Daily Telegraph,* Sept. 9, 1895.

71 Olive Ashton Harleston: In 1979, British historian Jeffrey P. Green traced the birth records of Olive Harleston: John Foster, Superintendent Registrar, Wigan, Lancashire, U.K., to Jeffrey Green, Crawley, West Sussex, July 18, 1979, and Walter Lloyd, Wigan, Lancashire, to Jeffrey P. Green, Crawley, West Sussex, U.K., Sept. 18, 1979, both in Edmund T. Jenkins Papers, Schomburg Center for Research in Black Culture, New York Public Library.

Chapter 6

75–78 Atlanta University: Edwin A. Harleston, Atlanta University transcript, 1900–1904; letters: "CSM," Atlanta, to Edwin A. Harleston, Feb. 6, 1903; "Little Bit," Louisville, KY, to Edwin A. Harleston, May 18 and 26, 1905; "Chihuahua," Oberlin, OH, to Edwin A. Harleston, Atlanta, Sept. 24, 1905; Inez Canty to Edwin A. Harleston, Feb. 2, 1910.

78–80 Segregation: Kirby, *Darkness at the Dawning*, 24–30, 103; Coleman Blease: Edgar, *South Carolina*, 468.

81–83 Teddy Harleston moves to Boston: George W. Cram, Cambridge, MA, to Edwin A. Harleston, May 11, 1905; School of the Museum of Fine Arts: a postcard Teddy wrote in 1923 explains, "I did not spend a year at Harvard. I was admitted but was advised to enter the Mus. Sch. Because H's work in Fine Arts was mainly theoretical and historical": Edwin A. Harleston, Boston, to Elise Forrest, Charleston, Dec. 3, 1923; Register of Pupils [1906], School of the Museum of Fine Arts [Boston]; *School of the Museum of Fine Arts, Thirty-first Annual Report for the Year Ending June 1, 1907* (Boston: Museum of Fine Arts, 1907); H. Winthrop Peirce, *History of the School of the Museum of Fine Arts, Boston, 1877–1927* (Boston: T. O. Metcalf, 1930); Alice Brooks, Boston, to Edwin Harleston, Apr. 11, 1911.

83–88 School of the Museum of Fine Arts, Boston: "Colored" hours: Charleston Museum Spring Calendar, 1928, John Bennett Papers, South Carolina Historical Society, Charleston; Henry Ossawa Tanner: Benjamin Brawley, *The Negro Genius* (New York: Dodd, Mead & Co., 1937; reprint, New York: Biblio and Tannen, 1966), 178 ff.; recipes for color: Edwin A. Harleston, diary/calendar 1909–10; first portrait commissions: F. B. Sanborn, Concord, MA, to Edwin A. Harleston, Boston, Apr. 12, 1912; Teddy Harleston in *The Sailor:* F. W. Coburn, *Boston Post,* clipping, ca. 1910.

Chapter 7

90–91 Inaugural parade: Chilton, *A Jazz Nursery,* 9; *Inaugural Addresses of the Presidents of the United States from George Washington 1789 to George Bush 1989* (Washington, DC: United States G.P.O., 1989).

91–94 Marriage of Ella Harleston and Daniel Jenkins: Jenkins's reputation: *Afro American Citizen* (Charleston), Jan. 17, 1900, Jenkins Orphanage vertical file, Charleston County Library; letter to Captain Harleston: D. J. Jenkins, Beldoc, SC, to E. G. Harleston, Charleston, SC, Oct. 4, 1912.

94–98 Life at Jenkins Orphanage: Children delivered by police: biographical data, Jenkins Orphanage Papers, Avery Research Center, Charleston; intake of new orphans: interview, John Dowling/Sarah Dowling/Mary Stiles, Feb. 7, 1979, in Edmund T. Jenkins Papers, Schomburg Center for Research in Black Culture, New York Public Library; "Cornbread and molasses": Rev. D. J. Jenkins, quoted in untitled pamphlet, Jenkins Orphanage, 1936, Edmund T. Jenkins Papers, Schomburg Center; escaped boys: Clarence Lunz, Judge of Probate, Charleston County, to Rev. D. J. Jenkins, May 2, 1934, Jenkins Orphanage Papers, Avery Research Center; punishment: Chilton, *A Jazz Nursery,* 14, 24; Ford Perry: D. J. Jenkins, Charleston, to Susie Littlejohn, n.d., Jenkins

Orphanage Papers, Avery Research Center; memoir of orphanage: Jabbo Smith, "Autobiographical material," box 1, Cladys "Jabbo" Smith Collection, New York Public Library for the Performing Arts.

98–101 Travels of Jenkins bands: Performance permit: William J. Wallins, Mayor, City of Yonkers, NY, July 21, 1921, Edmund T. Jenkins Collection, Schomburg Center for Research in Black Culture, New York Public Library; band venues: orphanage band member Amos White, untitled manuscript, p. 40, Edmund T. Jenkins Papers, Schomburg Center; girls' bands: author's interview with Harleston Fleming, Charleston, SC, August 1998.

101–105 Minstrels, coon songs, ragtime: Nathan Irvin Huggins, *Harlem Renaissance* (New York: Oxford, 1971), 248–80; Ted Gioia, *The History of Jazz* (New York: Oxford, 1997), 21 ff.; Jenkins band repertoire: Chilton, *A Jazz Nursery,* 16–21.

105–107 Trip to New York and U.K.: Band members: Chilton, *A Jazz Nursery,* 14–15; invitation to England: Jules Hurtig, Columbus, OH, to D. J. Jenkins, Charleston, Apr. 16 & 24, 1914, Edmund T. Jenkins Papers, Schomburg Center for Research in Black Culture, New York Public Library.

Chapter 8

108–109 Teddy Harleston in Boston: Calendar art: T. K. Gibson, Atlanta, to Edwin A. Harleston, Boston, Feb. 25, 1911 & Mar. 11, 1912; high school teacher: Mattie Marsh, Charleston, to Edwin A. Harleston, Boston, Mar. 30, 1912.

111–113 South Carolina when Teddy Harleston returns: Edgar, *South Carolina,* 450, 467–68; Newby, *Black Carolinians,* 120, 124–25, 134; Mamie Randall: Harleston family oral tradition; Charleston city directories, 1900–1925; U.S. Census, 1900 and 1910.

116–121 Elise Forrest: Party invitation: Elise Forrest, 97 Morris Street, Charleston, to Edwin A. Harleston, Charleston, Sept. 8, 1913; bazaar: Elise Forrest, Charleston, to Edwin A. Harleston, Charleston, ca. Jan. 1915; Brownie camera: Anon., "History of Kodak cameras," Eastman Kodak Company Home Page (Mar. 1999), July 19, 2000, http://www.kodak.com/global/en/consumer/products/techInfo/aa13/aa13.shtml; love letter: Elise Forrest, Fowler, SC, to Edwin A. Harleston, Charleston, Nov. 25, 1915; requesting film: Elise Forrest, Cades, SC, to Edwin A. Harleston, Charleston, Nov. 25, 1915 & Jan. 3, 1916; "saucy" pictures: Edwin A. Harleston, Atlanta, to Elise Forrest, New York, ca. Apr. 1919; mosquito: Elise Forrest, Charleston, to Edwin A. Harleston, Charleston, June 16, 1919; danced with men: Elise Forrest, Charleston, to Edwin A. Harleston, Oct. 9, 1916.

122 Embalming school: Edwin A. Harleston, 2 University Place, Atlanta, to Elise Forrest, New York, ca. Oct. 1919; Elise Forrest, Howard

Orphanage, Kings Park, Long Island, to Edwin A. Harleston, New York City, Jan. 11 & 28, 1917.

Chapter 9

124–26 Jenkins family in London: Ella Harleston Jenkins, London, to Edwin A. Harleston, Charleston, July 27, 1914; biography of Edmund Jenkins: Green, *Edmund Thornton Jenkins;* and Gwendolyn B. Bennett, "Edmund T. Jenkins: Musician," *Opportunity* magazine (Nov. 1925), 338–39.

126–30 Jenkins band at London exposition: Chilton, *A Jazz Nursery,* 16–21; visit of George V: Gerald Kiralfy, Anglo-American Exposition, London, to P. G. Daniels [assistant to Rev. Jenkins], July 13, 1914, Edmund T. Jenkins Papers, Schomburg Center for Research in Black Culture, New York Public Library; Coleman Blease: Daniel J. Jenkins, London, to Hon. Cole L. Blease, Gov. of South Carolina, Columbia, SC, July 23, 1914, Edmund T. Jenkins Papers, Schomburg Center.

130–31 Olive Harleston in Wigan: Walter Lloyd, Wigan, England, to Jeffrey Green, Crawley, England, Sept. 18, 1979; Laura E. Gilbert, Wigan, England, to Jeffrey Green, Sept. 4, 1979, Edmund Jenkins Papers, Schomburg Center for Research in Black Culture, New York Public Library.

132 "Fever heat over the war question": Ella Harleston Jenkins, Paris, to Edwin A. Harleston, Charleston, Aug. 1, 1914.

Chapter 10

137–39 Robert Harleston and Marie Forrest: Marie Forrest, Washington, DC, to Robert Harleston, Charleston, Oct. 16, 1911 & Apr. 24, 1912; Howard University: Marie Forrest, Washington, DC, to Robert Harleston, Charleston, Mar. 3, 1912.

140–43 Elise Forrest and Teddy Harleston in New York: Howard Orphanage: Bill Bleyer, "Kings Park: A Beacon for Those in Need," Long Island Historical Society Home Page, July 18, 2000, http://www.lihistory.com; "I had hysterics": Elise Forrest, Kings Park, Long Island, to Edwin A. Harleston, New York City, Jan. 11 & 28, 1917; Auguste Renouard: Habenstein, *American Funeral Directing,* 225–26, 310–11; Tom Reynolds, "The Embalmers' Supply Company, History" (Nov. 1997), July 18, 2000, http://www.embalmers.com; Donald M. Chase, "History of Colorado Funeral Enterprises by City or Town," Colorado Funeral Directors' Association Home Page, July 18, 2000, http://www.cofda.org/history/denver.htm; "sending you a curl": Elise Forrest, New York, to Edwin A. Harleston, Charleston, Mar. 1, 1917.

143–45 World War I: blacks enlisting in SC: Maurine Akua McDaniel, "Edwin Augustus Harleston, Portrait Painter, 1882–1931," dissertation, Emory University, 1994, 121; Teddy's attempted enlistment: E. B. Garey to Whom It May Concern, May 24, 1917; Notice of Classification for

E. A. Harleston, Oct. 7 & 12, 1918; Local Board for Division 1, Nov. 13, 1918; "What are you planning, Ted?": Elise Forrest, Asheville, NC, to Edwin A. Harleston, Charleston, Sept. 18, 1918.

145–48 Tuberculosis: Edwin A. Harleston, Charleston, to Elise Forrest, New York, July 20, 1920; Marie Forrest's weight: Marie Forrest, Asheville, NC, to Edwin A. Harleston, Charleston, Sept. 7, 1918; Gussie's shoes: Marie Forrest, Asheville, NC, to Edwin A. Harleston, Charleston, Sept. 17, 1918.

148–50 Gussie Harleston: Interviews with EHW, Atlanta, 1996–2000; Edwin A. Harleston, Charleston, to Elise Forrest, Washington, DC, Aug. 10, 1920; Edwin A. Harleston, Charleston, to Elise Harleston, Tuskegee, AL, Nov. 6, 1921; "perk as a beetle": Edwin A. Harleston, Charleston, to Elise Forrest, Sept. 5, 1920; "bloomer dresses": Edwin A. Harleston, Charleston, to Elise Harleston, Tuskegee, AL, Oct. 6, 1921; recital: Edwin A. Harleston, Charleston, to Elise Harleston, Tuskegee, AL, Oct. 22, 1921; Christmas: Edwin A. Harleston, Charleston, to Elise Forrest, New York, ca. Jan. 1920.

Chapter 11

151–52 Teddy resumes painting: McDaniel, "Edwin Augustus Harleston," 133–70; "thousand and one interests of our group": Edwin A. Harleston, Boston, to Elise Forrest, Charleston, Nov. 26, 1923.

152–55 Elise and photography school: "I think I can arrange it": Edwin A. Harleston, Cleveland, to Elise Forrest, New York, June 23, 1919; "the strain of marital relations": Marie F. Harleston, Summerville, SC, to Elise Forrest, New York, May 9, 1920; black photographers: Deborah Willis, *Reflections in Black: A History of Black Photographers, 1840 to the Present* (New York: W. W. Norton, 2000); Valencia Hollins Coar, *A Century of Black Photographs: 1840–1960* (Providence, RI: Rhode Island School of Design, 1983); Michael Francis Black Papers, Special Collections Library, Duke University, Durham, NC.

155–57 The *Mercury* riot: Letter to editor: Edwin A. Harleston, Charleston, to Editor, *News and Courier,* Charleston, Apr. 15, 1919; riot and lynchings: "All Quiet in Town After the Rioting," *Charleston News and Courier,* May 13, 1919; W. E. Burghardt Du Bois, *Dusk of Dawn: An Essay Toward an Autobiography of a Race Concept* (New York: Schocken Books, 1968), 264; David Levering Lewis, *When Harlem Was in Vogue* (New York: Penguin, 1981), 18; Tyler Stovall, *Paris Noir: African Americans in the City of Light* (Boston: Houghton-Mifflin, 1996), 27; "Negroes Deplore the Mob Spirit": photocopy of editorial from unidentified Charleston newspaper, ca. May 1917.

157–63 Elise at photography school: "so sweet and trim": Edwin A. Harleston, Atlanta, to Elise Forrest, New York, ca. Apr. 1919; "see . . . all the swells": Edwin A. Harleston, Charleston, to Elise Forrest, New York, Oct. 3, 1919; "$100 down": Elise Forrest, 241 Decatur St., Brooklyn, to Edwin A. Harleston, Charleston, ca. Sept. 1919; "Jews, Germans, and Irish": Elise Forrest, New York, to Edwin A. Harleston, Charleston, ca. Sept. 1919; "go to the Metropolitan Museum": Edwin A. Harleston, Charleston, to Elise Forrest, New York, ca. Jan. 1920; "her little soft body": Elise Forrest, New York, to Edwin A. Harleston, Charleston, Mar. 2, 1920; "Nice Jim Crow": Elise Forrest, 241 Decatur St., Brooklyn, to Edwin A. Harleston, Charleston, ca. Sept. 1919; "marry me": Elise Forrest, New York, to Edwin A. Harleston, Charleston, Jan. 25, 1920; "your promises are nil": Elise Forrest, New York, to Edwin A. Harleston, Charleston, Jan. 20, ca. 1920; "don't court the blues": Edwin A. Harleston to Elise Forrest, New York, ca. Oct. and ca. Nov. 1919; "Teddy boy—please hurry": Elise Forrest, New York, to Edwin A. Harleston, Charleston, Dec. 28, 1919; "poor little creature": Edwin A. Harleston, Charleston, to Elise Forrest, New York, ca. Jan. 1920; Edwin A. Harleston, Cleveland, to Elise Forrest, New York, June 23, 1919; "spoiled by the gentlemen": Elise Forrest, New York, to Edwin A. Harleston, Charleston, Mar. 8, 1920; "continued peevishness": Edwin A. Harleston, Charleston, to Elise Forrest, New York, ca. Jan. 1920; "it is good practice": Edwin A. Harleston, Atlanta, to Elise Forrest, New York, ca. Apr. 1919; Edwin A. Harleston, Charleston, to Elise Forrest, New York, Sept. 24, 1919.

163–67 Engagement and marriage: "of course, we will marry": Edwin A. Harleston, Charleston, to Elise Forrest, New York, ca. Jan. 1920; "my heart is bursting": Elise Forrest, New York, to Edwin A. Harleston, Charleston, Apr. 6 & July 4, 1920; "I salute you, the bride elect": Edwin A. Harleston, Charleston, to Elise Forrest, New York, July 7, 1920; "come get your bird": Elise Forrest, New York, to Edwin A. Harleston, Charleston, June 29, July 11, & ca. Aug. 1920; "my Christian duty": Elise Forrest, New York, to Edwin A. Harleston, Charleston, June 29, 1920; "face the preacher": Edwin A. Harleston, Charleston, to Elise Forrest, Washington, DC, Aug. 10, 1920; "you'd be utterly spoiled": Edwin A. Harleston, Charleston, to Elise Forrest, New York, Sept. 5, 1920; "Five days of Extreme Pleasure": Elise Forrest, misc. journal note, ca. Sept. 1920.

Chapter 12

169–72 Edmund Jenkins in London: Programme of Grand Orchestral Concert, May 18, 1916, Edmund T. Jenkins Collection, Center for Black Music Research, Columbia College, Chicago; conflict with the Reverend Mr. Jenkins: Edmund T. Jenkins, Paris, to Daniel Joseph Jenkins, Charleston, Apr. 20, 1926, Edmund Thornton Jenkins Collection, Schomburg Center for Research in Black Culture, New York Public Library.

173–76 "Hell Fighters": Reid Badger, *A Life in Ragtime: A Biography of James Reese Europe* (New York: Oxford University Press, 1995); Stovall, *Paris Noir,* 10–12; Stephen and Herbert Wright: Chilton, *A Jazz Nursery,* 23–24.

176–83 Tom Delaney: Jabbo Smith on running away: quoted in Whitney Balliett, "Starting at the Top," *The New Yorker,* Dec. 3, 1979, 116–25; and quoted in Owen Coyle, "Jabbo Smith—His Horn Is Hot Again," *Mississippi Rag,* July 1976; "Jazz-Me Blues": Lillyn Brown and Her Jazz-Bo Syncopaters, recorded May 9, 1921, in *The Complete Works of Esther Bigeous, Lillyn Brown, Albert Brown & the Remaining Titles of Ada Brown in Chronological Order (1921–28),* Document Records DOCD-5489.

183–88 Edmund Jenkins and jazz: Southern Syncopated Orchestra: Arthur Briggs, Paris, to Rainer Lotz, June 3, 1979, Edmund T. Jenkins Papers, Schomburg Center for Research in Black Culture, New York Public Library; Talbot-Darracq: Edmund T. Jenkins, New York, to W. H. Des Verney, Dec. 12, 1925; Invoice, B. Claridge, S.T.D. Motor Services, London, to E. T. Jenkins, 28 Torrington Sq., London, May 2, 1923, in Edmund Thornton Jenkins Collection, Schomburg Center; "The 'darkie' question": quoted in Green, *Edmund Thornton Jenkins,* 110.

Chapter 13

191–96 NAACP: Du Bois, *Dusk of Dawn,* 95; "Art Student": Richard Mickey, Charleston, to Board of Directors, NAACP, New York, Mar. 10, 1917, NAACP Papers, Charleston, SC, branch, Library of Congress, Manuscripts Division; seamstresses: letters from U.S. Representatives (Minnesota) Carl van Dyke, Harold Knutson, Clarence Miller, Thomas Schall, and Halvor Steenerson, and U.S. Senators (Minnesota) Frank Kellogg and Knute Nelson, June–July 1917, in NAACP Papers, Charleston, SC, branch, Library of Congress, Manuscripts Division; Petition, Committee of Colored Citizens, Charleston, May 10, 1917; murder and rape cases: Edwin A. Harleston, Charleston, to John R. Shillady, New York, Nov. 2, 1918, NAACP Papers, Charleston, SC, branch, Library of Congress, Manuscripts Division; teachers' campaign: "Awake," editorial, *The Crisis,* NAACP monthly (Apr. 1917); Richard Mickey, Charleston, to Board of Directors, NAACP, New York, Dec. 21, 1917, NAACP papers, Charleston, SC, branch, Library of Congress, Manuscripts Division; Richard Mickey, Charleston, to National Office, NAACP, New York, July 8, 1918; Petition, Edwin A. Harleston, et al., to Governor R. A. Cooper et al., Jan. 18, 1919; Resolution, Board of School Commissioners of the city of Charleston, SC, Feb. 1919; conference: Edwin A. Harleston, Cleveland, to Elise Forrest, New York, June 23, 1919; Edwin A. Harleston, speaking notes, "Cleveland 1919."

197–98 Elise Harleston at Tuskegee Institute: "ready to receive you": Edwin A. Harleston, Atlanta, to Elise F. Harleston, Charleston, Aug. 21, 1921; "expecting Mr. Battey": Elise F. Harleston, Tuskegee, AL, to Edwin A. Harleston, Charleston, Oct. 26, 1921; "you are rusty, eh?": Edwin A. Harleston, Charleston, to Elise F. Harleston, Tuskegee, AL, Oct. 6, 1921; "Get the floors painted": Elise F. Harleston, Tuskegee, AL, to Edwin A. Harleston, Charleston, Oct. 26, 1921.

200–205 The Harleston Studio: Home ferment: Edwin A. Harleston, Columbus, OH, to Elise F. Harleston, Charleston, n.d.; "Highest Artistic Standards": Harleston Studio advertisement, 1923; commissions: Will Houston, State College, Orangeburg, SC, to Edwin A. Harleston, Charleston, Feb. 12, 1923; Katherine Harleston Fleming, St. Augustine, FL, to Edwin A. Harleston, Charleston, June 3, 1923; "Black Swan Records has a booth": Edwin A. Harleston, Atlanta, to Elise F. Harleston, Charleston, Aug. 17, 1921; Il Bronzino: McDaniel, "Edwin Augustus Harleston," 170; "only $5": Elise F. Harleston, Charleston, to Edwin A. Harleston, Chicago, Aug. 5 & 19, 1925; "Trust by this time": Elise F. Harleston, Charleston, to Edwin A. Harleston, New York, Nov. 11, 1925.

205–208 Teddy in New York: New York Public Library: Ernestine Rose, New York Public Library, to Edwin A. Harleston, Charleston, July 9, 1923; "in the National Gallery": Edwin A. Harleston, Washington, DC, to Elise F. Harleston, Charleston, Oct. 31 & Nov. 3, 1923; "kissing her on 'them'": Edwin A. Harleston, Pittsburgh, PA, to Elise F. Harleston, Charleston, June 26, 1924; interview: Charles S. Johnson, National Urban League, New York, to Edwin A. Harleston, Oct. 26, 1923; "I am dreaming": Edwin A. Harleston, New York, to Elise F. Harleston, Charleston, Nov. 7, 1923; on *Runnin' Wild*: Edwin A. Harleston, New York, to Elise F. Harleston, Charleston, Nov. 7, 1923; "coming bliss": Elise F. Harleston, Charleston, to Edwin A. Harleston, New York, ca. Jan. 1924; "my warm little ricebird": Edwin A. Harleston, New York, to Elise F. Harleston, Charleston, Nov. 10, 1923; Edwin A. Harleston, St. Augustine, FL, to Elise F. Harleston, Charleston, Apr. 13, 1924; "The shirts that you had worn": Elise F. Harleston, Charleston, to Edwin A. Harleston, New York, Jan. 13 & 15, 1924; "I want money!": Edwin A. Harleston, Roxbury Hospital, Boston, to Elise F. Harleston, Thanksgiving Day, 1923.

Chapter 14

209–10 Jenkins house: Robert P. Stockton, "House Version of Regency Style," *Charleston Evening Post*, Feb. 5, 1990.

212–15 Edmund Jenkins and jazz in Paris: Green, *Edmund Thornton Jenkins,* 111–62; Stovall, *Paris Noir,* 50 passim.

215–16 Layland family: Laura Gilbert, Wigan, Lancashire, England, to Jeffrey Green, Sept. 4, 1979, Edmund Jenkins Papers, Schomburg Center for Research in Black Culture, New York Public Library.

217–21 Edmund Jenkins in America: Edwin A. Harleston, Canton, MA, to Elise F. Harleston, Charleston, Jan. 6, 1924; Elise F. Harleston, Charleston, to Edwin A. Harleston, New York, Jan. 13, 1924; "hunting Edmund": Edwin A. Harleston, to Elise F. Harleston, Charleston, ca. Jan. 1924.

221–24 Black Americans in Paris: Stovall, *Paris Noir,* 44–54; "valet who can cook": Edmund T. Jenkins, Paris, to Daniel Jenkins, Charleston, Apr. 20, 1926, Edmund Jenkins Papers, Schomburg Center for Research in Black Culture, New York Public Library.

Chapter 15

229–30 Surgery in Massachusetts: Edwin A. Harleston to Elise F. Harleston, Charleston, Nov. 10, 13, & 15, 1923.

231–32 Mamie Randall: Elise F. Harleston, Charleston, to Edwin A. Harleston, St. Augustine, FL, Aug. 4, 1929.

232–34 1923–24 travels: "*our* old home": Edwin A. Harleston, Canton, MA, to Elise F. Harleston, Charleston, Dec. 5, 1923; "3 cases of Reconcentrated Dioxin": Edwin A. Harleston, Canton, MA, to Elise F. Harleston, Charleston, Nov. 15, 1923; "three children buried": Elise F. Harleston, Charleston, to Edwin A. Harleston, Chicago, Aug. 19, 1925; "Thank God for Gussie": Edwin A. Harleston, Boston, to Elise F. Harleston, Charleston, Jan. 1, 1924; "childlessness": Edwin A. Harleston, Philadelphia, to Elise F. Harleston, Nov. 1, 1924; "will you let me have Doris?": Elise F. Harleston, Charleston, to Edwin A. Harleston, Chicago, ca. July 1924 & Aug. 4, 1924.

234–36 Teddy's art career: "Please give price": Solomon Johnson, *The Savannah Tribune,* Savannah, GA, to Edwin A. Harleston, Charleston, Mar. 20, 1924; *The Bible Student*: Inez Canty, Chicago, to Edwin A. Harleston, Charleston, May 6, 1924; McDaniel, "Edwin Augustus Harleston," 181; "a picture of President Twitchell Ware": George A. Towns, Atlanta, to Edwin A. Harleston, Charleston, June 9, 1924; "there is but one colored man": W. E. B. Du Bois, New York, to Edwin A. Harleston, Charleston, July 16, 1924, and W. E. B. Du Bois, New York, to Blanche W. Stubbs, July 16, 1924; "from $1000 to $2000 according to size": Edwin A. Harleston, Chicago, to Elise F. Harleston, Charleston, July 27, 1924.

236 Teddy Harleston as "colored artist": Meta Warrick Fuller, Framingham, MA, to Edwin A. Harleston, Charleston, Nov. 12, 1925; other black artists: *Negro Artists: An Illustrated Review of Their Achievements* (New York: Harmon Foundation, 1935), exhibition catalog, 41 ff.; Gary A. Reynolds and Beryl J. Wright, *Against the Odds: African-American Artists and the Harmon Foundation* (Newark, NJ: The Newark Museum, 1989), 150 ff; Charleston Renaissance: Martha R. Severens, *The Charleston Renaissance* (Spartanburg, SC; Saraland).

239–400 Washington, DC, job: one hundred dollars per painting: Will Houston, Orangeburg, SC, to Edwin A. Harleston, Charleston, Feb. 12, 1923; "You have worked for ten years": Elise F. Harleston, Charleston, to Edwin A. Harleston, Nashville, TN, Sept. 12, 1930; Elise F. Harleston, Charleston, to Edwin A. Harleston, Chicago, Aug. 5, 1925; "Pays nicely, though": Edwin A. Harleston, Pittsburgh, PA, to Elise F. Harleston, Charleston, June 26, 1924; "the superintendent at Washington": Edwin A. Harleston, Chicago, to Elise F. Harleston, Charleston, ca. July 1924.

240–43 Chicago: "Your gray and bald-headed husband": Edwin A. Harleston, Chicago, to Elise F. Harleston, Charleston, June 28, 1924; Edwin A. Harleston, Chicago, to Elise F. Harleston, Charleston, July 7, 1924; Edwin A. Harleston course registration, Art Institute of Chicago, June 27, 1924; Jesse Binga: Carl R. Osthaus, "The Rise and Fall of Jesse Binga, Black Financier," *Journal of Negro History* 58:1 (1973), 39–60; "just back from the Bingas'": Edwin A. Harleston, Chicago, to Elise F. Harleston, Charleston, July 13, 1924; Binga portrait: Edwin A. Harleston, Chicago, to Elise F. Harleston, Charleston, ca. July 1924; "the 'Great Mogul's' portrait": Elise F. Harleston, Charleston, to Edwin A. Harleston, Chicago, ca. July 1924; "not a bad sitter": Edwin A. Harleston, Chicago, to Elise F. Harleston, Charleston, ca. July 1924; "like something in *Vanity Fair*": Edwin A. Harleston, Chicago, to Elise F. Harleston, Charleston, July 7, 1924 & Aug. 10, 1924.

243 The couple's separation: "I wanted you so": Elise F. Harleston, Charleston, to Edwin A. Harleston, Chicago, July 17, 1925; "'love' me so I could go to sleep": Elise F. Harleston, Charleston, to Edwin A. Harleston, Chicago, ca. Aug. & Aug. 27, 1924; "dependable—my wife": Edwin A. Harleston, Chicago, to Elise F. Harleston, Charleston, July 7 & 13, 1924.

244–45 Calvin Coolidge: Benjamin E. Robinson, Boston, to William M. Butler, Chicago, Sept. 12, 1924; "the President is favorable": William M. Butler, Chicago, to Benjamin E. Robinson, Boston, Sept. 15 & 24, 1924; "the rendering of white folk": Edwin A. Harleston, Washington, DC, to Elise F. Harleston, Charleston, ca. Sept. 1924.

245–46 Edmund Ware: George A. Towns, Brooklyn, to Edwin A. Harleston, Charleston, Sept. 2, 1924; Edmund T. Ware, Montclair, NJ, to Edwin A. Harleston, Charleston, Jan. 1, 1925.

246–48 Pierre S. Du Pont: "pronounced shadows": Edwin A. Harleston, Philadelphia, to Elise F. Harleston, Charleston, ca. Oct. 1924; "DuPont day celebration": Edwin A. Harleston, New York, to Elise F. Harleston, Charleston, Nov. 10, 1924; Du Pont thank you: "Pierre S. Du Pont, Wilmington, DE, to Edwin A. Harleston, Charleston, Nov. 12, 1924; "a poor and barren sort of affair": Edwin A. Harleston, New York, to Elise F. Harleston, Charleston, Nov. 24, 1924; "success,

Success, *Success*": Katherine Harleston Fleming, St. Augustine, FL, to Edwin A. Harleston, Charleston, Dec. 29, 1924.

Chapter 16

250–57 Jabbo Smith: Chilton, *A Jazz Nursery;* Coyle, "Jabbo Smith"; Blake Green, *Datebook,* Sept. 6, 1981; Cladys "Jabbo" Smith Collection, New York Public Library for the Performing Arts; Balliett, "Starting at the Top," 116–25; Mike Joyce, "Interview with Jabbo Smith," *Cadence,* May 1982; M. W. Stearns, "Bubber Miley's Jungle Iron Choking & Sobbing Fades Dunn," *Downbeat,* Chicago, June 1937, 10; Dick Holbrook, "Let's Get Together on Jabbo Smith," in *Record Research,* 35 (June 1961), 7–8; "Louis Armstrong had gotten the better of the kid": Bill Crow, *Jazz Anecdotes* (New York: Oxford University Press, 1990), 91.

257–60 Tommy Benford: Whitney Balliett, "Profiles: New York Drummers," *The New Yorker,* Nov. 5, 1979, 52–76.

260–64 *Porgy:* James M. Hutchisson, *DuBose Heyward: A Charleston Gentleman and the World of Porgy and Bess* (Jackson, MS: University Press of Mississippi, 2000), 50–85; from the John Bennett Papers, South Carolina Historical Society, Charleston: "Sunday budgets," Aug. 14, Sept. 11, & Oct. 9, 1927, & Jan. 15, 1928; DuBose Heyward, Hotel Manger, New York, to John Bennett, Charleston, Sept. 6, 1927; Mrs. John Bennett, Charleston, to DuBose Heyward, New York, ca. Sept. 1927.

Chapter 17

266–68 Passing for white: Interviews with EHW, Oct. 3 & Dec. 22, 1998 & Mar. 27, 1999; Williamson, *New People,* 100.

268–72 Harlem: Lewis, *When Harlem Was in Vogue;* Steven Watson, *The Harlem Renaissance: Hub of African-American Culture, 1920–1930* (New York: Pantheon, 1995), 88, 131, 136, 142, 144; Teddy Harleston in New York: "Lots of N.Y. jazz on now": Edwin A. Harleston, New York, to Elise F. Harleston, Charleston, Oct. 14, 1925; "I . . . couldn't do the 'Charleston'": Edwin A. Harleston, Chicago, to Elise F. Harleston, Charleston, Aug. 2, 1925; "I saw Alethia": Edwin A. Harleston, New York, to Elise F. Harleston, Charleston, Oct. 10, 1925; "do the Georgia grind": Lewis, *When Harlem Was in Vogue,* 106; "How is it": W. E. B. Du Bois, "Criteria of Negro Art," *The Crisis,* 32:6 (Oct. 1926), 290–97, quoted in Lewis, *When Harlem Was in Vogue,* 177; "I am doing the exhibitions": Edwin A. Harleston, New York, to Elise F. Harleston, Charleston, Nov. 4, 1925.

273–74 Julia Peterkin: Elise F. Harleston, Charleston, to Edwin A. Harleston, ca. 1928.

274–75 Successes stopped: Edwin A. Harleston, New York, to Elise F. Harleston, Charleston, Nov. 10, 1924; "Well cheer up, my love": Edwin A. Harleston, Philadelphia, to Elise F. Harleston, Charleston,

Nov. 1, 1924; "It does come a little, eh my partner?": Edwin A. Harleston, Chicago, to Elise F. Harleston, Charleston, Aug. 12, 1926.

276–78 Charleston Museum: "Captain was stopped . . . by Mr. John Bennett": Elise F. Harleston, Charleston, to Edwin A. Harleston, Oct. 28, 1925; "Unforeseen circumstances": Laura M. Bragg, Charleston to Edwin A. Harleston, Charleston, Apr. 22, 1926; "I should have been . . . better prepared": Edwin A. Harleston, Charleston, to Laura M. Bragg, Charleston, May 7, 1926; "those who 'aspire to paint'": Elise F. Harleston, Charleston, to Edwin A. Harleston, St. Augustine, FL, Sept. 14, 1929.

278–79 Harmon Foundation: Edwin A. Harleston, Charleston, to George Haynes, New York, July 27, 1926; Mary Beattie Brady, Harmon Foundation, to George Haynes, New York, Nov. 5, 1926; report of Francis Jones, painter, Nov. 6, 1926; memorandum, conference with Miss Brady, Dr. Lindsay, Mr. Simons, Dr. Haynes, Nov. 9, 1926; and memorandum, conversation of George Haynes with Wm. Boring, Nov. 29, 1926, all in Harmon Foundation Papers, Library of Congress, Manuscripts Division; "The judges": William E. Harmon, New York, to Edwin A. Harleston, Charleston, Dec. 9, 1926; "I repeat, I love you": Elise F. Harleston, Charleston, to Edwin A. Harleston, Nashville, Sept. 12, 1930.

280–82 Teddy's disappointment: "there are several 'prospects'": Edwin A. Harleston, St. Augustine, FL, to Elise F. Harleston, Charleston, Sept. 4, 1929; Yolande Du Bois: W. E. B. Du Bois, New York, to Edwin A. Harleston, Charleston, Aug. 4, 1928; character in a novel: DuBose Heyward, *Mamba's Daughters* (New York: Literary Guild, 1929), 216, 220, 224.

282–83 "The Building of a Picture": J. B. Randolph, Claflin College, Orangeburg, SC, to Edwin A. Harleston, Charleston, Jan. 19, Feb. 1, & Feb. 14, 1929; "one hundred brown college girls": Sue Bailey, New York, to Edwin A. Harleston, Charleston, May 9, 1930; "All the fixin's": Edwin A. Harleston, King's Mountain, NC, to Elise F. Harleston, Charleston, June 11, 1930.

283–85 Aaron Douglas: "Can you stay . . . until about the 15th . . . ?": Aaron Douglas, Nashville, TN, to Edwin A. Harleston, Charleston, July 12, 1930; "cordial and affable fellow": Edwin A. Harleston, Nashville, TN, to Elise F. Harleston, Charleston, Aug. 3, 1930; "a little short of cosmic": Aaron Douglas, New York, to Edwin A. Harleston, Charleston, ca. Oct. 1930.

287–88 Condolences: Ada Hinton, Canton, MA, to Elise F. Harleston, Charleston, May 12, 1931; J. B. Randolph, Orangeburg, SC, to Elise F. Harleston, Charleston, May 12, 1931; Helen Baker, Bronx, NY, to Elise F. Harleston, Charleston, May 15, 1931; W. E. B. Du Bois, New York, to E. C. Mickey, Charleston, May 12, 1931.

Chapter 18

289–93 Elise after Teddy Harleston's death: Helen Chisholm, Saybrook, CT, to Elise F. Harleston, Charleston, Aug. 29, Oct. 12, & Oct. 28, 1931; H. Chisholm, South Byfield, MA, to Elise F. Harleston, Charleston, Sept. 8, 1931; life insurance: Edwin A. Harleston, Chicago, to Elise F. Harleston, Charleston, Aug. 1, 1925; "I . . . must return the pictures": Helen Chisholm, Saybrook, CT, to Elise F. Harleston, Charleston, Nov. 30 & Dec. 17, 1931; Atlanta School of Social Work: Nellie Towns, Atlanta, to Elise F. Harleston, Charleston, Jan. 8, 1932; Helen Chisholm, Saybrook, CT, to Elise F. Harleston, Charleston, Jan. 13, 1932; Mame Dugas, Charlotte, NC, to Elise F. Harleston, Charleston, Feb. 15, 1932.

294–95 Tennis court: Sylvia Harleston, Charleston, to Elise Harleston, Atlanta, Aug. 4, 1932.

296–97 Elise Harleston and Jack Wheeler: Elvira Forrest, Charleston [to Elise Harleston Wheeler], Aug. 17, 1932.

298–300 Edwina and Sylvia Harleston: "It's trimmed with lace": Edwina (Gussie) Harleston, Charleston, to Elise Harleston Wheeler, Baltimore, 1932 & Mar. 9, 1934; Sylvia Harleston: Edwina (Gussie) Harleston, letter fragment, n.d.

300–301 Fire at Jenkins Orphanage: *Charleston News and Courier*, Sept. 24, 1932; Mar. 18, 22, 23, 29, & May 31, 1933.

303–304 Talladega: Edwina (Gussie) Harleston, transcript, Talladega College, 1934–39; Edwina (Gussie) Harleston, Talladega, AL, to Elise Harleston Wheeler, Baltimore, Oct. 20, 1934, & Mar. 18, 1935.

Chapter 19

305–307 Jenkins Orphanage investigation: *Charleston News and Courier*, Mar. 18, 22, 23, 29, & 31, 1933; Bennett diary: "Sunday budget," Nov. 2, 1928, John Bennett Papers, South Carolina Historical Society, Charleston; Emily Bullitt, Bureau of Social Welfare, City of Charleston, to Edwin A. Harleston, Charleston, June 27, 1928; *Time* magazine: "Jenkins Bands," *Time*, Aug. 26, 1935; *Charleston News and Courier*, Aug. 25, 1935; "housing project for the whites": Edwina (Gussie) Harleston, Talladega, AL, to Elise Harleston Wheeler, Baltimore, ca. spring 1938; "I plead guilty": Rev. D. J. Jenkins, draft of letter to the editor of the *Charleston Evening Post*, Feb. 3, 1936, Jenkins Orphanage vertical file, South Carolina Historical Society, Charleston.

307–13 Edwina Harleston at Talladega: Edwina (Gussie) Harleston, Talladega, AL, to Elise Harleston Wheeler, Baltimore, Mar. 6 & 18, Apr. 20, May 6 & 21, Nov. 6, and Dec. 20, 1935, Mar. 12, 1937, and ca. spring 1938; Ogunquit: Edwina (Gussie) Harleston, Ogunquit, ME, to Elise Harleston Wheeler, Baltimore, May 21, June 10, June 27,

Aug. 3, and Sept. 3, 1935; "I hate her": Edwina (Gussie) Harleston, Talladega, AL, to Elise Harleston Wheeler, Baltimore, ca. May 1937.

314–18 Freddie Green and Cat Anderson: Thomas Johnson, "'Count Basie's Left Hand' Comes to the Caroliniana," *Carolina Columns,* bulletin of the South Caroliniana Society (Columbia, SC), Fall 1999, 4 passim; Donald Clarke, *Wishing on the Moon: The Life and Times of Billie Holiday* (New York: Viking, 1994); John S. Wilson, "Count Basie, 79, Dead," *New York Times,* Apr. 27, 1984; Helen and Stanley Dance, "Interview with Freddie Green, Aug. 9, 1977," North By South: Charleston to Harlem, The Great Migrations, Will Scott and Peter Rutkoff, eds., http://www.northbysouth.org/1998/index.htm (Oct. 23, 2000); Chilton, *A Jazz Nursery,* 52–60; John Chilton, *Who's Who of Jazz: Storyville to Swing Street* (New York: Time-Life, 1978).

318–19 Death of Rev. D. J. Jenkins: Certificate of Death, Daniel Joseph Jenkins, South Carolina Bureau of Vital Statistics, July 30, 1937, Jenkins Orphanage Papers, Avery Research Center, Charleston; *Charleston News and Courier,* July 31, Aug. 1, 3, & 4, 1937; Eloise C. Jenkins, fiduciary, Aug. 7, 1937; Final Account of Eloise C. Jenkins, as Administratrix, Nov. 15, 1938, Charleston County Probate Court.

319 Northwestern University: Edwina (Gussie) Harleston, transcript, Northwestern University, 1940–44.

321–23 Jenkins Orphanage relocation: *Charleston News and Courier,* May 23, June 1 & Sept. 12, 1938, and Oct. 5, 1939.

Chapter 20

327 50 Ashe Street and 34 Magazine Street: Susan Harleston Singleton to Clara Pinckney Brunson, conveyance, Apr. 19, 1941, Register of Mesne Conveyance, Charleston County, T42-105; Orphan Aid Society to Housing Authority of the City of Charleston, conveyance, Sept. 22, 1938, Register of Mesne Conveyance, Charleston County, V40-11.

327–33 Edwina Whitlock in Gary: Edwina Harleston Whitlock, Gary, IN, to Elise Harleston Wheeler, Baltimore, May 28, 1955.

339 Social Caseworker I: Edwina H. Whitlock, "Notice of Eligibility, County of Los Angeles Civil Service Commission," Oct. 19, 1964.

Chapter 21

345–46 Eloise Harleston Jenkins and Olive Jenkins: "Wealthy Widows," *Ebony,* Aug. 1949; "she is even now paying the price": Olive Jenkins Campbell, Yonkers, NY, to Edwina Fleming, Charleston, Aug. 23, 1974.

347 Ritz Sandwich Shop: Charleston City Directory, 1940, 1942.

349–50 Jabbo Smith: A. L. Seiman to Lorraine Gordon, Jan. 18, 1991, Cladys "Jabbo" Smith Collection, New York Public Library for the Performing Arts.

Permissions and Photography Credits

The author is grateful to the following for permission to quote material as noted below:

Edward B. Marks Music, for lines from "Jazz-Me Blues," words and music by Tom Delaney, copyright 1921, Edward B. Marks Music Company; copyright renewed. Used by permission. All rights reserved. Lines from "Georgia Stockade Blues," words and music by Tom Delaney, copyright 1925, Edward B. Marks Music Company; copyright renewed. Used by permission. All rights reserved.

Jomo Zimbabwe, for lines from "Levee Lounge Lizard," words and music by Edmund T. Jenkins (1926), copyright 1990, Jomo Zimbabwe. Used by permission. All rights reserved.

Unless noted below, photographs are courtesy of Edwina Harleston Whitlock.

Section One

Page 1, top: photograph by the author, 2001; page 1, bottom: "A Philosophic Cock," James Akin, hand-colored aquatint, early 1800s, American Antiquarian Society, Worcester, MA.; page 2, top and bottom: Courtesy of Audrey Battiste, Atlanta, GA; page 5, top: South Carolina Historical Society, Charleston, SC; page 6, bottom: The Charleston Museum, Charleston, SC; page 7, top Avery Research Center, College of Charleston, Charleston, SC; page 7, bottom: The Charleston Museum; page 8, top: South Carolina Historical Society; page 8, bottom: Avery Research Center; page 9, bottom: Historic American Buildings Survey, Prints and Photographs Division, Library of Congress; page 10, top: The Charleston Museum; page 10, bottom, right: Courtesy of Jeffrey Green, Center for Black Music Research, Columbia College, Chicago; page 11, top: Courtesy of Barbara Braithwaite, Yonkers, NY; page 12: Courtesy of Mae Gentry, Atlanta, GA, photograph by Joseph Ivansco; page 13, bottom: photograph by Joseph Ivansco.

Section Two

Page 3, top: photograph by Joseph Ivansco; page 4, top: Delaware Public Archives, Dover, DE; page 4, bottom: Gibbes Museum of Art/Carolina Art Association, Charleston, SC; page 6, top: Avery Research Center; page 7, top: South Carolina Historial Society; page 7, bottom: Avery Research Center; page 8, top: Courtesy of Jeffrey Green, Center for Black Music Research; page 9: Avery Research Center; page 10, top: South Carolina Historical Society; page 10, bottom: Avery Research Center; page 11, middle: Center for Black Music Research; page 11, bottom: South Carolina Library, University of South Carolina, Columbia, SC; page 12, top: New York Public Library for the Performing Arts.

Acknowledgments

Edwina Harleston Whitlock, amanuensis of the Harleston family, eighty-five years old at the publication of the book, made this account possible. Beginning in 1960 and for several decades, she acted as unpaid archivist and obdurate collector of data and documents, interview leads, letters, paintings, postcards, and memorabilia of her family, and made herself a bridge to the past. In 1998, she invited me to tell her family's story however I chose. A profound appreciation goes to Edwina Whitlock.

Thank you also to Henry Ferris, editor at William Morrow/HarperCollins, who saw the strength in this story. Much gratitude to him for handling everything with care. I'm grateful to Nichole Green, my hardworking researcher, who organized sources and put her hands on additional pictures, memoirs, manuscripts, vital records, and critical histories. The assistance that came from Edwina Whitlock's daughters, including the hospitality and friendliness of Sylvia Whitlock, and the kindnesses and favors of Mae Gentry, helped immeasurably.

A word of thanks to people who allowed interviews and provided personal papers and photographs: Audrey Battiste, Barbara Braithwaite, Sarah Dowling, Harleston Fleming, Lillian Hern, Albert and Mary Hinton, Felder Hutchinson, Maude Jenkins, Jomo Zimbabwe, and others. And additional gratitude to the extended Harleston family, in the South and across the States, whose collective stories—most of which I haven't been able to tell—could fill more pages.

Redge and Jane Hanes of North Carolina were unusually generous, loaning me a house where I disappeared with the manuscript, an act of kindness that allowed me to finish the job. Thanks to my wife, Elizabeth, who tolerated my vanishing, and to her mother, Katherine Guckenberger, who gave good advice. Elizabeth was obliged to get a dog in my absence, but didn't give up on me.

Index